# Advanced Materials for Biological and Biomedical Applications

Materials Research Society
Symposium Proceedings Volume 1569

# Advanced Materials for Biological and Biomedical Applications

Symposium held April 1–5, 2013, San Francisco, California U.S.A.


**EDITORS**

**Michelle Oyen**
Cambridge University
Cambridge, United Kingdom

**Andreas Lendlein**
Helmholtz-Zentrum Geesthacht
Teltow, Germany

**William T. Pennington**
Clemson University
Clemson, South Carolina, U.S.A.

**Lia Stanciu**
Purdue University
W. Lafayette, Indiana, U.S.A

**Sonke Svenson**
Drug Delivery Solutions LLC
Arlington, Massachusetts, U.S.A.

Co-edited by

**Marc Behl**
Helmholtz-Zentrum Geesthacht
Teltow, Germany



MRS®

Materials Research Society
Warrendale, Pennsylvania

CAMBRIDGE
UNIVERSITY PRESS

**CAMBRIDGE**
**UNIVERSITY PRESS**

Shaftesbury Road, Cambridge CB2 8EA, United Kingdom

One Liberty Plaza, 20th Floor, New York, NY 10006, USA

477 Williamstown Road, Port Melbourne, VIC 3207, Australia

314–321, 3rd Floor, Plot 3, Splendor Forum, Jasola District Centre, New Delhi – 110025, India

103 Penang Road, #05–06/07, Visioncrest Commercial, Singapore 238467

Cambridge University Press is part of Cambridge University Press & Assessment, a department of the University of Cambridge.

We share the University's mission to contribute to society through the pursuit of education, learning and research at the highest international levels of excellence.

www.cambridge.org
Information on this title: www.cambridge.org/9781605115467

Materials Research Society
506 Keystone Drive, Warrendale, PA 15086
http://www.mrs.org



First published 2014

CODEN: MRSPDH

*A catalogue record for this publication is available from the British Library*

ISBN 978-1-605-11546-7 Hardback

# CONTENTS

## *INTERFACE OF (MULTIFUNCTIONAL) BIOMATERIALS*

## *RESPONSIVE MULTIFUNCTIONAL BIOMATERIALS*



*Invited Paper

*Invited Paper

# PREFACE

This MRS Proceedings volume pulls together several symposia on related topics in the area of biomaterials, all of which were held April 1–5 at the 2013 MRS Spring Meeting in San Francisco, California:

Symposium LL, "Hybrid Inorganic-Biological Materials,"

Symposium MM, "New Tools for Cancer Using Nanomaterials, Nanostructures, and Nanodevices,"

Symposium NN, "Multifunctional Biomaterials," and

Symposium QQ, "Conjugated Polymers in Sensing and Biomedical Applications."

Research in the field of biomaterials is progressing rapidly. Biomaterials that can mimic functions of natural tissue and can provide, receive, and respond to signals from their environment are the key topic of fundamental as well as translational research. These signals include interactions with synthetic molecules, biological species, and physical stimuli. Recently the concept of functions independent of the materials basis turned out as a promising pathway. Of particular interest are multifunctional materials. This symposium proceedings volume represents recent advances in materials sciences for biomedical applications, including inorganic, hybrid, and polymeric biomaterials. The papers are divided into five sections: (1) Multifunctional Biomaterials, (2) Interface of (Multifunctional) Biomaterials, (3) Responsive Multifunctional Biomaterials, (4) Particulate Biomaterials, and (5) Multifunctional Biomaterials in Sensors and Applications. Each paper of this volume reflects one piece of the puzzle of the exciting recent developments occurring in the broad field of biomaterials such as stimuli-sensitive polymer composite materials, smart or active materials, efficient drug delivery vehicles, and novel sensor applications. We hope that these papers convey the breadth of exciting advancements happening in the area of biomaterials.

Michelle Oyen
Andreas Lendlein
William T. Pennington
Lia Stanciu
Sonke Svenson
Marc Behl

November 2013

# Acknowledgments

The papers published in this volume result from three MRS Spring 2013 symposia—LL, MM, NN, and QQ. We sincerely thank all of the oral and poster presenters of the symposia who contributed to this proceedings volume. In addition we thank the reviewers of these manuscripts for their valuable feedback to the editors and to the authors. Furthermore, we greatly acknowledge the organizational help of the Materials Research Society and its staff, particularly the Publications staff for guiding us smoothly through the submission/review process. The organizers of Symposium NN thank the Biomedical Polymers Laboratory of the Soochow University, Suzhou, China, the Jiangsu Key Laboratory of Advanced Functional Polymer Design and Application (Soochow University), Suzhou, China, and Aldrich Materials Science for their generous financial support.

# Materials Research Society Symposium Proceedings

Volume 1536 – Film Silicon Science and Technology, 2013, P. Stradins, A. Boukai, F. Finger, T. Matsui, N. Wyrsch, ISBN 978-1-60511-513-9

Volume 1537E – Organic and Hybrid Photovoltaic Materials and Devices, 2013, S.W. Tsang, ISBN 978-1-60511-514-6

Volume 1538 – Compound Semiconductors: Thin-Film Photovoltaics, LEDs, and Smart Energy Controls, 2013, M. Al-Jassim, C. Heske, T. Li, M. Mastro, C. Nan, S. Niki, W. Shafarman, S. Siebentritt, Q. Wang, ISBN 978-1-60511-515-3

Volume 1539E – From Molecules to Materials—Pathways to Artificial Photosynthesis, 2013, J.-H. Guo, ISBN 978-1-60511-516-0

Volume 1540E – Materials and Integration Challenges for Energy Generation and Storage in Mobile Electronic Devices, 2013, M. Chhowalla, S. Mhaisalkar, A. Nathan, G. Amaratunga, ISBN 978-1-60511-517-7

Volume 1541E – Materials for Vehicular and Grid Energy Storage, 2013, J. Kim, ISBN 978-1-60511-518-4

Volume 1542E – Electrochemical Interfaces for Energy Storage and Conversion—Fundamental Insights from Experiments to Computations, 2013, J. Cabana, ISBN 978-1-60511-519-1

Volume 1543 – Nanoscale Thermoelectric Materials, Thermal and Electrical Transport, and Applications to Solid-State Cooling and Power Generation, 2013, S.P. Beckman, H. Böttner, Y. Chalopin, C. Dames, P.A. Greaney, P. Hopkins, B. Li, T. Mori, T. Nishimatsu, K. Pipe, R. Venkatasubramanian, ISBN 978-1-60511-520-7

Volume 1544E – *In-Situ* Characterization Methods in Energy Materials Research, 2014, J.D. Baniecki, P.C. McIntyre, G. Eres, A.A. Talin, A. Klein, ISBN 978-1-60511-521-4

Volume 1545E – Materials for Sustainable Development, 2013, R. Pellenq, ISBN 978-1-60511-522-1

Volume 1546E – Nanoparticle Manufacturing, Functionalization, Assembly and Integration, 2013, H. Fan, T. Hyeon, Z. Tang, Y. Yin, ISBN 978-1-60511-523-8

Volume 1547 – Solution Synthesis of Inorganic Functional Materials—Films, Nanoparticles and Nanocomposites, 2013, M. Jain, Q.X. Jia, T. Puig, H. Kozuka, ISBN 978-1-60511-524-5

Volume 1548E – Nanomaterials in the Subnanometer-Size Range, 2013, J.S. Martinez, ISBN 978-1-60511-525-2

Volume 1549 – Carbon Functional Nanomaterials, Graphene and Related 2D-Layered Systems, 2013, P.M. Ajayan, J.A. Garrido, K. Haenen, S. Kar, A. Kaul, C.J. Lee, J.A. Robinson, J.T. Robinson, I.D. Sharp, S. Talapatra, R. Tenne, M. Terrones, A.L. Elias, M. Paranjape, N. Kharche, ISBN 978-1-60511-526-9

Volume 1550E – Surfaces of Nanoscale Semiconductors, 2013, M.A. Filler, W.A. Tisdale, E.A. Weiss, R. Rurali, ISBN 978-1-60511-527-6

Volume 1551 – Nanostructured Semiconductors and Nanotechnology, 2013, I. Berbezier, J-N. Aqua, J. Floro, A. Kuznetsov, ISBN 978-1-60511-528-3

Volume 1552 – Nanostructured Metal Oxides for Advanced Applications, 2013, A. Vomiero, F. Rosei, X.W. Sun, J.R. Morante, ISBN 978-1-60511-529-0

Volume 1553E – Electrical Contacts to Nanomaterials and Nanodevices, 2013, F. Léonard, C. Lavoie, Y. Huang, K. Kavanagh, ISBN 978-1-60511-530-6

Volume 1554E – Measurements of Atomic Arrangements and Local Vibrations in Nanostructured Materials, 2013, A. Borisevich, ISBN 978-1-60511-531-3

Volume 1556E – Piezoelectric Nanogenerators and Piezotronics, 2013, X. Wang, C. Falconi, S-W. Kim, H.A. Sodano, ISBN 978-1-60511-533-7

Volume 1557E – Advances in Scanning Probe Microscopy for Imaging Functionality on the Nanoscale, 2013, S. Jesse, H.K. Wickramasinghe, F.J. Giessibl, R. Garcia, ISBN 978-1-60511-534-4

Volume 1558E – Nanotechnology and Sustainability, 2013, L. Vayssieres, S. Mathur, N.T.K. Thanh, Y. Tachibana, ISBN 978-1-60511-535-1

Volume 1559E – Advanced Interconnects for Micro- and Nanoelectronics—Materials, Processes and Reliability, 2013, E. Kondoh, M.R. Baklanov, J.D. Bielefeld, V. Jousseaume, S. Ogawa, ISBN 978-1-60511-536-8

Volume 1560E – Evolutions in Planarization—Equipment, Materials, Techniques and Applications, 2013, C. Borst, D. Canaperi, T. Doi, J. Sorooshian, ISBN 978-1-60511-537-5

Volume 1561E – Gate Stack Technology for End-of-Roadmap Devices in Logic, Power and Memory, 2013, S. Banerjee, ISBN 978-1-60511-538-2

# MATERIALS RESEARCH SOCIETY SYMPOSIUM PROCEEDINGS

Volume 1562E – Emerging Materials and Devices for Future Nonvolatile Memories, 2013, Y. Fujisaki, P. Dimitrakis, D. Chu, D. Worledge, ISBN 978-1-60511-539-9
Volume 1563E – Phase-Change Materials for Memory, Reconfigurable Electronics, and Cognitive Applications, 2013, R. Calarco, P. Fons, B.J. Kooi, M. Salinga, ISBN 978-1-60511-540-5
Volume 1564E – Single-Dopant Semiconductor Optoelectronics, 2014, M.E. Flatté, D.D. Awschalom, P.M. Koenraad, ISBN 978-1-60511-541-2
Volume 1565E – Materials for High-Performance Photonics II, 2013, T.M. Cooper, S.R. Flom, M. Bockstaller, C. Lopes, ISBN 978-1-60511-542-9
Volume 1566E – Resonant Optics in Metallic and Dielectric Structures—Fundamentals and Applications, 2013, L. Cao, N. Engheta, J. Munday, S. Zhang, ISBN 978-1-60511-543-6
Volume 1567E – Fundamental Processes in Organic Electronics, 2013, A.J. Moule, ISBN 978-1-60511-544-3
Volume 1568E – Charge and Spin Transport in Organic Semiconductor Materials, 2013, H. Sirringhaus, J. Takeya, A. Facchetti, M. Wohlgenannt, ISBN 978-1-60511-545-0
Volume 1569 – Advanced Materials for Biological and Biomedical Applications, 2013, M. Oyen, A. Lendlein, W.T. Pennington, L. Stanciu, S. Svenson, ISBN 978-1-60511-546-7
Volume 1570E – Adaptive Soft Matter through Molecular Networks, 2013, R. Ulijn, N. Gianneschi, R. Naik, J. van Esch, ISBN 978-1-60511-547-4
Volume 1571E – Lanthanide Nanomaterials for Imaging, Sensing and Optoelectronics, 2013, H. He, Z-N. Chen, N. Robertson, ISBN 978-1-60511-548-1
Volume 1572E – Bioelectronics—Materials, Interfaces and Applications, 2013, A. Noy, N. Ashkenasy, C.F. Blanford, A. Takshi, ISBN 978-1-60511-549-8
Volume 1574E – Plasma and Low-Energy Ion-Beam-Assisted Processing and Synthesis of Energy-Related Materials, 2013, G. Abrasonis, ISBN 978-1-60511-551-1
Volume 1575E – Materials Applications of Ionic Liquids, 2013, D. Jiang, ISBN 978-1-60511-552-8
Volume 1576E – Nuclear Radiation Detection Materials, 2014, A. Burger, M. Fiederle, L. Franks, D.L. Perry, ISBN 978-1-60511-553-5
Volume 1577E – Oxide Thin Films and Heterostructures for Advanced Information and Energy Technologies, 2013, G. Herranz, H-N. Lee, J. Kreisel, H. Ohta, ISBN 978-1-60511-554-2
Volume 1578E – Titanium Dioxide—Fundamentals and Applications, 2013, A. Selloni , ISBN 978-1-60511-555-9
Volume 1579E– Superconducting Materials—From Basic Science to Deployment, 2013, Q. Li, K. Sato, L. Cooley, B. Holzapfel, ISBN 978-1-60511-556-6
Volume 1580E – Size-Dependent and Coupled Properties of Materials, 2013, B.G. Clark, D. Kiener, G.M. Pharr, A.S. Schneider, ISBN 978-1-60511-557-3
Volume 1581E – Novel Functionality by Reversible Phase Transformation, 2013, R.D. James, S. Fähler, A. Planes, I. Takeuchi, ISBN 978-1-60511-558-0
Volume 1582E – Extreme Environments—A Route to Novel Materials, 2013, A. Goncharov, ISBN 978-1-60511-559-7
Volume 1583E – Materials Education—Toward a Lab-to-Classroom Initiative, 2013, E.M. Campo, C.C. Broadbridge, K. Hollar, C. Constantin, ISBN 978-1-60511-560-3

**Prior Materials Research Symposium Proceedings available by contacting Materials Research Society**

**Multifunctional Biomaterials**

**Mater. Res. Soc. Symp. Proc. Vol. 1569 © 2013 Materials Research Society**
**DOI: 10.1557/opl.2013.837**

# Tailoring of Mechanical Properties of Diisocyanate Crosslinked Gelatin-Based Hydrogels

Axel T. Neffe[1,2,3], Tim Gebauer[1,2,3], and Andreas Lendlein[1,2,3]
[1]Institute of Biomaterial Science and Berlin-Brandenburg Centre for Regenerative Therapies, Helmholtz-Zentrum Geesthacht, Kantstrasse 55, 14513Teltow, Germany.
[2]Department of Chemistry, University of Potsdam, Potsdam, Germany
[3]Helmholtz Virtual Institute "Multifunctional Biomaterials for Medicine", Teltow and Berlin, Germany

## ABSTRACT

Polymer network formation is an important tool for tailoring mechanical properties of polymeric materials. One option to synthesize a network is the addition of bivalent crosslinkers reacting with functional groups present in a polymer. In case of polymer network syntheses based on biopolymers, performing such a crosslinking reaction in water is sometimes necessary in view of the solubility of the biopolymer, such as gelatin, and can be beneficial to avoid potential contamination of the formed material with organic solvents in view of applications in biomedicine. In the case of applying diisocyanates for the crosslinking in water, it is necessary to show that the low molecular weight bifunctional crosslinker has fully reacted, while tailoring of the mechanical properties of the resulting hydrogels is possible despite the complex reaction mechanism. Here, the formation of gelatin-based hydrogel networks with the diisocyanates 2,4-toluene diisocyanate, 1,4-butane diisocyanate, and isophorone diisocyanate is presented. It is shown that extensive washing of materials is required to ensure full conversion of the diisocyanates. The use of different diisocyanates gives hydrogels covering a large range of Young's moduli (12-450 kPa). The elongations at break (up to 83%) as well as the maximum tensile strengths (up to 410 kPa) of the hydrogels described here are much higher than for lysine diisocyanate ethyl ester crosslinked gelatin reported before. Rheological investigations suggest that the network formation in some cases is due to physical interactions and entanglements rather than covalent crosslink formation.

## INTRODUCTION

Tailoring of polymer properties to the requirements of a specific application can e.g. be performed by adjusting comonomer types and ratios,[1] polymer architecture, molecular weight, and sometimes processing enhancing or reducing the formation of crystallites. Polymer network formation by introduction of covalent or physical netpoints has been shown to be particularly effective in controlling the mechanical properties of a material right after synthesis, but also during the degradation phase.[2] Biopolymer-based networks have attracted growing interest as being from regrowing resources and inherently display functionalities useful e.g. for application in biomedicine, such as degradability and adhesion sites for cells.[3] However, tailoring of their material properties is challenging in view of the high tendency for self-organization due to the high number of functional groups such as hydrogen bond donors (-OH, $-NH_2$), hydrogen bond acceptors (e.g. –OH, $-NH_2$, C=O), (hetero)aromatic rings etc. Such self-organization is often thermodynamically favored though kinetically hindered, leading to changes in material properties over time. A typical example are gelatin-based materials, in which triple helicalization of single chains is observed at temperatures <35 °C,[4] resulting in the formation of physical netpoints, which leads to gelation.[5] It was shown that such triple helicalization of chains can be avoided when performing a crosslinking of gelatin at temperatures above $T_{gel}$, by either covalent crosslinking with L-lysine diisocyanate ethyl ester (LDI) [6] or alternatively when introducing defined physical netpoints by functionalizing gelatin with amino acid derived aromatic groups.[7] In this way, the material properties of the formed hydrogels can be controlled by the amount of crosslinks introduced.

Here, it was investigated if in the crosslinking of gelatin with varying amounts of 2,4-toluene diisocyanate (TDI), 1,4-butane diisocyanate (BDI), or isophorone diisocyanate (IPDI) in water similar relationships between diisocyanate excess and mechanical properties as for LDI are observed, or if the different reactivities and solubilities of the diisocyanates lead in view of the several possible and competing reaction pathways to other effects. Furthermore, the goal was to broaden the accessible range of mechanical properties by varying the crosslinker type. In the following, the gelatin-based networks crosslinked with the different diisocyanates types at different ratios is compared to networks crosslinked with LDI.

## EXPERIMENTAL DETAILS

### Materials

Gelatin (type A from porcine skin, 200 bloom), poly(ethylene glycol)-*block*-poly(propylene glycol)-*block*-poly(ethylene glycol) (PEPE, $M_n$ = 14 600 g·mol$^{-1}$, average composition $PEG_{141}$-$PPG_{44}$-$PEG_{141}$), 2,4-toluene diisocyanate (TDI), 1,4-butane diisocyanate (BDI), and isophorone diisocyanate (IPDI) were purchased from Sigma-Aldrich Chemie GmbH (Steinheim, Germany) and used without further purification. L-lysine diisocyanate ethyl ester (LDI) was purchased from Shanghai Infine Chemical Co., Ltd. (Shanghai, China) and distilled under vacuum before usage.

### Hydrogel Synthesis

In a typical synthesis, 10 g gelatin was dissolved in 90 mL of double distilled water at 48 °C in a 250 mL round bottom flask under stirring. 1 g PEPE was added and the mixture vigorously stirred for one hour at 48 °C. An 3-, 5-, or 8-fold molar excess of BDI, TDI, LDI or IPDI compared to the number of amino groups in the gelatin ($n_{amino} = 3.167 \times 10^{-4}$ mol / $g_{gelatin}$) was added to the mixture stirring continued for 0.5, 1.5, 6, or 30 minutes (for BDI, TDI, LDI or IPDI, respectively, according to their prior determined reaction kinetics). The viscous reaction mixture was casted into polystyrene petri dish (94 mm diameter). The petri dish was capped and left standing at room temperature overnight. The formed hydrogel film was removed from the dish and washed with double distilled water for 3 days and subsequently dried at room temperature. The evaluation of unreacted isocyanate groups was performed on samples, in which the diisocyanates were reacted with glycine instead of gelatin, while keeping all other reaction conditions constant.

**Water Uptake (*H*)**

Dry hydrogel samples were immersed in water at room temperature for 24 h. The swollen samples were carefully blotted from each side with lab tissue paper to remove excess water. The water uptake was determined using the following equation:

$$H = \frac{W_s - W_d}{W_s} \cdot 100,$$

with $W_s$ and $W_d$ being the weights of the swollen and dry samples, respectively. For each composition, three replicates were used for the determination of *H*.

**Tensile Tests**

Mechanical tests were performed at 37 °C on a tensile tester (Zwick Z2.5, Zwick GmbH, Ulm, Germany) equipped with an water tank with temperature control. Film samples of standard dimensions (ISO 527-2/1BB), equilibrated in water at 37 °C, were tested with 0.02 N pre-force and 5 mm·min$^{-1}$ crosshead speed, until break. Young's moduli (E) of the different gelatin films were calculated from the linear region, generally around 0.5 – 5 % strain of the tensile stress–strain plot.

**Rheological Analysis**

Storage modulus *G'*, loss modulus *G''*, and viscosity $|\eta^*|$ of swollen hydrogel networks were determined with a Rheotec Haake MARS II rheometer (Thermo Fischer Scientific, Reichenthal, Germany) using samples of 20 mm diameter and a 20 mm plate-plate measurement geometry with Peltier element for temperature control. Measurements were performed in temperature-sweep mode in the linear viscoelastic region applying constant force and constant frequency. Heating rate was set to 2 K·min$^{-1}$ for all measurements. Evaporation of water and drying of samples during the temperature sweep was suppressed by covering the sensor plates with a Teflon solvent trap and using a solvent reservoir.

## RESULTS AND DISCUSSION

The formation of hydrogel networks could be observed with all crosslinker types applied and with all ratios added. In test reactions, it was investigated whether the diisocyanates would completely react, either with the nucleophilic -$NH_2$ and -OH groups present in gelatin or with water. The reaction with water would lead to decarboxylation of an isocyanate group and formation of a new amine, which can then engage in an addition reaction with isocyanates.[6] This reaction leads to the formation of oligomeric crosslinks between gelatin chains as well as oligoureas grafted on gelatin chains, in both cases contributing to the mechanical properties of the resulting hydrogels. However, initially unreacted isocyanate groups are undesired as later reaction might change material properties over time, and might result in toxic effects when applied in a biological environment. Figure 1 shows the IR spectra of IPDI reacted in different amounts with glycine. The picture given is representative also for the other crosslinkers. Directly after synthesis, large amounts of unreacted diisocyanates can be detected in the IR spectra through the absorption band at app. 2250 cm$^{-1}$. This band increased with increasing amount of diisocyanate added in the reaction, and despite using water as solvent, not all isocyanate groups had reacted. Extensive washing drastically reduces the band indicative for the isocyanate, and the reaction is completed after three days of washing. Therefore, hydrogels synthesized were washed for the same amount of time before mechanical testing.

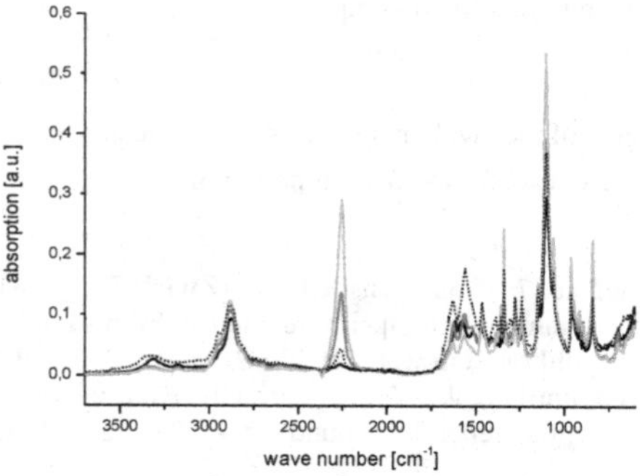


**Figure 1.** IR spectra of the reaction mixture of IPDI in 3- (black), 5- (dark grey) and 8-fold (light grey) excess with glycine directly after the reaction, as well as after extensive washing (8-fold excess, dotted). Immediately after crosslinking, a large number of unreacted isocyanate groups (band at app. 2250 $cm^{-1}$) are still present in the samples, which may change material properties through further crosslinking. The unreacted isocyanate groups can be drastically reduced by washing.

Results from the tensile tests and water uptake studies are summarized in Table 1. While for LDI, as described before, with an increase of diisocyanate excess an increase in Young's modulus $E$ and maximum tensile strength $\sigma_{max}$ as well as a decrease in elongation at break $\varepsilon_b$ and water uptake $H$ was observed, these relationships were not observed for all other diisocyanates. In the BDI samples, $E$ was comparable to the LDI samples, however, $\varepsilon_b$ as well as $\sigma_{max}$ were higher than for LDI, and H did not change much with the amount of BDI added in the reaction. For the IPDI crosslinked samples, values for $E$ were higher than macroscopic Young's moduli reported for tissues like skeletal muscle (100 kPa) or spinal cord (89 kPa)[8], but in the same range as hydrogels e.g. based on *N*-vinylpyrrolidone and methylmethacrylate used for cell encapsulation.[9] One has to keep in mind that the mechanical properties of the gels rely on the network density, to which covalent as well as physical netpoints contribute. The covalent netpoints are formed through links between gelatin chains by single crosslinker moieties or oligoureas, while physical netpoints can potentially be formed through gelatin chains engaging in triple helices or interaction of grafted oligoureas. $E$ and $\varepsilon_b$ relate differently to covalent and physical netpoints. While $E$ is measured for low elongations and therefore both physical and covalent netpoints fully contribute, $\varepsilon_b$ is generally larger for materials with a lower amount of covalent crosslinks. If during the tensile tests the physical netpoints break first, while the covalent netpoints remain, the larger $\varepsilon_b$ could be related to a relatively higher number of physical netpoints in BDI crosslinked films compared to LDI crosslinked films. The TDI and IPDI containing hydrogels displayed much higher $E$ values as well as $\sigma_{max}$ compared to LDI crosslinked films. A potential explanation is here an increased amount of the number of crosslinks or shorter connection between the chains. That $H$ is not changing strongly for one crosslinker type with increasing amount can be related to the point that after a certain number of netpoints is formed, further crosslinking does not decrease the swellability of the network anymore, or that the increasing absolute number of netpoints is counteracted by increasing distance between the netpoints (i.e. longer oligourea chains are formed).

**Table 1.** Summary of the mechanical properties of the hydrogels determined in tensile tests and water uptake at 37 °C, as well as rheological properties of the gels depending on the type and amount of diisocyanate added to the gelatin.

| Crosslinker | -NCO ex.[a] | $E$ [kPa][b] | $\varepsilon_b$ [%][b] | $\sigma_{max}$ [kPa][b] | $H$ [wt.%][b] | $G'$ [Pa][c] | $G''$ [Pa][c] | $\|\eta^*\|$ [Pa·s][c] |
|---|---|---|---|---|---|---|---|---|
| BDI | 3 | 12 ± 5 | 52 ±26 | 10 ± 0 | 752 ± 22 | 2263 | 362 | 405 |
| | 5 | 15 ± 5 | 82 ± 39 | 10 ± 0 | 672 ± 11 | 2756 | 114 | 4390 |
| | 8 | 51 ± 16 | 83 ± 9 | 40 ± 20 | 752 ± 10 | 3892 | 692 | 6291 |
| TDI | 3 | 93 ± 22 | 38 ± 11 | 40 ± 10 | 871 ± 17 | n.d. | n.d. | n.d. |
| | 5 | 100 ± 20 | 45 ± 9 | 50 ± 10 | 842 ± 50 | 3742 | 359 | 2992 |
| | 8 | 160 ± 62 | 34 ± 7 | 60 ± 20 | 891 ± 55 | 5549 | 415 | 16103 |
| IPDI | 3 | 289 ±34 | 51 ± 3 | 180 ± 20 | 383 ± 27 | 13824 | 1500 | 44260 |
| | 5 | 449 ± 45 | 71 ± 4 | 410 ± 50 | 312 ± 4 | 13673 | 941 | 14541 |
| | 8 | 370 ± 54 | 39 ± 4 | 160 ± 20 | 328 ± 23 | 19644 | 2247 | 6294 |
| LDI | 3 | 13 ± 3 | 43 ± 25 | 7 ± 3 | 942 ± 113 | 4309 | 1416 | 7219 |
| | 5 | 35 ± 9 | 31 ± 19 | 13 ± 8 | 725 ± 20 | 5155 | 431 | 8233 |
| | 8 | 55 ± 11 | 30 ± 12 | 19 ± 11 | 438 ± 36 | 1503 | 135 | 2402 |

[a] molar excess of the amount of diisocyanates compared to the amount of amino groups in gelatin determined in a TNBS assay. [b] determined at 37 °C. [c] determined in the linear region at 25 °C. No errors are given, as the rheological data were determined in single experiments. Please compare to Figure 3; for TDI samples, values drastically decreased upon heating to 37 °C. n.d.: not determinable.

In Figure 2, stress-strain curves of the samples G10_INCO5 and G10_BNCO5 are displayed, which are representative for all samples investigated. Notably, a nearly linear behavior of stress and strain was observed, though commonly for rubber-like polymer networks sigmoidal behavior is recorded.[10] The results suggest that the samples broke before the region of plastic deformation was reached.

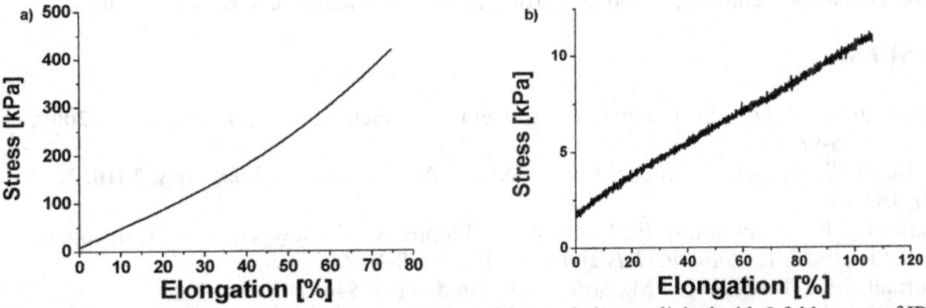


**Figure 2.** Curves from the tensile test of hydrogels formed from gelatin crosslinked with 5-fold excess of IPDI (a) or BDI (b), which are representative for the whole group of materials discussed in this paper.

The rheological behavior of the samples is described in Table 1 and Figure 3. Notably, *G'* increased in the linear region with diisocyanate excess for IPDI, BDI, and TDI crosslinked samples, but not for the LDI crosslinked samples. Interesting is the change of storage modulus *G'* with the temperature (Figure 3). Here, it was observed that only the hydrogels formed through reaction with IPDI or BDI did show constant mechanical properties over a wide range of temperatures, while TDI crosslinked gels showed a gel-sol transition at ca. 35 °C, which can be explained in a system with physical netpoints being cleaved at increased temperatures.

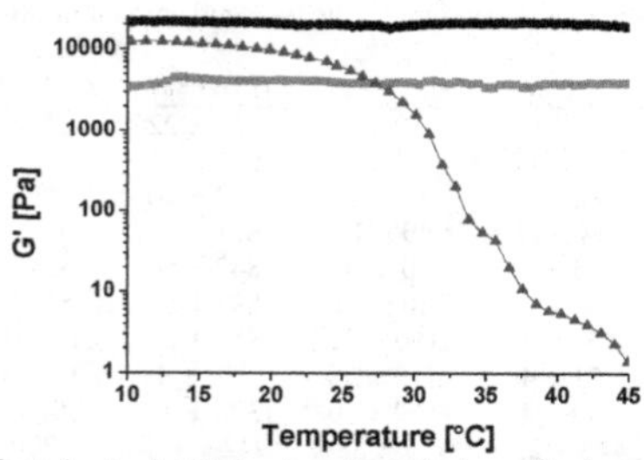


**Figure 3.** Temperature dependent rheological behavior of gelatin-based hydrogels crosslinked with 8-fold excess of IPDI (black), BDI (grey), or TDI (dark grey) as shown for *G'*.

## CONCLUSIONS

Polymer hydrogel networks based on gelatin and a small diisocyanate acting as crosslinker were successfully synthesized in water. Despite the high reactivity of the isocyanate group towards water, extensive washing and prolonged contact to the solvent is required for completing the reaction. By choosing different diisocyanate types and ratios between amino groups of gelatin and the diisocyanate, it is possible to adjust the mechanical properties, e.g. Young's moduli, $\varepsilon_b$, and $\sigma_{max}$, of the hydrogels over a wide range. The complex reaction mechanism of diisocyanates with gelatin in water, potentially leading to direct crosslinks and grafting reactions as well as the putative formation of gelatin triple helices gives rise to materials with different network architectures and properties, as shown in tensile and rheological tests.

## REFERENCES


1. Lendlein, A.; Zotzmann, J.; Feng, Y.; Alteheld, A.; Kelch, S., *Biomacromolecules* **2009,** *10* (4), 975-982.
2. Neffe, A. T.; Tronci, G.; Alteheld, A.; Lendlein, A., *Macromol. Chem. Phys.* **2010,** *211* (2), 182-194.
3. Hennessy, K. M.; Pollot, B. E.; Clem, W. C.; Phipps, M. C.; Sawyer, A. A.; Culpepper, B. K.; Bellis, S. L., *Biomaterials* **2009,** *30* (10), 1898-1909.
4. Gornall, J. L.; Terentjev, E. M., *Soft Matter* **2008,** *4* (3), 544-549.
5. Bigi, A.; Panzavolta, S.; Rubini, K., *Biomaterials* **2004,** *25* (25), 5675-5680.
6. Tronci, G.; Neffe, A. T.; Pierce, B. F.; Lendlein, A., *J. Mater. Chem.* **2010,** *20*, 8875.
7. Zaupa, A.; Neffe, A. T.; Pierce, B. F.; Nochel, U.; Lendlein, A., *Biomacromolecules* **2011,** *12* (1), 75-81.
8. Levental, I.; Georges, P. C.; Janmey, P. A., *Soft Matter*, **2007**, *3*, 299-306.
9. Jen, A.C.; Wake, M.C.; Mikos, A.G., *Biotech. Bioeng.* **1996**, *50*, 357-364.
10. Meissner, B., *Polymer* **2000,** *41* (21), 7827-7841.

Mater. Res. Soc. Symp. Proc. Vol. 1569 © 2013 Materials Research Society
DOI: 10.1557/opl.2013.831

# Immunological investigations of oligoethylene glycols functionalized with desaminotyrosine and desaminotyrosyltyrosine

Konstanze K. Julich-Gruner[1,#], Toralf Roch[1,#], Nan Ma[1], Axel T. Neffe[1], Andreas Lendlein[1]

[1]*Institute of Biomaterial Science and Berlin-Brandenburg Centre for Regenerative Therapies, Helmholtz-Zentrum Geesthacht, Kantstrasse 55, 14513 Teltow, Germany*
[#] Equally contributing authors

## ABSTRACT

Biomaterials require thorough *in vitro* testing before being applied *in vivo*. Unwanted contaminations of biomaterials but also their intrinsic properties can cause uncontrolled immune response leading to severe consequences for the patient. Therefore, immunological evaluation of materials for biomedical applications is mandatory before entering clinical application. In order to introduce physical netpoints, the aromatic compounds desaminotyrosine (DAT) and desaminotyrosyl-tyrosine (DATT) were successfully used to functionalize linear and star-shaped oligoethylene glycol (lOEG and sOEG) as previously described. The materials showed properties of surfactants and have potential to be used for solubilization of lipophilic drugs in water. Furthermore, the materials are susceptible for $H_2O_2$ degradation as determined by MALDI-ToF MS analyses. This is important for potential *in vivo* applications, as macrophages can release reactive oxygen species (ROS) under inflammatory conditions. As it is known that surfactant solutions of high concentration can lead to cell lysis, the effects of OEG-DAT(T) solutions on murine RAW macrophages were investigated. Even at highest OEG-DAT(T) concentration of 1000 $\mu g{\cdot}mL^{-1}$ the viability of the RAW cells was not significantly impaired. Additionally, the polymers were incubated with whole human blood and the production of inflammatory cytokines such as the tumor necrosis factor (TNF)-α and interleukin (IL)-6 was determined. Only at high concentrations, the OEG-DAT(T) solution induced low levels of TNF-α and IL-6, indicating that a mild inflammatory reaction could be expected when such high OEG-DAT(T) concentrations are applied *in vivo*. Similarly, the OEG-DAT(T) solution did not induce ROS in monocytes and neutrophils after incubation with whole human blood. Conclusively, the data presented here demonstrate that OEG-DAT(T) do not lead to a substantial activation of the innate immune mechanisms and could therefore be investigated for solubilizing pharmaceutical agents.

## INTRODUCTION

When designing a material for biomedical application such as drug delivery, the functionality of the biomaterial needs to be established and fully characterized. Moreover, the possible biological effects induced by the materials need to be studied carefully before application *in vivo*. Inappropriate materials preparation could result in microbial contaminations such as endotoxins (lipopolysaccharides (LPS)), which are components of the cell wall of gram-negative bacteria and can lead to the unspecific immune activation inducing fever or in the worst case multi organ failure [1]. Furthermore, other microbial products from fungi, gram-positive bacteria or viral pathogens can as well induce similar immune response. These contaminations can be already introduced by the starting materials or during material syntheses [2]. Immune cells such as macrophages express pattern-recognition receptors (PRR), which qualifies them as tool to investigate material bound microbial contamination, but also material intrinsic properties that may induce immune activation.

Previously, the aromatic compounds desaminotyrosine (DAT) and desaminotyrosyltyrosine (DATT) have been successfully used to functionalize gelatin in order to form physically crosslinked networks [3]. However, in gelatin, triple helix formation can occur and may induce additional physical netpoints influencing the overall mechanical properties of the hydrogel. Therefore, linear and star-shaped oligoethylene glycols (lOEG and sOEG) were chosen for DAT(T) functionalization, because their backbones allow no additional interactions (Figure 1) [4]. This previous study showed that the functionalized OEGs did not form gels, as proven by rheological measurements. In general, the compounds in aqueous solution had a higher loss modulus G'' compared to storage modulus G' indicating properties of a liquid than a gel. Due to the structure of the macromolecule with a large hydrophilic core and two hydrophobic end groups, the polymers showed properties of a surfactant and have potential to be used to solubilize lipophilic drugs in water. As high concentrations of surfactants can induce cell lysis we hypothesized that a similar effect may occur for the OEG-DAT(T). Here, the cellular response of macrophages and the immunological evaluation of these materials are shown. When tested with whole human blood the production of inflammatory cytokines such as IL-6 and TNF-α was investigated. Additionally, the potential of the materials to induce ROS in distinct immune cells was evaluated, because ROS, which are generated during inflammatory reactions, could lead to a degradation of the OEGs.

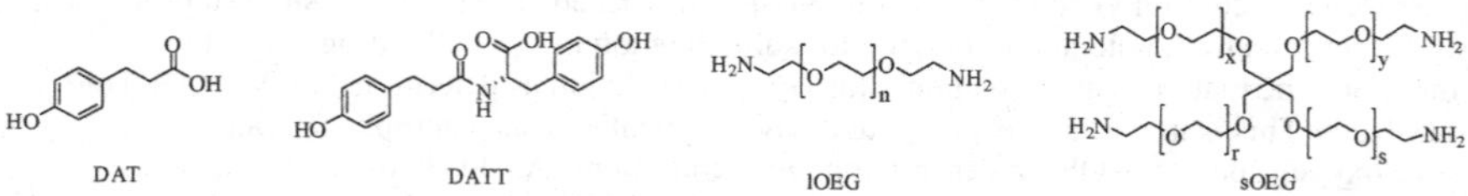


**Figure 1** Chemical structures of desaminotyrosine (DAT) and desaminotyrosyltyrosine (DATT) used in this work for end group functionalization of linear (lOEG) and star-shaped oligoethylene glycols (sOEG).

## EXPERIMENTAL DETAILS

### Syntheses of DAT(T)-functionalized oligoethylene glycols (OEG)

lOEG, (3 kDa) and sOEG (5 kDa) with amino groups as endgroups were functionalized with desaminotyrosine (DAT) or desaminotyrosyl-tyrosine (DATT) by activation of the acid using 1-ethyl-3-(3-dimethylaminopropyl) carbodiimide (EDC) as previously described [4]. The compounds were characterized by $^1$H-NMR (DRX 500 Avance spectrometer, Bruker, Rheinstetten, Germany; software Topspin version 1.3), IR (8400S FTIR spectrometer, Shimadzu, Duisburg, Germany), and MALDI-ToF mass spectrometry (ultrafleXtreme MALDI-TOF, Bruker, Bremen, Germany and dihydroxybenzoic acid as matrix). For the degradation studies, solutions of sOEG-DAT(T) (5 kDa) were treated with $H_2O_2$. 100 µL polymer solution (10 wt%) were mixed with 50 µL PBS buffer and 50 µL $H_2O_2$ (1 wt%). After 5 min the reaction was quenched with $FeSO_4$. The solutions were centrifuged and the supernatants were analyzed by MALDI-ToF.

### Immunological Assays

The RAW-Blue cell culture and activation were performed as previously described [5]. In Brief, $5x10^5$ RAW-Blue™ cells (InvivoGen, San Diego, USA) were cultured for 24 hours with increasing concentrations of the OEG-based materials. After incubation, cell culture supernatants were harvested and directly incubated for 3 hours with the QuantiBlue™ solution.
To determine the percentage of dead RAW cells, 1 µg·mL$^{-1}$ 4',6-diamidino-2-phenylindole

(DAPI) was added immediately prior analysis, which was performed with the MACSQuant flow cytometer (Milteniy Biotec). For all samples, aggregated cells were excluded via FCS-A/FCS-H scatter parameter as previously described.[6] Data were analyzed using FlowJo software (Tree Star, Ashland, USA).

The study was performed according to the criteria of the Nordkem-workshop [7, 8]. Blood was obtained from apparently healthy volunteers as previously described [5]. The study protocol received an approval by the institutional committee of the Charité Berlin.

The cytokine secretion by cells of the human blood was quantified in cell culture supernatants using Bio-Plex suspension multiplex array systems (Bio-Plex 200®, BioRad, Germany) according to the manufacturer's instructions. IL-6 and TNF-α microbead assays were purchased from BioRad. ROS were determined in whole humane blood using the Phagoburst® assay kit (ORPEGEN Pharma, Heidelberg, Germany) as previously described [5].

The evaluation of oxidative burst activity was performed by flow cytometry using the MACSQuant® analyzer. Flow cytometric data were analysed using FlowJo software (Tree Star, Ashland, USA). For statistical analysis GraphPad Prism (La Jolla, CA 92037, USA) was used.

## RESULTS AND DISCUSSION

The different OEGs were successfully functionalized with DAT or DATT by coupling of the carbodiimide activated acids to the amino endgroups of OEGs. The compounds were characterized by $^1$H-NMR, IR, and MALDI-ToF mass spectrometry. The degree of functionalization was determined from the $^1$H-NMR spectra by relating the integrals from the aromatic protons of DAT (δ = 6.6 pm) or the α-CH group of DATT (δ = 4.4 pm) to the signal of the $CH_2$ group of the repeating unit of OEG (δ = 3.5 ppm). The results are summarized in Table 1. The lOEG-DATT and the sOEG-DATT had a very high degree of functionalization (d.f.), whereas in the case of DAT in average only 2 or arms of 4 were functionalized.

**Table 1** Degree of functionalization (d.f.) and number average molecular weight ($M_n$) for the functionalized OEGs.

| | *d.f.* [%]* | *$M_n$** [Da]* |
|---|---|---|
| lOEG-DATT | 92 | 3618 (3110)# |
| sOEG-DAT | 61 | 5882 (5340)# |
| sOEG-DATT | 87 | 6430 (5340)# |

* calculated from $^1$H-NMR
** determined by MALDI-ToF
# $M_n$ of unfunctionalized OEG

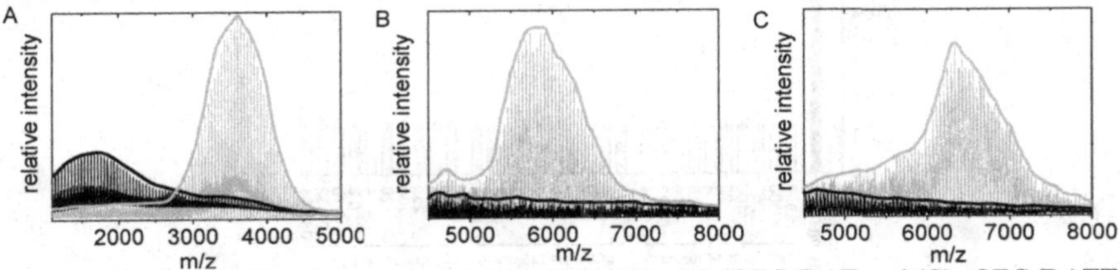


**Figure 2** MALDI-ToF mass spectra of (A) lOEG-DATT, (B) sOEG-DAT and (C) sOEG-DATT before treatment with $H_2O_2$ (grey line) and after the reaction (black line). The lines are added as guide to the eye.

As previously described, OEG can be degraded by hydrogen peroxide [9]. Here, it was investigated whether lower $H_2O_2$ concentration (0.25 wt%) can also mediate degradation. Even after a reaction time of five minutes the mass of functionalized OEGs was strongly reduced

(Figure 2). For the sOEG the degradation was even faster and observed degraded masses are lower compared to linear OEG. One reason could be that the different architectures of the sOEG allows the preferential cleavage at the branching points.

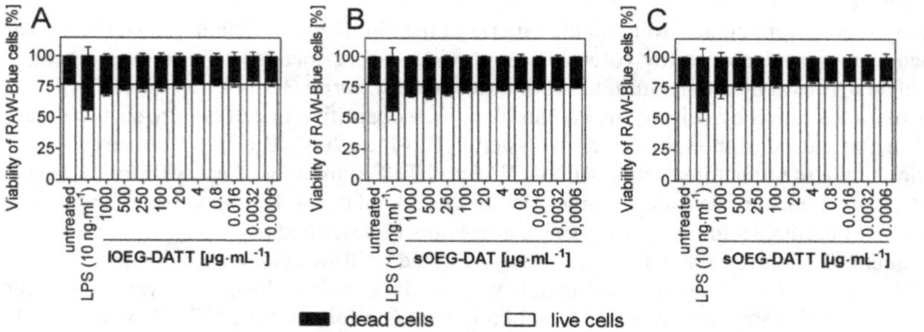


**Figure 3** Survival of RAW cells. RAW cells were treated for 24 hours with (A) lOEG-DATT, (B) sOEG-DAT and (C) sOEG-DATT. The average of two independent experiments are shown (mean ± range).

Aqueous solutions of the functionalized OEGs have a lower surface tension than water and lOEG-DATT has a critical micelle concentration of 0.1 mg·mL$^{-1}$. Thus, the OEGs showed the properties of surfactants, indicating applicability for solubilizing lipophilic drugs. To investigate the potential of the materials for application *in vivo*, a first step is to check whether these materials are cell compatible and do not induce unwanted immune responses. The incubation of RAW macrophages with OEG solutions revealed that the viability was not substantially influenced by the different OEG (Figure 3). Even at very high concentrations (1 mg·mL$^{-1}$) most of the cells were viable, whereas RAW cells treated with 10 ng·mL$^{-1}$ LPS showed substantial apoptosis, which is a physiological process to avoid uncontrolled immune reactions.[10]

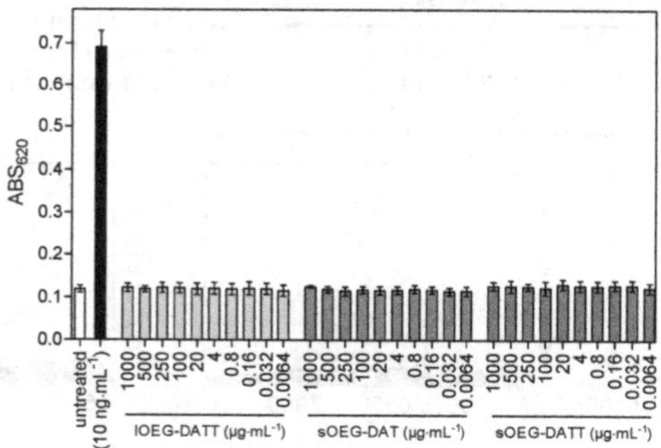


**Figure 4** RAW-Blue$^{TM}$ cell-based assay for detection of microbial contamination. The average of two independent experiments analyzed in triplicates are shown (mean ± standard deviation).

The RAW-Blue$^{TM}$ cell line is genetically engineered to secrete an alkaline phosphatase when getting in contact with microbial products, such as LPS, which induced a strong activation of the

RAW-Blue™ cells. However, none of the tested materials showed macrophage activation, indicating that the OEG were neither contaminated with microbial products nor have intrinsic immuno-stimulatory effects (Figure 4).
In the next step, it was investigated whether lOEG-DATT or sOEG-DAT(T) induce the secretion of pro-inflammatory cytokines in human whole blood (Figure 5). It was observed that only high concentration of OEG led to the production of low amounts of IL-6 and TNF-α. The cytokine secretion seems to be higher in sOEG-DATT treated blood compared to lOEG-DATT, which might be explained, by the fact that more arms are functionalized in the star-shaped OEG than in the linear OEG.

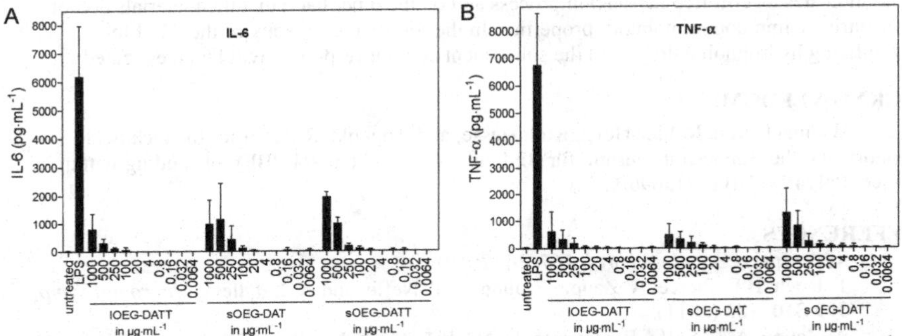


**Figure 5** Cytokine secretion in human whole blood, (A) of IL-6 and (B) of TNF-α. LPS was used at 10 ng·ml$^{-1}$. The average of four individual donors are shown (mean ± standard deviation).

The biomaterial-induced ROS production could lead to the degradation of the OEG, but will also enhance inflammation and tissue damage [11]. In whole human blood, ROS can be produced by neutrophils and monocytes in response to bacteria such as *Escherichia coli* (Figure 6). When human blood was incubated with the different OEG neither neutrophils nor monocytes generated ROS, indicating that no severe systemic consequence are expected when these OEG would be applied *in vivo*.

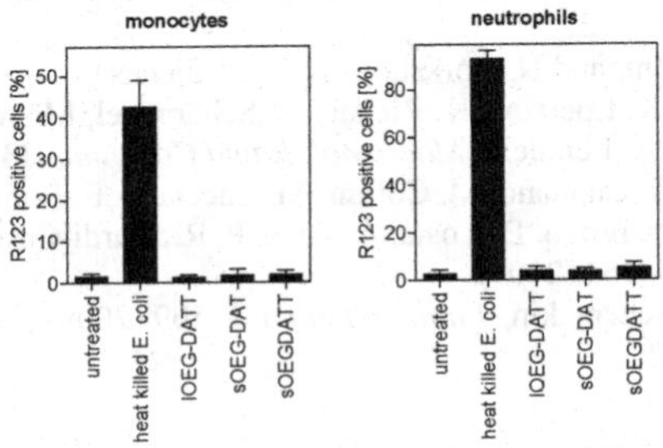


**Figure 6** ROS production of whole blood monocytes and neutrophils after incubation with 1 mg·mL$^{-1}$ DAT(T) functionalized OEG. The average of four individual donors are shown (mean ± range).

## CONCLUSIONS

Linear and star-shaped OEGs were functionalized with DAT or DATT. Since these materials showed the properties of surfactants, their potential application could be the solubilization of lipophilic drugs in water. Therefore, these materials have to be evaluated biologically *in vitro*. Direct cell studies were performed. RAW macrophage stayed viable after incubation with the OEGs and did not show activation. At a concentration of 0.1 mg·mL$^{-1}$ correlating to the CMC of the material, no substantial inflammatory cytokine response and ROS generation could be detected in human blood.

Therefore, all investigated OEG can be considered as immuno-compatible, which on the one hand indicates a germ free fabrication process and on the other hand that the materials do not bear intrinsic immuno-stimulatory properties. In the future, the capacity of the OEG for solubilizing hydrophobic drugs and the subsequent cellular responses will be investigated.

## ACKNOWLEDGMENTS

We thank Anja Müller-Heyn, Ruth Hesse, and Angelika Ritschel for the technical support, and the Bundesministerium für Bildung und Forschung (BMBF) for funding within project Poly4Bio BB (0315696A).

## REFERENCES

1. S.M. Opal, *Contrib Nephrol*, **167**, 14 (2010).
2. T. Roch, B.F. Pierce, A. Zaupa, F. Jung, A.T. Neffe, and A. Lendlein, *Macromol Symp*, **309-310**, 182 (2011).
3. A. Zaupa, A.T. Neffe, B.F. Pierce, U. Nöchel, and A. Lendlein, *Biomacromolecules*, **12**, 75 (2011).
4. K.K. Julich-Gruner, A.T. Neffe, and A. Lendlein, *J. Appl. Biomater. Function. Mater.*, **10**, 170 (2012).
5. T. Roch, J. Cui, K. Kratz, A. Lendlein, and F. Jung, *Clin Hemorheol Microcirc*, **50**, 131 (2012).
6. C. Wischke, S. Mathew, T. Roch, M. Frentsch, and A. Lendlein, *J Control Release*, **164**, 299 (2012).
7. B. Berg, H. Solberg, J. Nilsson, and N. Tryding, *John Wiley & Sons Chichester (UK)*, 55 (1981).
8. R. Gräsbeck, T. Alström, and N. workshop, Wiley, Chichester.
9. M. Lange, S. Braune, K. Luetzow, K. Richau, N. Scharnagel, M. Weinhart, A.T. Neffe, F. Jung, R. Haag, and A. Lendlein, *Macromol. Rapid Commun.*, **33**, 1487 (2012).
10. I. Zanoni, R. Ostuni, G. Capuano, M. Collini, M. Caccia, A.E. Ronchi, M. Rocchetti, F. Mingozzi, M. Foti, G. Chirico, B. Costa, A. Zaza, P. Ricciardi-Castagnoli, and F. Granucci, *Nature*, **460**, 264 (2009).
11. M.T. Quinn and I.A. Schepetkin, *J Innate Immun*, **1**, 509 (2009).

**Mater. Res. Soc. Symp. Proc. Vol. 1569 © 2013 Materials Research Society**
**DOI: 10.1557/opl.2013.839**

# Influence of diisocyanate reactivity and water solubility on the formation and the mechanical properties of gelatin-based networks in water

Tim Gebauer[1, 2, 3], Axel T. Neffe[1, 2, 3], and Andreas Lendlein[1, 2, 3]
[1]Institute of Biomaterial Science and Berlin-Brandenburg Centre for Regenerative Therapies, Helmholtz-Zentrum Geesthacht, Kantstraße 55, 14513 Teltow, Germany
[2]Institute of Chemistry, University of Potsdam, 14476 Potsdam-Golm, Germany
[3]Helmholtz Virtual Institute – Multifunctional Biomaterials for Medicine, Teltow and Berlin, Germany

## ABSTRACT

Gelatin can be covalently crosslinked in aqueous solution by application of diisocyanates like L-lysine diisocyanate ethyl ester in order to form hydrogels. Reaction of isocyanate groups with water is however a limiting factor in hydrogel network formation and can strongly influence the outcome of the crosslinking process. Here, diisocyanates with different water solubility and reactivity were applied for the formation of gelatin-based hydrogel networks and the mechanical properties of the hydrogels were investigated to gain a better understanding of starting material/ hydrogel property relations. L-Lysin diisocyanate ethyl ester (LDI), 2,4-toluene diisocyanate (TDI), 1,4-butane diisocyanate (BDI), and isophorone diisocyanate (IPDI) were selected, having different solubility in water ranging from $10^{-4}$ to $10^{-2}$ mol·L$^{-1}$. BDI and LDI were estimated to have average reactive isocyanates groups, whereas TDI is highly reactive and IPDI has low reactivity. Formed hydrogels showed different morphologies and were partially very inhomogeneous. Gelation time (1 to 50 minutes), water uptake (300 to 900 wt.-%), and mechanical properties determined by tensile tests (E-moduli 35 to 370 kPa) and rheology (Shear moduli 4.5 to 19.5 kPa) showed that high water solubility as well as high reactivity leads to the formation of poorly crosslinked or inhomogeneous materials. Nevertheless, diisocyanates with lower solubility in water and low reactivity are able to form stable, homogeneous hydrogel networks with gelatin in water.

## INTRODUCTION

Mechanical properties of hydrogels are generally influenced by the degree of crosslinking of the polymer network [1]. In biomaterials research, crosslinking of biopolymers with isocyanates is a common approach to form covalently crosslinked multifunctional hydrogel systems with tailorable mechanical properties [2, 3, 4, 5]. Unfortunately these crosslinking reactions are typically carried out in organic solvents, which can remain entrapped inside the formed polymer networks and thus increase their potential toxicity.

In order to exclude toxic solvents gelatin-based hydrogels were established, which were covalently crosslinked with diisocyanates in water [6, 7]. These reactions provide hydrogels with tailorable mechanical properties and degradation behavior. Crosslinking reactions were performed using L-lysine diisocyanate ethyl ester (LDI) in water with the application of a surfactant that confers improved solubility for the rather hydrophobic diisocyanate.

Although isocyanates preferably react with primary amines to urea derivatives, reaction with water, resulting in the formation of free amines and carbon dioxide and thus allowing further side reactions, must be considered because of the high $H_2O$ concentration [8, 9]. Investigation of the crosslinking mechanism of LDI in water clearly showed the formation of mono- and oligomeric crosslinks as well as grafted species [6]. Additional reactions which may occur include formation of biuret adducts, oligomerization of isocyanates, or amide bond formation [9]. The application of highly reactive species like isocyanates under aqueous conditions is rather uncommon. The crosslinking reaction of gelatin in water with differently reactive and water soluble diisocyanates should lead to a more fundamental understanding, how the diisocyanate crosslinkers influence the mechanical properties and which kind is preferable for network formation in water.

In this work the influence of diisocyanates with different water solubility and inherent reactivity on the mechanical properties of gelatin-based hydrogels were investigated. The choice of crosslinking reagent should give a general overview how changes in reaction behavior influence network formation. In general diisocyanates react readily with nucleophiles like amines or hydroxyl groups. Activation of the isocyanate group, and thus higher reactivity, can occur through electron withdrawing groups at the nitrogen atom. The opposite effect is achieved by electron donating groups, whereas the isocyanate group is deactivated and has lower reactivity. Therefore, L-Lysin diisocyanate ethyl ester (LDI), 2,4-toluene diisocyanate (TDI), 1,4-butane diisocyanate (BDI), and isophorone diisocyanate (IPDI) were chosen, ranging in water solubility from ~ $10^{-4}$ mol·$L^{-1}$ to $10^{-2}$ mol·$L^{-1}$ and having differently activated isocyanate groups .Here, TDI is highly reactive due to activation of the isocyanates by the possibility to delocate electrons into the aromatic ring. In contrast, the aliphatic ring in IPDI with its methyl side groups has a +I-effect that deactivates the isocyanates [10]. LDI and BDI do not have significant activation or deactivation of their isocyanate groups, but different solubilities (Table I).

In the following, gelatin-based hydrogel networks crosslinked with four different diisocyanates were prepared. The gelation times, the structure of the different films, and their mechanical properties are compared, to give an insight into potential selection criteria for diisocyanate crosslinkers in hydrogel formation.

## EXPERIMENTAL DETAILS

### Materials

Gelatin (from porcine skin, type A, 200 bloom), poly(ethylene glycol)-*block*-poly(propylene glycol)-*block*-poly(ethylene glycol) (Pluronic® F-108, short PEPE, average $M_n$ ~ 14,600 g/mol), 2,4-toluene diisocyanate (TDI), 1,4-butane diisocyanate (BDI), and isophorone diisocyanate (IPDI) were purchased from Sigma-Aldrich Chemie GmbH (Steinheim, Germany) and used without further purification. L-lysine diisocyanate ethyl ester (LDI) was purchased from Shanghai Infine Chemical Co., Ltd. (Shanghai, China) and distilled under vacuum before usage.

### Determination of Average Gelation Time & Synthesis of Hydrogels

10 g gelatin were dissolved in 90 mL of double distilled water at 48 °C in a 250 mL round bottom flask under constant magnetic stirring using a conical magnetic stirr bar of 35 mm length. 1 g PEPE was added to each gelatin solution and vigorously stirred for one hour at 48 °C. An 8-fold excess of diisocyanate to the number of amino groups in the gelatin ($n_{amino} = 3.167 \cdot 10^{-4}$ mol per $g_{gelatin}$) was added to the mixture at 48 °C under constant stirring at 500 rpm and the time was taken until the magnetic stir bar was stopped due to gelation. The experiment was repeated 3 times for each diisocyanate.

For the synthesis of hydrogel films, the reaction procedure was repeated. After addition of the diisocyanate the mixture was then stirred for 75% of previously determined gelation time. The viscous reaction mixtures were casted into polystyrene petri dishes (94 mm diameter). The petri dishes were capped and left standing at room temperature overnight. The formed hydrogel films were removed from the dishes and washed in double distilled water for 3 days, exchanging water daily. Washed films were dried in the air at room temperature.

**Water Uptake (*H*), and degree of swelling (*Q*)**

Dry hydrogel samples (6 mm diameter discs) were weighed and then immersed in water at room temperature for 24 h. The swollen samples were carefully blotted from each side with lab tissue paper to remove excess water. The mass change was used to determine $H$ using the following equation, $H = (W_s\text{-}W_d)/W_s \times 100$, whereby $W_s$ and $W_d$ are the weights of the swollen and dry samples, respectively. Degree of swelling $Q$ was calculated from $H$ using equation (2). $\rho_p$ and $\rho_s$ are the densities of the dry hydrogel and the solvent (here $H_2O$). For each composition, three replicas were used for the determination of $H$ and $Q$.

$$H = \frac{(W_s - W_d)}{W_s} \cdot 100 \quad \text{(Eq. 1)} \qquad Q = \left(1 + \frac{\rho_P}{\rho_s}\left(\frac{W_s}{W_d} - 1\right)\right) \cdot 100 \quad \text{(Eq. 2)}$$

**Tensile Tests**

Mechanical tests were performed at 37 °C on a tensile tester (Zwick Z2.5, Zwick GmbH, Ulm, Germany) equipped with an water tank with temperature control. Film samples of standard dimensions (ISO 527-2/1BB), equilibrated in water at 37 °C, were tested with 0.02 N pre-force and 5 mm·min$^{-1}$ crosshead speed, until break. E-moduli of the different gelatin films were calculated from the linear region, generally between 0.5 – 5% elongation ($\varepsilon$) in the stress–strain plot.

**Rheological Analysis**

Storage modulus $G'$ and loss modulus $G''$ of swollen hydrogel networks were investigated on a Rheotec Haake MARS II rheometer (Thermo Fischer Scientific, Reichenthal, Germany) using samples of 20 mm diameter and a 20 mm plate-plate measurement geometry with Peltier element for temperature control. Measurements were performed in temperature-sweep mode in the linear viscoelastic region applying constant force and constant frequency. Heating rate was set to 2 K·min$^{-1}$ for all measurements. Evaporation of water and drying of samples during the temperature sweep was suppressed by covering the sensor plates with a Teflon solvent trap and using a solvent reservoir.

## RESULTS AND DISCUSSION

In order to determine initial reaction conditions for the formation of diisocyanate crosslinked gelatin-based hydrogel networks and to get an overview how the reaction kinetics for the different crosslinkers are, network formation with the different crosslinkers keeping all other parameters constant were performed and the time until gelation of the reaction mixture was determined (Table I). The gelatin concentration was set to 10 wt.-%, as in this concentration reaction with LDI and hexamethylene diisocyanate gave stable gels. Evaluation of the results showed that reactivity and water solubility both have a strong influence on the speed of the network formation. In comparison BDI has a lower expected reactivity than TDI, but has an order of magnitude higher solubility in water. Both crosslinkers react fast and lead to gelation within a short time frame. On the other hand, LDI has solubility comparable to TDI, but is not as reactive, resulting in 4 to 5 times longer gelation time. Finally IPDI has the lowest solubility and the lowest reactivity of the four diisocyanates and indeed reacts 20 to 25 times slower than BDI or TDI and still 4 to 5 times slower than LDI. Reaction with the surfactant, which is potentially possible, was not observed in test reactions (data not shown).

**Table I:** Solubility values, reactivity, and gelation time

| Diisocyanate | Solubility (at 25 °C in $H_2O$) | Relative Reactivity[a] | Gelation time |
|---|---|---|---|
| LDI | $8.6 \cdot 10^{-4}$ mol·L$^{-1}$ | average | 6 – 15 min |
| BDI | $9.2 \cdot 10^{-3}$ mol·L$^{-1}$ | average | 1 – 2 min |
| TDI | $1.2 \cdot 10^{-3}$ mol·L$^{-1}$ | high | 2 – 3 min |
| IPDI | $1.1 \cdot 10^{-4}$ mol·L$^{-1}$ | low | 40 – 50 min |

[a]Reactivity in comparison to other diisocyanates with regards to activation of the isocyanate group through mesomeric or inductive effects. An increase in reactivity is likely connected to faster reaction as well as a higher number of reaction pathways.

With these results, the reaction time for the hydrogel films that can be used for mechanical testing could be chosen. To be able to handle the reaction mixture before gelation and to cast films, stirring time was set to approximately 75% of the lowest determined reaction time. So in the following synthesis of hydrogel films, TDI was stirred for 90 s, BDI for 45 s, LDI for 360 s and IPDI for half an hour.

All reactions resulted in the formation of hydrogel films. Films prepared from BDI, LDI and IPDI looked homogeneous on the macroscopic level, while films prepared from TDI showed a large amount of sphere-like inhomogenicities that were distributed throughout the bulk material and especially located on the bottom side in the petri dish (Figure 1).

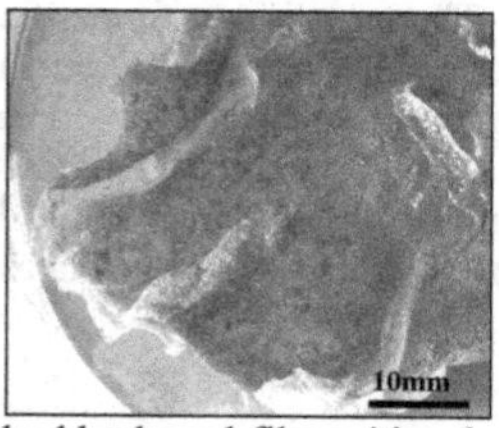


**Figure 1:** TDI crosslinked hydrogel film with spherical inhomogenicities

The water uptake of the hydrogels correlated with the gelation time of the crosslinking reaction (Table II). Diisocyanate crosslinkers with fast gelation time (BDI and TDI) have the highest water uptake (700 to 900 wt.-% respectively), whereas TDI film samples showed a strong tendency to curl while swelling. This could hint on a severe crosslinking gradient and inhomogeneous distribution of the crosslinker within the samples. LDI has reduced water uptake of around 400 wt.-% and the slowest reacting IPDI the lowest value of 300 wt.-% water uptake. This leads to the conclusion that the fast reacting diisocyanates can only achieve low crosslinking of the gelatin chains. Especially the highly inhomogeneous TDI-crosslinked hydrogel shows the highest water uptake. Here, it is possible that unwanted side reactions have a much stronger influence leading to less crosslinking and gradually changing crosslinking throughout the hydrogel film. In contrast, crosslinkers with lower reactivity and solubility in water (LDI and especially IPDI) seem to give well crosslinked networks.

**Table II:** Water uptake (*H*), degree of swelling (*Q*), and shear modulus (*G*) of crosslinked hydrogels at 25 °C. Young's Modulus (*E*), maximum stress ($\sigma_{max}$), and elongation at break ($\varepsilon_b$) form tensile tests at 37 °C.

| Crosslinker | $H^a$ [wt.-%] | $Q^a$ [vol.-%] | *E* [kPa] | $\sigma_{max}$ [kPa] | $\varepsilon_b$ [%] | *G* [kPa] |
|---|---|---|---|---|---|---|
| TDI | 891 ± 55 | 1172 ± 67 | 160 ± 62 | 60 ± 20 | 34 ± 7 | 5.5 ± 0.2 |
| BDI | 714 ± 10 | 1037 ± 13 | 51 ± 16 | 40 ± 20 | 83 ± 9 | 4.5 ± 0.2 |
| LDI | 438 ± 36 | 630 ± 43 | 55 ± 9 | 19 ± 11 | 30 ± 12 | 11.7 ± 0.2 |
| IPDI | 328 ± 23 | 501 ± 23 | 370 ± 54 | 160 ± 20 | 39 ± 4 | 19.5 ± 0.2 |

[a]*H* and *Q* are stated as average of three samples, error is given as maximum deviation.

Investigation of elastic moduli of the formed network showed a high tensile strength for the slow reacting IPDI (Table II), which fits to the low water uptake. During tensile tests it became also obvious that TDI crosslinked samples were gradually inhomogenous from the top to the bottom of the film. TDI samples had a soft top cover and a quite stiff bottom part where also most of the previously mentioned sphere-like inhomogenicities were located. This could explain the higher E-modulus of 160 kPa. Here the higher crosslinked bottom part dominates in the tensile test while the soft top is not contributing. At the same time the less crosslinked part mainly determines the high water uptake. As the tensile tests were performed at 37 °C, BDI crosslinked samples showed a severe softening in swollen state including partial dissolution of material. Also $\varepsilon_b$ is highest for BDI crosslinked hydrogels, which can be interpreted by assuming that BDI samples contain much less covalent crosslinks than the other samples and are mainly stabilized by physical netpoints. During tensile testing, physical crosslinks break first, increasing the mobility of polymer chains under stress while covalent crosslinks remain and stabilize the network, resulting in larger elongation before break [11]. This also explains the rather high water uptake of BDI samples. However, differences in swelling of the samples partially resulting in orientation of polymer chains has to be taken into account as well. LDI crosslinked samples had the lowest elastic modulus and also the lowest maximum stress that was applied during measurements. Nevertheless, low water uptake of the LDI samples lead to the conclusion that these networks are well crosslinked. The trend in elastic strength observed in tensile tests could also be found in rheological measurements and is represented by the shear modulus *G*.

## CONCLUSIONS

Formation of covalently crosslinked gelatin-based hydrogel networks with diisocyanates in water is strongly influenced by the water solubility and reactivity of the used diisocyanates. Observation of the morphology of formed hydrogels as well as investigation of mechanical properties showed that either relatively high water solubility or high reactivity of the applied diisocyanates leads to inhomogeneous or poorly crosslinked networks, while diisocyanates with lower solubility and/or low reactivity are able to form stable hydrogels. Such behavior has been

reported before for crosslinked polyacrylamide hydrogels, where delayed crosslinker/polymer contact with highly reactive crosslinkers led to heterogeneous networks [12, 13].
BDI is the best soluble of the studied diisocyanates with average relative reactivity. BDI crosslinked gelatin networks showed high water uptake at room temperature and strong softening as well as partial dissolution of the material at 37 °C. This leads to the conclusion that water solubility mainly supports the reaction of the isocyanate group with the water molecules, decreasing the potential to react with amino groups of gelatin.
TDI has the highest reactivity of the diisocyanates and has mediocre solubility in water. It was only forming highly inhomogeneous networks with an apparent crosslinking gradient throughout the hydrogel. The high water uptake seemed to be caused by the low crosslinked top side of the films, although the elastic modulus was the second highest caused by the highly crosslinked bottom. This again highlights the inhomogeneous reaction of the TDI with gelatin. Here it is probable that highly reactive isocyanates seem to force oligomer formation and thus also decreasing the potential to evenly react with gelatin.
LDI has as well average solubility as average reactivity. Networks formed by LDI crosslinking had lower water uptake and low E-modulus but were homogeneous and stable at 37 °C.
IPDI has the lowest solubility and the lowest reactivity of the investigated diisocyanates. Material formed by the reaction of IPDI with gelatin showed lowest water uptake and highest stiffness (highest E-modulus). Both results lead to the conclusion that slower reaction of isocyanates with water and also high hydrophobicity support crosslinking of the gelatin chains and allow better distribution of reactive species in the solution during network formation.
The results suggest that for homogeneous films with high crosslinking density diisocyanates with low solubility and low reactivity are beneficial. It would furthermore be interesting to investigate if it is possible to design a diisocyanate with high water solubility but low reactivity, e.g. through steric hindrance of the isocyanate groups. In this way the addition of a surfactant to the reaction could be avoided.

**REFERENCES**


1. L.E. Nielsen, *J. Macromol. Sci. - Revs. Macromol. Chem.*, **1969**, *3*, 69
2. S. Van Vlierberghe, P. Dubruel, E. Schacht, *Biomacromol.*, **2011**, *12*, 1387
3. L.H.H. Olde Damink, P.J. Dijkstra, M.J.A. van Luyn, P.B. van Wachem, P. Nieuwenhuis, J. Feijen, *J. Mater. Sci.: Mat. Med.*, **1995**, *6*, 429
4. P.J. Nowatzki, D.A. Tirrell, *Biomaterials*, **2004**, *25*, 1261
5. G.A. Digenis, T.B. Gold, V.P. Shah, *J. Pharm. Sci.*, **1994**, *83*, 915
6. G. Tronci, A.T. Neffe, B.F. Pierce, A. Lendlein, *J. Mater. Chem.*, **2010**, *20*, 8875
7. B.F. Pierce, E. Pittermann, N. Ma, T. Gebauer, A.T. Neffe, M. Hölscher, F. Jung, A. Lendlein, *Macromol. Biosci.*, **2012**, *12*, 312
8. R.G. Arnold, J.A. Nelson, J.J. Verbanc, *Chem. Revs.*, **1957**, *57*, 47
9. H. Ulrich, *J. Polymer Sci.: Macromol. Rev.*, **1976**, *11*, 93
10. M. Ionescu, *Chemistry and Technology of Polyols for Polyurethanes*, Repra Technology Ltd., Shawsbury, UK, **2005**
11. A.T. Neffe, G. Tronci, A. Alteheld, A. Lendlein, *Macromol. Chem. Phys.*, **2010**, *211*, 182
12. N. Orakdogen, O. Okay, *Polymer Bulletin*, **2006**, *57*, 631
13. S. Abdurrahmanoglu, O. Okay, *J. Macromol. Sci., Part A: Pure and Applied Chem.*, **2008**, *45*, 769

**Mater. Res. Soc. Symp. Proc. Vol. 1569 © 2013 Materials Research Society**
**DOI: 10.1557/opl.2013.830**

# The influence of the co-monomer ratio of poly[acrylonitrile-*co*-(*N*-vinylpyrrolidone)]s on primary human monocyte-derived dendritic cells

Toralf Roch, Marc Behl, Michael Zierke, Benjamin F. Pierce, Karl Kratz, Thomas Weigel, Nan Ma, and Andreas Lendlein
*Institute of Biomaterial Science and Berlin-Brandenburg Center for Regenerative Therapies (BCRT), Helmholtz-Zentrum Geesthacht, Kantstraße 55, 14513 Teltow, Germany*

## ABSTRACT

A major goal in the field of regenerative medicine is to improve our understanding of how biomaterial properties affect cells of the immune system. Systematic variation of defined chemical properties could help to understand which factors determine and modulate cellular responses. A series of copolymers poly[acrylonitrile-*co*-(*N*-vinylpyrrolidone)]s (P(AN-*co*-NVP)) served as model system, in which increasing hydrophilicity was adjusted by increasing the content related to the NVP based repeating units ($n_{NVP}$) (0, 4.6, 11.8, 22.3, and 29.4 mol%). The influence of increasing $n_{NVP}$ contents on cellular response of human primary monocyte derived dendritic cells (DC), which play a key role in the initiation of immune responses, was investigated. It was shown using the LAL-Test as well as a macrophage-based assay, that the materials were free of endotoxins and other microbial contaminations, which could otherwise bias the readout of the DC experiments. The increasing $n_{NVP}$ content led to a slightly increased cell death of DC, whereas the activation status of DC was not systematically altered by the different P(AN-*co*-NVP)s as demonstrated by the expression of co-stimulatory molecule and cytokine secretion. Similarly, under inflammatory conditions mimicked by the addition of lipopolysaccharides (LPS), neither the expression of co-stimulatory molecules nor the release of cytokines was influenced by the different copolymers. Conclusively, our data showed that this class of copolymers does not substantially influence the viability and the activation status of DC.

## INTRODUCTION

Antigens and microbial products can activate DC to initiate immune responses [1]. The manipulation of the DC response is of high therapeutic interest since DC can induce anti-tumor responses, enhance the efficacy of vaccines, and could suppress auto-immune reaction [2]. Biomaterial properties such as chemical composition or surface roughness were described to influence the phenotype and activation status of DC [3, 4]. However, it is still challenging to deduce, which material properties are the most important in affecting DC phenotype. Therefore investigations, in which distinct chemical and physical materials properties are systematically altered will shed light on DC-biomaterial interactions [5]. Therefore, P(AN-*co*-NVP) copolymers with increasing $n_{NVP}$ content (4.6, 11.8, 22.3, and 29.4 mol%) showing a systematic hydrophilicity, which was documented by decreasing contact angles and an increased water uptake, were prepared. In the following, abbreviated sample IDs were used, for example $n_{NVP}$ 4.6 mol% refers to as with $n_{NVP}$4.6. Since the P(AN-*co*-NVP) copolymers with higher $n_{NVP}$ content inhibited the growth of fibroblast and showed a reduced adherence of endothelial cells as well as macrophages [6, 7], we hypothesize that materials with higher hydrophilicity levels also have an influence on the cellular response of DC. Here, P(AN-*co*-NVP) copolymers with $n_{NVP}$ contents of 0, 4.6, 11.8, 23.3, and 29.4 mol% were synthesized and their effects on the viability, the expression of co-stimulatory molecules, and the secretion of inflammatory mediators by DC after cultivating on the different P(AN-*co*-NVP) copolymers were explored. The data presented here should improve our understanding how systematic increases in $n_{NVP}$ contents and the associated increase in hydrophilicity influence DC material interactions.

## EXPERIMENTAL DETAILS

### Material synthesis

AN (Aldrich, >99% purity) was distilled prior usage. NVP (Aldrich; >99% purity) was purified by use of inhibitor remover (Aldrich). The copolymers were synthesized by solvent free emulsion polymerization in technical scale of 1 kg batch size using KMBS/KPS as radical starter. The reaction was quenched by the addition of hydrochloric acid (HCl), the precipitated polymer was filtered off, washed with 40 L of water (Millipore grade) and afterwards dried at 30 °C at 2 mbar until constant weight was achieved. The $n_{NVP}$ content of the obtained copolymers was determined by elemental analysis.

### Processing of test specimen

The obtained raw copolymer materials were homogenized by milling and dried at 60 °C in vacuum before processing into disc shaped test specimen with a diameter of Ø 13 mm and a thickness of around 1 mm. The processing was realized by a sintering method utilizing a custom made tool. All copolymers test specimen were prepared in a two-step process: first sintering at temperatures in the range from 125 °C to 140 °C under a load of 2000-2500 kg for 150 to 180 s and subsequently at 60 °C for 120 to 150 seconds with the same load. The disc shaped samples were sterilized with ethylene oxide (45 °C, exposure time: 180 min) prior to biological tests or surface characterization.

### Determination of microbial contaminations

The material eluates were prepared according to ISO 10993-12 as previously described [8]. The eluates were analyzed for LPS contaminations using the LAL test (Lonza, Cologne, Germany), which was performed according to manufacturer instructions. The RAW-Blue cell culture and activation were performed as previously described [8]. In brief, $5x10^5$ RAW-Blue™ cells (InvivoGen, San Diego, USA) were cultured for 24 hours in 24 well plates, in which the different P(AN-*co*-NVP) samples were inserted. After incubation, cell culture supernatants were harvested and incubated for 3 hours with the QuantiBlue™ solution.

### Generation of peripheral blood mononuclear cell (PBMC) derived DC, co-stimulatory molecule and cytokine expression analysis

DC generation, determination of co-stimulatory molecule, and cytokine expression were performed as previously described [9]. In brief, the monocytes isolation KitII (Miltenyi Biotec, Bergisch-Gladbach, Germany) was used to isolate monocytes from PBMC by depleting the CD14 negative cells. Routinely, more than 95% of the purified cells were CD14 positive. Monocytes were differentiated into DC by the addition of 10 µg/mL interleukin(IL)-4 (Miltenyi Biotec, Bergisch-Gladbach, Germany) and 100 $\mu g \cdot mL^{-1}$ granulocyte macrophage - colony stimulating factor (R&D Systems, Minneapolis, USA). DC were harvested after six days and $1x10^6$ cells were seeded in 24 well plates together with the different material samples (Ø 13 mm) for 24 hours with or without 1 $\mu g \cdot mL^{-1}$ lipopolysaccharides (LPS) (LPS from *E.coli* O111:B4, Axxora GmbH, Lörrach, Germany). The differentiation of monocytes into DC was monitored by CD209 expression and resulted routinely in more than 95% CD209-positive cells. Prior staining of surface molecules, Fc-receptors were blocked with FcR-Blocking Reagent (Miltenyi Biotec) for 15 min on ice to avoid unspecific antibody binding. After washing, DC were stained for 10 min on ice with anti-CD209-APC (clone DCN47.5, Miltenyi Biotec) to identify DC and to determine the activation status with either anti-CD86-FITC (clone FM95, Miltenyi Biotec) and anti-HLA-DR-PE (clone AC122, Miltenyi Biotec) or anti-CD40-FITC (clone 5C3, BD Pharmingen), and anti-CD80-PE (clone L307.4, BD Pharmingen). To discriminate live and dead cells, 1 $\mu g \cdot mL^{-1}$ 4',6-diamidino-2-phenylindole (DAPI) was added immediately prior analysis, which was performed with the MACSQuant flow cytometer (Miltenyi Biotec). For all samples, aggregated cells were excluded via FCS-A/FCS-H scatter parameters as previously described [9]. Data were analyzed using FlowJo software (Tree Star, Ashland, USA). The cytokine secretion by DC was quantified in cell culture supernatants using Bio-Plex suspension multiplex array systems (Bio-Plex 200®, BioRad, Germany) according to the manufacturer's instructions. Microbead assays were purchased from BioRad.

**Statistical analysis**

For all DC experiments the average of two independent experiments, each with three individual donors (n=6), is shown. Data were plotted and statistically analyzed using GraphPad Prism (La Jolla, CA92037, USA). For all samples mean values and the standard error of the mean are given unless otherwise indicated. Multiple sample tests were performed using one-way analysis of variance (ANOVA). Tukey-adjusted p-values lower 0.05 were considered as significant.

## RESULTS and DISCUSSION

**Materials**

The P(AN-*co*-NVP)s were prepared using free radical polymerization under aqueous conditions and processed into chips (Figure 1).

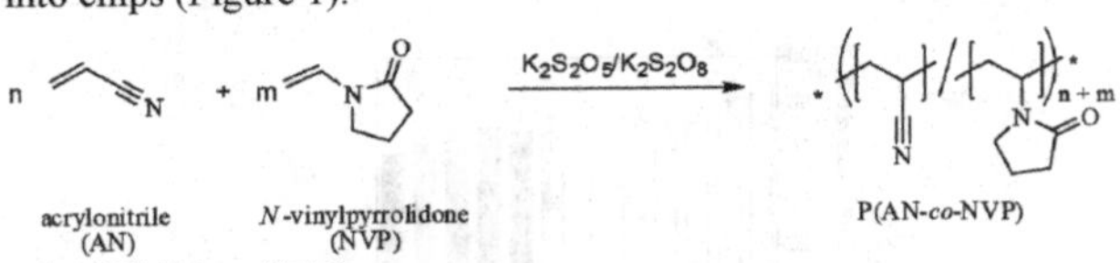


**Figure 1:** Synthesis of P(AN-*co*-NVP)

With increasing the $n_{NVP}$ content from 4.6 mol% to 29.4 mol% the P(AN-*co*-NVP) showed decreasing contact angles ($\theta_{Adv}$) from 65.2 ± 2.8° to 57.2 ± 2.6° with similar hysteresis values and an increasing volumetric degree of swelling from 102 ± 0.3 vol% to 128 ± 1.3 vol%, while the surface roughness ($R_q$) was slightly reduced from 0.06 µm to 0.04 µm, indicating that higher $n_{NVP}$ contents correlate with higher hydrophilicity All materials were sterilized with ethylene oxide, which did not lead to significant changes of the contact angles, the roughness levels, the glass transition temperature ($T_g$) and the mass loss at (TGA at 500 °C) (data not shown).

**Activation of murin macrophages**

According to the LAL-test the eluates of all samples showed endotoxin levels below 0.06 $EU \cdot mL^{-1}$, which is well below 0.5 $EU \cdot mL^{-1}$, the limit set by the U.S. Food and Drug Administration [10]. Since the LAL-test only detects soluble LPS, a cell-based assay using RAW-Blue™ macrophages was performed to detect material bound endotoxins and other microbial contamination. After engagement of pattern recognition receptors by microbial products, the RAW-Blue™ cells secrete an alkaline phosphatase, which is proportional to the degree of contamination and can be quantified in a color change reaction. In agreement with the LAL-test, none of the investigated P(AN-*co*-NVP) samples induced a RAW-Blue™ cells response, while LPS, which served as positive control, induced a substantial cellular response

(Figure 2). The data indicate a germ free manufacturing process of the different polymers.

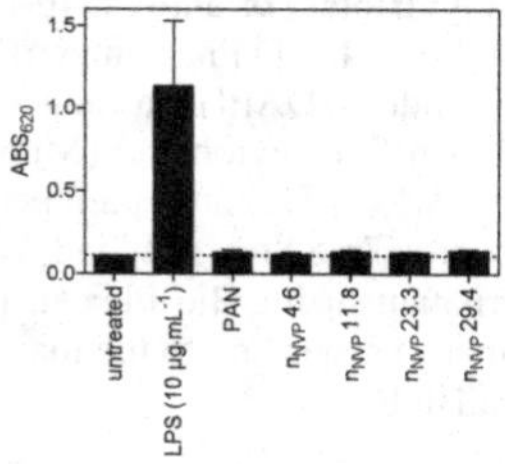


**Figure 2:** Raw-Blue™ cell-based assay to determine material-bound microbial contaminations. The average of two independent experiments measured in triplicates is shown (mean ± range).

## Survival and activation status of dendritic cells

Since increased hydrophilicity could have an influence on the viability of DC, the cell were cultured on the different P(AN-*co*-NVP) samples with increasing $n_{NVP}$ contents and DAPI incorporation was analyzed by flow cytometry. The viability of DC decreased with increasing $n_{NVP}$ amounts. However, the frequency of live cells was never below 60%. The LPS treatment reduced the DC survival, which is an expected physiological apoptotic process described to maintain self-tolerance and to prevent autoimmune diseases [11]. However, the different copolymers had no significant effect on the survival of LPS-activated DC (Figure 3).

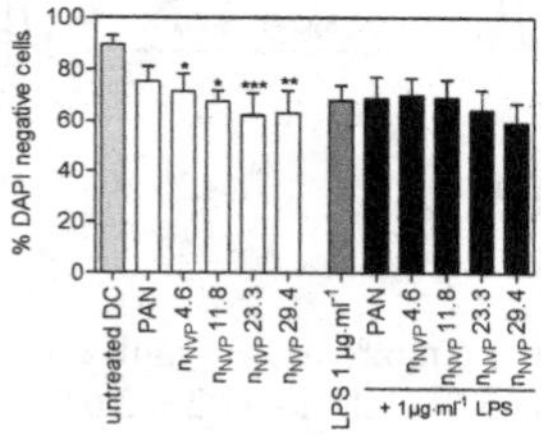


**Figure 3:** Survival of DC cultured on different P(AN-*co*-NVP) samples with or without LPS. Dead cells were defined as DAPI positive. Significant differences compared to untreated DC are indicated by *, (*for $p<0.05$, **for $p<0.05$, and *** for $p<0.001$).

The increased cell death of DC led to the assumption that the cells get activated by the materials with higher NVP-contents, because activation of DC would subsequently lead to apoptosis.

The activation status of DC was determined by the expression of co-stimulatory molecules such as CD40, CD80, and CD86 as well as by the expression level of MHCII (HLA-DR). The quantification of the flow cytometric data revealed that the expression of CD40, CD80, was not altered by the different P(AN-*co*-NVP) and was comparable to DC cultured without the copolymer samples. The expression of CD40 and CD80 remained unchanged when DC were activated with LPS and cultured together with the different P(AN-*co*-NVP) copolymers (Figure 4A,B). The expression levels of CD86 and HLA-DR were significantly increased only when DC were cultured on P(AN-*co*-NVP) on $n_{NVP}$4.6. However, a systematic increase of the DC responses after cultivation on the different P(AN-*co*-NVP) samples could not be observed, indicating the $n_{NVP}4.6$ reflects the optimal ratio to induce the expression of CD86 and HLA-DR on DC (Figure 4C,D). The data indicate that the increased cell death (Figure 3**Figure 3:** Survival of DC cultured on different P(AN-*co*-NVP) samples with or without LPS. Dead cells were defined as DAPI positive. Significant differences compared to untreated DC are indicated by *, (*for p<0.05, **for p<0.05, and *** for p<0.001).

) mediated by the P(AN-*co*-NVP) cannot be attributed to the material induced activation of DC. In future studies it will be elucidated whether the material-induced cell death is an apoptotic or necrotic process.

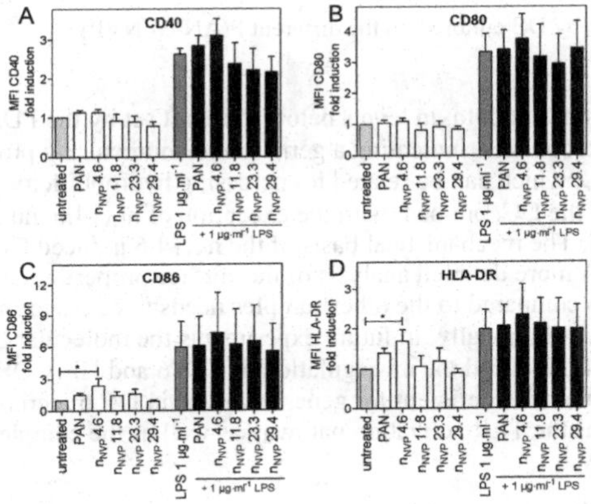


**Figure 4:** Expression of co-stimulatory molecules by DC after cultivation on P(AN-*co*-NVP) copolymers. The fold changes of the median fluorescence intensities of the respective activation marker were plotted. (*for p<0.05).

The cytokine profile of DC can guide the subsequent response into $T_H1$ or $T_H2$ direction [12]. Using the multiplex profiling the expression levels of the inflammatory mediators IL-1β, IL-1ra, IL-6, IL-8, IL-9, IL-10, IL-12(p70), IL-17, eotaxin, G-CFS, GM-CSF, IFN-γ, IP-10, MCP-1, MIP-1α, MIP-1β, PDGF-BB, Rantes, and TNF-α were determined after DC were cultured on the different P(AN-*co*-NVP) copolymers. The expression levels of all analyzed factors were not significantly influenced by the different copolymers, neither under homeostatic conditions nor under inflammatory condition simulated by the addition of LPS (data not shown, Figure 5). The expression of IL-12(p70) as $T_H1$-driving cytokine, of TNF-α as very potent pro-inflammatory cytokine and MIP-1β as chemokine are shown representatively in Figure 5. Although not significant, the MIP-1β production was increased for all samples and the highest levels were secreted by DC cultured with $n_{NVP}23.3$. An increased MIP-1β expression was recently described for DC, which have been cultured on polystyrene and poly(ether imide), indicating that distinct polymeric surfaces selectively induce MIP-1β, but no other inflammatory mediator [4]. The functional consequences of the increased MIP-1β production will be investigated in future studies.

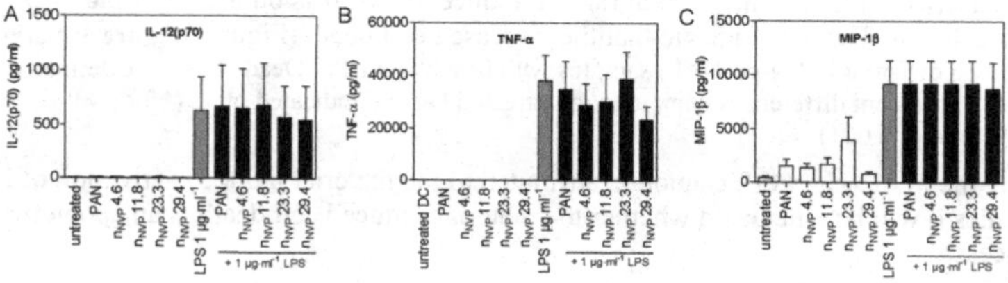


**Figure 5:** Cytokine secretion by DC cultured on the different P(AN-*co*-NVP).

## CONCLUSIONS

All tested P(AN-co-NVP)s had endotoxin levels below the limit set by the FDA and did not activate RAW-Blue™ macrophages, indicating a germ free manufacturing process. The increasing amounts of $n_{NVP}$, which can be related to increasing hydrophilicity, slightly diminished the viability of the DC, but had, with the exception of $n_{NVP}$4.6, no influence on the activation status of the DC. The mechanistical basis of the $n_{NVP}$4.6 induced DC activation remains to be elucidated. A more detailed analysis of the surface properties after swelling in cell culture medium of $n_{NVP}$4.6 compared to the other samples needs to be performed to anticipate why $n_{NVP}$4.6 activates DC. Additionally, in future experiments the molecular biological mechanisms of how $n_{NVP}$ 4.6 induced the up-regulation of CD86 and HLA-DR will be addressed by using dendritic cells from mice deficient for genes involved in DC material-mediated activation pathways such as the TLR activation pathway as well as the complement receptor or integrin activation pathway.

## ACKNOWLEDGMENTS

We thank Anja Müller-Heyn, for the technical support and the Bundesministerium für Bildung und Forschung (BMBF) for funding within project 0315696A „Poly4bio BB".

## REFERENCES

1. C. Munz, R.M. Steinman, and S. Fujii, *J Exp Med*, **202**, 203 (2005).
2. R.M. Steinman and J. Banchereau, *Nature*, **449**, 419 (2007).
3. P.M. Kou and J.E. Babensee, *J Biomed Mater Res A*, **96**, 239 (2011).
4. T. Roch, K. Kratz, N. Ma, F. Jung, and A. Lendlein, *Clin Hemorheol Micro*, **DOI 10.3233/CH-131699** (2013).
5. V.P. Shastri and A. Lendlein, *Mrs Bull*, **35**, 571 (2010).
6. G. Boese, C. Trimpert, W. Albrecht, G. Malsch, T. Groth, and A. Lendlein, *Tissue Eng*, **13**, 2995 (2007).
7. L.S. Wan, Z.K. Xu, X.J. Huang, X.D. Huang, and K. Yao, *Acta Biomater*, **3**, 183 (2007).
8. T. Roch, J. Cui, K. Kratz, A. Lendlein, and F. Jung, *Clin Hemorheol Microcirc*, **50**, 131 (2012).
9. C. Wischke, S. Mathew, T. Roch, M. Frentsch, and A. Lendlein, *J Control Release*, **164**, 299 (2012).
10. M.B. Gorbet and M.V. Sefton, *Biomaterials*, **26**, 6811 (2005).
11. I. Zanoni, R. Ostuni, G. Capuano, M. Collini, M. Caccia, A.E. Ronchi, M. Rocchetti, F. Mingozzi, M. Foti, G. Chirico, B. Costa, A. Zaza, P. Ricciardi-Castagnoli, and F. Granucci, *Nature*, **460**, 264 (2009).
12. M.L. Kapsenberg, *Nat Rev Immunol*, **3**, 984 (2003).

Mater. Res. Soc. Symp. Proc. Vol. 1569 © 2013 Materials Research Society
DOI: 10.1557/opl.2013.796

# The Synthesis and Characterization of Nano-hydroxyapatite (nHAP)-*g*-poly(lactide-*co*-glycolide)-*g*-collagen Polymer for Tissue Engineering Scaffolds

Didarul Bhuiyan[a], Michael J. Jablonsky[b], John Middleton[a] and Rina Tannenbaum[a,c]

[a]Department of Biomedical Engineering and [b]Department of Chemistry, [c]UAB Comprehensive Cancer Center and the School of Medicine, University of Alabama at Birmingham, Birmingham, AL 35924

## ABSTRACT

Bone grafts, commonly performed to augment bone regeneration from autologous or alleogeneic sources, carry an enormous cost, estimated at upwards of 21 billion dollars per year. Hydroxyapatite (HAP) bio-ceramic has been widely used in clinic as a bone graft substitute material due to its biocompatibility and the similarity of its structure and composition to bone mineral. However, its applications are limited due to its lack of strength and toughness. Researchers have attempted to overcome these issues by combining HAP bio-ceramics into resorbable polymers to improve their mechanical properties. However, poor bonding between the HAP and the polymer caused separation at the polymer-filler interface. To overcome this, short chains of polymers were grafted directly from the hydroxyl groups on the surface of nanocrystalline HAP. Collagens, being the most abundant proteins in the body, and having suitable properties such as biodegradability, bioabsorbability with low antigenicity, high affinity to water, and the ability to interact with cells through integrin recognition, makes them a very promising candidate for the modification of the polymer surface. In this study, a novel method of synthesizing nano-hydroxyapatite (nHAP)-*g*-poly(lactide-*co*-glycolide)-*g*-collagen polymer was introduced. The synthesis process was carried out in several steps. First, poly (lactide-*co*-glycolide) (PLGA) polymer was directly grafted onto the hydroxyl group of the surface of n-HAP particles by ring-opening polymerization, and subsequently coupled with succinic anhydride. In order to activate the co-polymer for collagen attachment, the carboxyl end group obtained from succinic anhydride was reacted with N-hydroxysuccinimide (NHS) in the presence of dicyclohexylcarbodiimide (DCC) as the cross-linking agent. Finally, the activated co-polymer was attached to calf skin collagen type I, in hydrochloric acid/phosphate buffer solution and the precipitated co-polymer with attached collagen was isolated. The synthesis was monitored by $^1$H NMR and FTIR spectroscopies and the products after each step were characterized by thermal analysis (TGA and DSC). These composite materials will be tested as potential scaffolds for tissue engineering applications.

## INTRODUCTION

Bone grafting is a standard surgical practice, involving the repair and replacement of deficient bone caused by congenital disorders, traumatic injury, or bone tumors.[1,2] More than half a million bone graft procedures are performed each year in the United States, fact which makes bone grafts the second most frequently transplanted material.[3-5] Autogenous bone grafts have less probability of being rejected by the immune system, but availability may in some cases be limited. Ongoing research is aimed at the development of a scaffolds made of synthetic or natural biomaterials.[6-17] However, there are currently no synthetic bone graft substitute materials available that have similar or superior biological and mechanical properties compared to bone. Hydroxyapatite (HAP) bio-ceramic is currently used in clinic as a bone graft substitute material, but due to its fragility, this material is only suitable for repairing small fracture or non load bearing bone defects because of its low mechanical strength and weak fatigue resistance.[8,11-12] Compounding HAP with resorbable polymers such as poly(L-lactide) (PLLA) should have resulted in improved mechanical properties,[9-17] but the material phase-separated due to the lack of interfacial bonding between the two components. Therefore, fabrication of synthetic bone graft substitutes that possess the desired properties and maintain their homogeneous morphology would require the modification and bioactivation of the polymer surface, such as the tethering of collagen.[6] In this study, synthesis of a novel nano-hydroxyapatite (nHAP)-*g*-poly(lactide-*co*-glycolide)-*g*-collagen block co-polymer was introduced. The synthesis process was carried out in several steps. First, poly (lactide-*co*-glycolide) (PLGA) polymer was directly grafted onto the hydroxyl group on the surface of n-HAP particles by ring-opening polymerization, and subsequently coupled with succinic anhydride. In order to activate the co-polymer for collagen attachment, the carboxyl end group obtained from succinic anhydride was reacted with N-hydroxysuccinimide (NHS) in the presence of dicyclohexylcarbodiimide (DCC) as the cross-linking agent. Finally, the activated co-polymer was attached to calf skin collagen type I, in hydrochloric acid/phosphate buffer solution and the precipitated co-polymer with attached collagen was isolated. The synthesis was monitored by $^1$H NMR and FTIR spectroscopies and the products after each step were characterized by thermal analysis (TGA and DSC).

## EXPERIMENTAL DETAILS

D,L-lactide and Glycolide monomer were obtained from Ortec, Inc. (Easley, S.C.). Hydroxyapatite Nanopowder (<200 nm) was purchased from Sigma Aldrich. Lyophilized calf skin collagen type I ($M_w$ 1,000,000 g/mol) was purchased from Elastin, Inc. (Owensville, Missouri). Previously dried 0.84 g nano-hydroxyapatite (nHAP) was mixed with 39.42 g D,L-Lactide and 10.58 g Glycolide monomers in a glass reactor placed in an oil bath at 150 °C. The molten monomer and nHAP mixture was sonicated with a sonicator probe for 5 min, following by the addition of 0.05 g stannous octoate catalyst. Samples were collected at during the polymerization process at different time intervals and analyzed by $^1$H NMR spectroscopy. After 2 h of polymerization, 0.2 g succinic anhydride was added to the reactor. After 4.5 h of polymerization, the reactor was taken out of the oil bath and the nHAP-PLGA co-polymer was collected. 10 g of the co-polymer was dissolved in 200 mL dried methylene chloride and 0.39 g NHS and 0.09 g DCC was added. The reaction was allowed to continue for 20 h under nitrogen atmosphere. The polymer solution was then dissolved in ethyl acetate and precipitated in anhydrous diethyl ether. The precipitated polymer was collected, dried and stored at 4 °C. 160 mg calf skin collagen type I was dissolved in 223.8 mL of 1 mM HCl in an ice bath and subsequently diluted with 300 mL of 50 mM phosphate buffer (pH 7.4). 0.96 g nHAP-PLGA-NHS polymer was dissolved in 26.6 mL anhydrous DMF and added drop wise into the collagen solution in the ice bath. The precipitated co-polymer sample was collected every 15 minutes for TGA experiments. After 3 h of reaction, the n-HAP-*g*-poly(lactide-*co*-glycolide)-*g*-collagen block co-polymer was filtered, dried and stored at 4 °C. NMR samples were prepared by dissolving the co-polymers in $CDCl_3$, purchased from Fisher Scientific. $^1$H NMR spectra were measured on a Bruker DPX-300 spectrometer at 300 MHz. The $^1$H spectra for samples collected during the polymerization at various time intervals minutes were used to determine the ratio of monomer to polymer in the samples, and subsequently the kinetic behavior. The TGA experiments were conducted at a heating rate of 20 °C·min$^{-1}$ from room temperature to 600 °C in an unsealed platinum pan under nitrogen atmosphere. The equipment used was the 2950 Thermogravimetric Analyzer manufactured by Du Pont Instruments.

## RESULTS AND DISCUSSION

The conversion of D,L lactide and Glycolide monomers to PLGA polymer was monitored by $^1$H NMR spectroscopy. Figure 1 shows the evolution of the polymer formation by examining the $^1$H NMR at various time intervals for two chemical shift ranges, 1.3 – 1.7 ppm corresponding to $CH_3$ groups (A) and 4.6 – 5.3 ppm corresponding to $CH_2$ and CH groups (B). The chemical shift values for the $CH_3$, $CH_2$, and CH functional groups for D,L lactide and Glycolide monomers and the PLGA polymer are summarized in the table contained in Figure 1. We can see that the peaks at 4.99 and 5.08 ppm (corresponding to $CH_2$ of the Glycolide and CH of the D,L-lactide, respectively) decrease while the peaks in the 4.7 – 4.9 and 5.2 – 5.3 ppm range (corresponding to $CH_2$ and CH of the PLGA co-polymer, respectively) increase with time. After purification and activation with NHS and DCC, the monomer peaks disappear completely and only the peaks corresponding to the PLGA polymer can be observed.

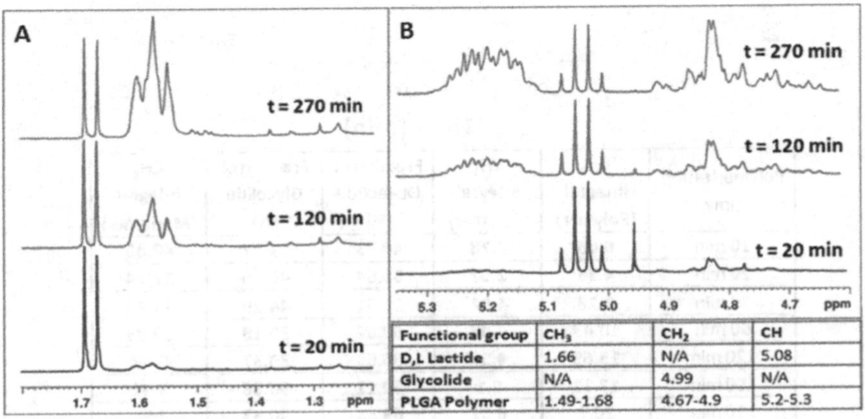


| Functional group | $CH_3$ | $CH_2$ | CH |
|---|---|---|---|
| D,L Lactide | 1.66 | N/A | 5.08 |
| Glycolide | N/A | 4.99 | N/A |
| PLGA Polymer | 1.49-1.68 | 4.67-4.9 | 5.2-5.3 |

**Figure 1:** Stack plot of $^1$H NMR spectra for the 1.2-1.8 ppm range (A) and the 4.6-5.4 ppm range (B), for the samples collected at different time intervals during the polymerization process.
**Table 1:** The chemical shift values for the $CH_3$, $CH_2$, and CH functional groups

By integrating the area under each of the pertinent peaks, the ratio between D,L-lactide and Glycolide monomers within the PLGA co-polymers could be obtained, indicating the progress of the reaction. The fractions of D,L-lactide and Glycolide at different time interval during the polymerization process are summarized in Table 2. The fraction of the Glycolide monomer in the PLGA polymer was higher at the beginning of synthesis but decreased with the progression of the polymerization process together with the increase of the D,L-lactide monomer fraction. At the end, after the purification and activation with NHS and DCC, the ratio of D,L-lactide and Glycolide monomers in the polymer was 73:27, which is very close to the initial target composition of 75-25 in the PLGA polymer. This indicates that the uptake of monomers into the growing polymer chain during polymerization is asymmetric with respect to both monomers, most likely leading to the formation of monomer blocks in the polymer, thus generating a polymer with mixed random and block architecture.[18] Based on the data in the table associated with Figure 2 and plotted in Figure 2, the initial polymerization proceeded with the random uptake of monomers, i.e. comparable fractions of both monomers in the growing polymer chain, generating a random copolymer. However, in the later stages of the polymerization, there was a preferential uptake of the D,L lactide monomer, most like generating a block architecture.

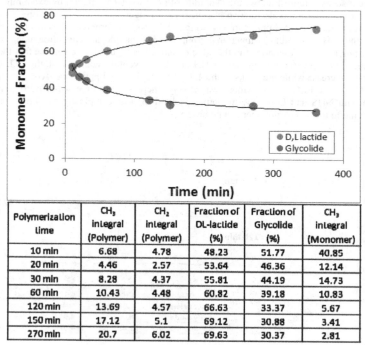

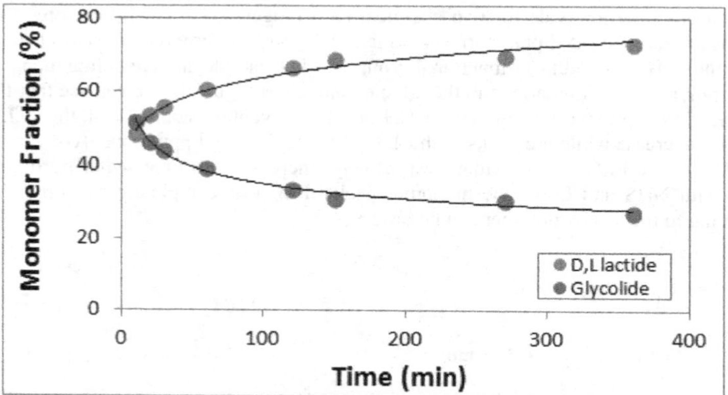


| Polymerization time | $CH_3$ integral (Polymer) | $CH_2$ integral (Polymer) | Fraction of DL-lactide (%) | Fraction of Glycolide (%) | $CH_3$ integral (Monomer) |
|---|---|---|---|---|---|
| 10 min | 6.68 | 4.78 | 48.23 | 51.77 | 40.85 |
| 20 min | 4.46 | 2.57 | 53.64 | 46.36 | 12.14 |
| 30 min | 8.28 | 4.37 | 55.81 | 44.19 | 14.73 |
| 60 min | 10.43 | 4.48 | 60.82 | 39.18 | 10.83 |
| 120 min | 13.69 | 4.57 | 66.63 | 33.37 | 5.67 |
| 150 min | 17.12 | 5.1 | 69.12 | 30.88 | 3.41 |
| 270 min | 20.7 | 6.02 | 69.63 | 30.37 | 2.81 |

**Figure 2:** The fraction of the monomers measured and calculated at various reaction times representing the uptake of these monomers in the growing polymer chain. **Table 2:** Summary of integration results obtained from $^1$H NMR spectra for each functional group

By comparing the peaks of the $CH_3$ functional group of the D,L-lactide monomer and those of the PLGA polymer, the monomer to polymer concentration ratio was obtained. This ratio decreased non-linearly as a function of reaction time. Fitting the data to a kinetic expression indicated that the synthesis of the PLGA co-polymer followed second order kinetics with a calculated rate constant of $4.83x10^{-4}$ $s^{-1}$.

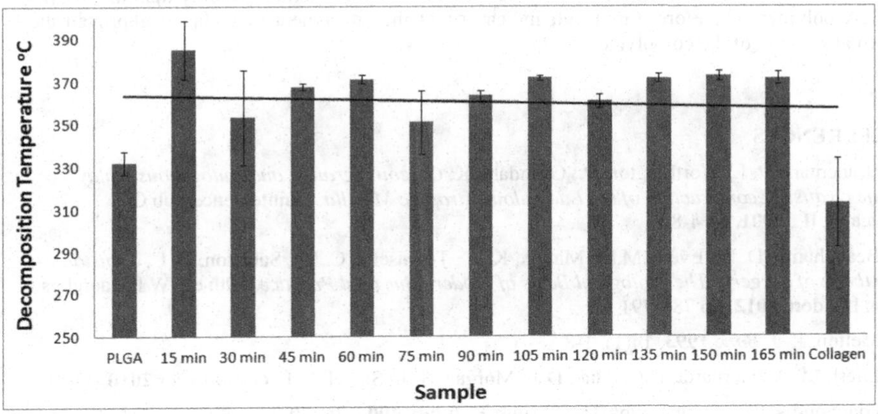


**Figure 3:** Average decomposition temperature for the co-polymer samples at different time intervals of collagen addition during synthesis process. The solid line represents the average decomposition temperature obtained by a linear fit of the data.

The thermal stability of the nHAP-PLGA-collagen tri-block co-polymer was measured by thermo-gravimetric analysis (TGA). The TGA decomposition profiles of samples collected at different stages during the synthesis process were compared to the decomposition profile of PLGA and collagen, and are shown in Figure 3. The nHAP-PLGA co-polymer exhibited a decomposition temperatures below 330 °C. However, after adding the collagen, the decomposition temperatures measured at different time intervals increased significantly. The decomposition temperature of the collagen-containing copolymers increased to around 370 °C immediately upon the addition of the collagen, and it remained constant during the reaction process. Given the fact that the decomposition temperature of calf skin collagen type I alone was only around 310 °C, it is clear that tethering the collagen to the PLGA copolymer has had a profound effect on the thermal stability of the hybrid material, beyond what would have been expected by assuming a simple composite effect. It is very likely the attachment of the collagen to the PLGA polymer has intimately altered the morphology of the copolymer, which in turn, has enhanced the thermal properties compared to the nHAP-PLGA copolymer and the collagen itself. This result is consistent with the result from DSC experiments.

## CONCLUSIONS

In the study, PLGA polymer was successfully grafted on the surface of nano particles of HAP. Chemical linkage was formed between nHAP particles and PLGA polymer and verified with $^1$H NMR analysis. Results from $^1$H NMR spectroscopy demonstrated that the synthesis reaction followed second order kinetics. The nHAP grafted PLGA polymer was subsequently attached to Collagen type I, and cross-linked co-polymer was obtained. This linkage was confirmed by

FTIR, DSC and TGA analysis. The TGA analysis shows that the samples behaved as a single material and amount of impurity within the samples were negligible. TGA results also reveal that the novel nHAP-PLGA-Collagen co-polymer possesses higher thermal stability than the nHAP-PLGA polymer. Therefore, this result indicates that the attachment of collagen improved the thermal stability of the co-polymer.

**REFERENCES**


1. Branemark, P. I.; Worthington, P.; Grondahl, K. *Osseointegration and Autogenous Onlay Bone Grafts: Reconstruction of the Edentulous Atrophic Maxilla*; Quintessence Pub Co.: Chicago, IL, **2001**, pp 4-8.

2. Beauchamp, D. R.; Evers, M.B.; Mattox, K. L.; Townsend, C. M.; Sabiston, D. C. *Sabiston Textbook of Surgery: The Biological Basis of Modern Surgical Practice,* 19th ed. W B Saunders Co.: London, **2012**, pp 783-791

3. Melton, L. J. *Bone* **1993**, 14(1), 1-8

4. Flierl, M. A.; Gerhardt, D.C.; Hak, D.J.; Morgan, S. J.; Stahel, P. F. *Orthopedics* **2010**, 33(2)

5. Giannoudis, P.V.; Dinopoulos, H., Tsiridis, E. *Injury* **2005.** 36, 20-27.

6. Porjazoska, A.; Yilmaz, O.K.; Baysal, K.; Cvetkovska, M.; Sirvanci, S.; Ercan, F.; Baysal, B. M. *J. Biomater. Sci. Polymer Edn.* **2006**, 17(3), 323-340

7. Liao, S.; Wang, W.; Uo, M.; Ohkawa, S.; Akasaka, T.; Tamura, K.; Cui, F.; Watari, F. *Biomaterials* **2005**, 26, 7564-7571

8. Zhang, P.; Hong, Z.; Yu, T.; Chen, X.; Jing, X. *Biomaterials* **2009**, 30, 58-70

9. Furukawa, T.; Matsusue, Y.; Yasunaga, T.; Shikinami, Y.; Okuno, M.; Nakamura, T. *Biomaterials* **2000**, 21, 889-898

10. Hasegawa, S.; Ishii, S.; Tamura, J.; Furukawa, T.; Neo, M.; Matsusue, Y.; Shikinami, Y.; Okuno, M.; Nakamura, T. *Biomaterials* **2006**, 27, 1327-1332.

11. Shikinami, Y.; Okuno, M. *Biomaterials* **1999**, 20, 859-877

12. Shikinami, Y.; Okuno, M. *Biomaterials* **2001**, 22, 3197-3211

13. Verheyen, C. C.; Wijn, J. R.; Blitterswijk, C. A.; Groot, K. *J Biomed Mater Res* **1992**, 26, 1277-1296

14. Verheyen, C. C.; Klein, C. P.; Bleick-Hogerovorst, J.M.; Wolke, J.G.; Blitterswijk, C. A.; Groot, K. *J Mater Sci: Mat in Med.* **1993**, 4(1), 58-65

15. Neffe, A.T.; Loebus, A.; Zaupa, A.; Stoetzel, C.; Mueller, F.A.; Lendlein, A. *Acta. Biomater.* **2011**, 7, 1693-1701

16. Ignjatovic, N.; Tomic, S.; Dakic, M.; Miljkovic, M.; Plavsic, M.; Uskokovic, D. *Biomaterials* **1999**, 20, 809-816

17. Deng, X.; Hao, J.; Wang, C. *Biomaterials* **2001**, 22, 2867-2873

18. Edgecombe, B.D.; Stein, J. A.; Frechet, J. M. *Macromolecules*, **1998**, 31, 1292-1304

Mater. Res. Soc. Symp. Proc. Vol. 1569 © 2013 Materials Research Society
DOI: 10.1557/opl.2013.765

# Marine Adhesive Containing Nanocomposite Hydrogel with Enhanced Materials and Bioadhesive Properties

Yuan Liu, Hao Zhan, Sarah Skelton and Bruce P. Lee*
Department of Biomedical Engineering, Michigan Technological University, Houghton, MI 49931, U. S. A.

## ABSTRACT

4-arm poly(ethylene glycol) end-capped with mimics of adhesive moiety found in mussel adhesive protein, dopamine, was combined with a biocompatible nano-silicate, Laponite, in creating a nanocomposite hydrogel with improved materials and adhesive properties. Dopamine's ability to form both irreversible covalent (cohesive and interfacial) and reversible physical (with Laponite) crosslinks was exploited in creating an injectable tissue adhesive. Incorporation of Laponite did not interfere with the curing of the adhesive. In some instances, increasing Laponite content reduced gelation time as dopamine-Laponite bond reduced the required number of covalent bonds needed for network formation. Incorporation of Laponite also increased compressive materials properties (e.g., max strength, energy to failure, etc.) of the nanocomposite without compromising its compliance as strain at failure was also increased. From lap shear adhesion test using wetted pericardium as the substrate, incorporating Laponite increased work of adhesion by 5 fold over that of control. Strong, physical bonds formed between dopamine and Laponite increased bulk materials properties, which contributed to the enhanced adhesive properties.

## INTRODUCTION

Rapid and effective wound closure remains an important goal of virtually all modern endoscopic and conventional surgical procedures and is essential for restoration of the structure and function of injured tissues. While the discontinuity in soft tissues is traditionally secured with mechanical perforating devices (e.g., sutures, tacks, and staples), these devices are also a source of tissue trauma, neural irritation and persistent pain [1, 2]. They are also not suitable for complicated procedures, such as stopping leaks of bodily fluids and air in blood vessels and tissues with rather low cohesive properties (e.g., lung, spleen, and kidney). Tissue adhesives can potentially simplify complex procedures, reduce surgery time, and minimize trauma. However, existing tissue adhesives are hampered by poor adhesive strength and poor biocompatibility [3].

A biologically inspired approach is described here in developing tissue adhesive with improved materials and adhesive properties. 3,4-Dihydroxyphenylalanine (DOPA), an amino acid found in large abundance in mussel adhesive proteins, is responsible for adhesion and rapid curing of these proteins [4]. DOPA is a unique and versatile adhesive molecule capable of binding to both organic and inorganic surfaces through either irreversible covalent or strong reversible bonds [5]. DOPA-modified PEG hydrogel have previously demonstrated improved adhesive strength over fibrin glue [6].

Hydrogels are considered a good candidate for adhesive biomaterials because of their excellent biocompatibility and ease in tailoring their compositions and physical properties for desired applications [7]. However, classic hydrogels are fragile with narrow elastic range. One approach in enhancing the fracture resistance of hydrogels is to incorporate inorganic nanoparticles, which form weak physical bonds (e.g., hydrogen bonding, electrostatic interaction) with the polymer network to increase the effective crosslinking density of the nanocomposite [8]. Previously, we exploited the strong physical interaction between network bound dopamine and a nano-silicate, Laponite, in forming a fracture resistant nanocomposite hydrogel [9].

Here, we constructed an injectable nanocomposite hydrogel as potential tissue adhesives based on a 4-arm poly(ethylene glycol) end-capped with dopamine (PEG-D4) and Laponite. The effect of Laponite concentration on the gelation time, materials properties, and adhesive properties of the nanocomposite were determined.

## EXPERIMENT

Dopamine and sodium periodate ($NaIO_4$) were obtained from Acros Organics (Geel, Belgium), and Laponite XLG (Laponite) was a gift from Southern Clay Products, Inc. (Austin, TX). PEG-D4 (10k Da PEG) was prepared as previously described [10]. PEG-D4 and $NaIO_4$ precursor solutions (with or without Laponite) in phosphate buffered saline (PBS) were mixed and cast into hydrogels via a dual-syringe system. The final concentrations of PEG-D4 and Laponite were kept at 150 mg/ml and 0–2 wt%, respectively, while $NaIO_4$ was kept at 0.1–1.5 molar ratio relative to dopamine. Gelation time of hydrogels was determined when the precursor solution ceased to flow in a tilted vial [11]. Hydrogel discs (diameter=15mm; thickness=2mm) equilibrated in PBS (pH=7.4) were compressed until they fully fractured at a rate of 1.8 mm/min using a materials testing system (8872 Instron, Norwood, MA). Hydrogels were subjected to lap shear adhesion test using wetted bovine pericardium (2.5cm × 5cm strips) as the substrate with an overlap of 2.5cm × 1cm [12, 13].

## DISCUSSION

### Design of the adhesive hydrogel

PEG-D4 nanocomposite hydrogels were prepared as shown in Scheme 1. PEG, a hydrophilic, biocompatible polymer used in numerous FDA approved products (e.g., CoSeal®, Baxter, Inc.), was employed as the major structural component for this tissue adhesive hydrogel. PEG is end-capped with dopamine, which impart the inert PEG chain with the ability for interfacial binding to tissue surface in the presence of moisture, self-polymerization induced by a mild oxidant, and strong physical bond to Laponite surface. $NaIO_4$ is used as a mild oxidant to initiate curing of the adhesive and the oxidant is reduced to form inert iodide in the process. PEG is linked to dopamine via a glutaric acid linker connected to PEG using an ester linkage, which is hydrolysable under slightly basic physiological environment (pH=7.4). Nanocomposite hydrogels lost 70 % of dry mass over 2 months when incubated in PBS at 37 °C.

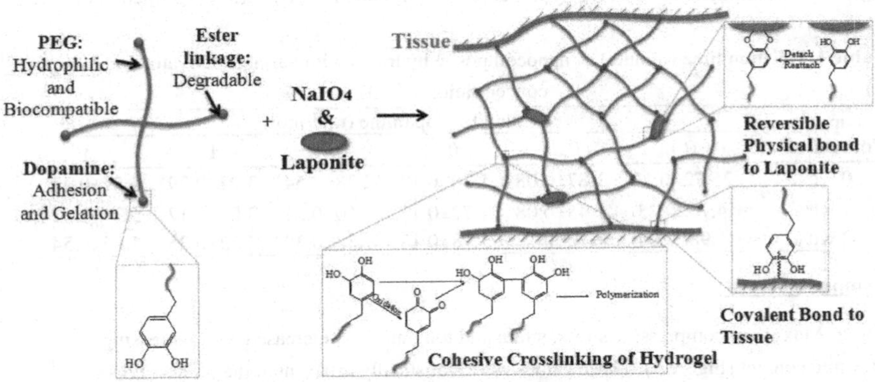


**Scheme 1.** Schematic representation of the formation of PEG-D4 nanocomposite hydrogel and its adhesion to biological tissue. Dopamine is capable of forming three types of crosslinks in this system: reversible bonding between dopamine and Laponite (**A**), covalent crosslinking between dopamine molecules resulting in curing of the adhesive (**B**), and interfacial covalent crosslinking between dopamine and functional groups (e.g., $-NH_2$) found on tissue surface (**C**).

### Gelation time

Mixing PEG-D4 with $NaIO_4$ with or without Laponite resulted in formation of cured hydrogel (Figure 1). For all formulations, gelation was the fastest at a $NaIO_4$:dopamine ratio between 0.25 and 0.75 (Table 1). Similar, $IO_4$ concentration dependence has been previously reported as the oxidized quinone reacts with unoxidized catechol groups to achieve rapid network formation [11]. Incorporation of Laponite into PEG-D4 hydrogels did not affect the gelation rate at $NaIO_4$:dopamine ratio greater than 0.25. At $NaIO_4$:dopamine ratio of 0.1, increasing Laponite from 0 to 2 wt% reduced the gelation time from over 25 min to 2 min. Dopamine formed strong interfacial bond with Laponite, which may have reduced the number of chemical crosslinks needed for hydrogel network formation. All subsequent tests utilized a $NaIO_4$:dopamine molar ratio of 0.5, given that an acceptable curing time (less than 2 min) was achieved using this $IO_4$ concentration.

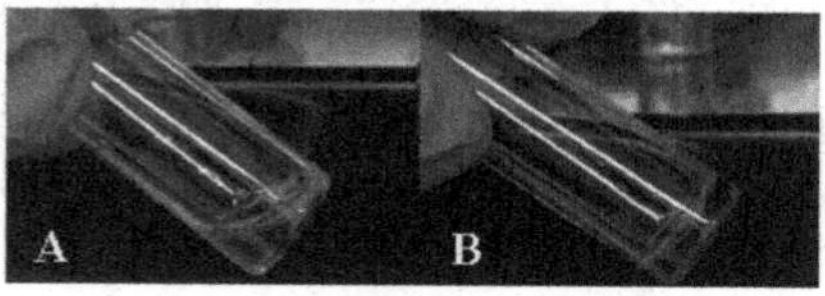


**Figure 1.** Photographs of adhesive hydrogel before (**A**) and after (**B**) curing in a tilted vial.

**Table 1.** Gelation time (minute) of nanocomposite hydrogel with variable activator and Laponite concentration (n = 3)

| Laponite Concentration | $NaIO_4$: Dopamine (mol/mol) | | | | | |
|---|---|---|---|---|---|---|
| | 0.1 | 0.25 | 0.5 | 0.75 | 1 | 1.5 |
| 0 wt% | 25.72±0.17 | 1.87±0.08 | 1.20±0.17 | 2.38±0.54 | 3.22±0.29 | 4.57±0.42 |
| 1 wt% | 4.70±0.25 | 1.45±0.08 | 1.72±0.17 | 2.60±0.21 | 3.08±0.17 | 5.33±1.52 |
| 2 wt% | 1.98±0.04 | 1.70±0.17 | 1.78±0.17 | 1.82±0.33 | 3.72±0.25 | 5.05±0.54 |

**Compression test**

Maximum compressive stress, strain and toughness all increased with increasing Laponite content (Figure 2). These values were statistically different at the highest Laponite content (2 wt%) tested. In the presence of external stress, breaking of reversible dopamine-Laponite bonds minimized structural damage to the irreversibly crosslinked network. As such, a higher strength and energy were necessary to fracture the more compliant nanocomposite hydrogel when compared to the Laponite-free PEG-D4. However, the elastic modulus was independent of Laponite concentration, indicating that the crosslinking density of these nanocomposites remained constant. Similarly, equilibrium water content (another measure of crosslinking density) for all formulations was also unaffected (data not shown). Given that PEG arm length is constant, these observations indicate that the interaction between Laponite and PEG did not contribute to the enhanced materials properties.

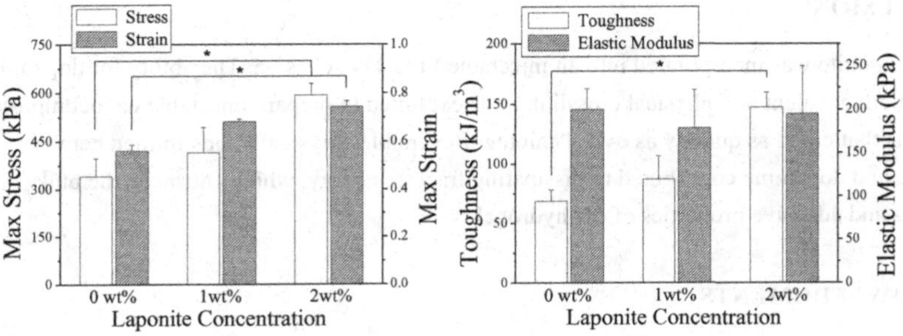


**Figure. 2** Results from compression test for PEG-D4 hydrogels with $NaIO_4$:dopamine molar ratio of 0.5 containing up to 2 wt% Laponite. *$p < 0.05$ (by one-way ANOVA test).

**Lap shear adhesion test**

Similar to the compression testing, incorporation of Lapontie improved the adhesive properties of the nancomposite hydrogel (Figure 3). Although the maximum adhesive strength was unaffected, measured work of adhesion for the nanocomposite hydrogel was over 5 times higher when compared to that of Laponite-free PEG-D4. As shown in Scheme 1, dopamine undergoes three competing crosslinking mechanisms to form the tissue adhesive. Incorporation of Laponite did not interfere with dopamine to form interfacial covalent bonds with the tissue surface. An effective tissue adhesive requires a balance of both elevated interfacial and cohesive properties. Incorporating Laponite resulted in nanocomposite hydrogels with elevated toughness, which directly translated to increased work of adhesion. Elevated fracture energy was required to disrupt the adhesive hydrogel in order to separate the adhesive joint.

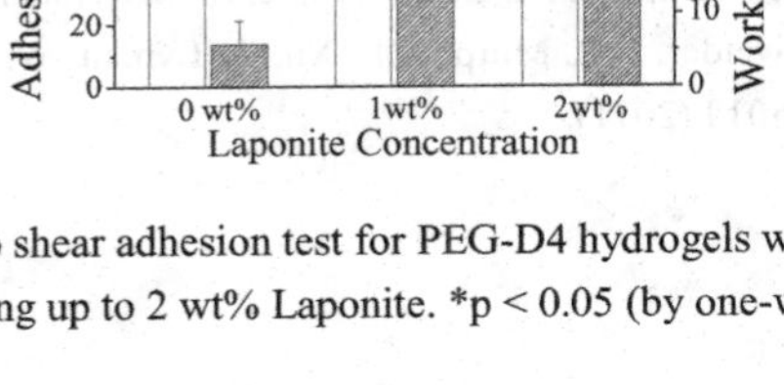

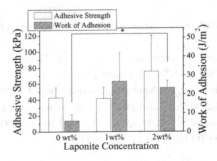


**Figure. 3** Results from lap shear adhesion test for PEG-D4 hydrogels with $NaIO_4$:dopamine molar ratio of 0.5 containing up to 2 wt% Laponite. *$p < 0.05$ (by one-way ANOVA test).

## CONCLUSIONS

Laponite was incorporated into an injectable PEG-D4 adhesive. The ability for dopamine to form both covalent and physical crosslinks was exploited to prepare injectable nanocomposite hydrogels that cured as quickly as over 1 minute. Reversible, physical bonds formed between Laponite and dopamine contributed to dissipating fracture energy, which enhanced the bulk materials and adhesive properties of the hydrogel.


## ACKNOWLEDGMENTS

This work was supported by start-up fund and Research Excellence Fund – Research Seed Grant (R01290) provided by Michigan Technological University.


## REFERENCES


1. M. Hidalgo, M. J. Castillo, J. L. Eymar and A. Hidalgo, *Hernia* **9**, 242 (2005).
2. E. Stark, K. Oestreich, K. Wendl, B. Rumstadt and E. Hagmüller, *Surg. Endosc.* **13**, 878 (1999).
3. Y. Ikada, In *Wound Closure Biomaterials and Devices*, edited by C. C. Chu, J. A. Von Fraunhofer and H. P. Greisler, (CRC press, 1997) pp. 317.
4. B. P. Lee, P. B. Messersmith, J. N. Israelachvili, and J. H. Waite, *Annu. Rev. Mater. Res.* **41**, 99 (2011).
5. H. Lee, N. F. Scherer and P. B. Messersmith, *Proc. Natl. Acad. Sci.* **103**, 12999 (2006).
6. S. A. Burke, M. Ritter-Jones, B. P. Lee, P. B. Messersmith, *Biomed. Mater.* **2**, 203 (2007).
7. B. D. Ratner and S. J. Bryant, *Annu. Rev. Biomed. Eng.* **6**, 41 (2004).
8. K. Haraguchi and T. Takehisa, *Adv. Mater.* **14**, 1120 (2002).
9. S. Skelton, M. Bostwick, K. O'Connor, S. Konst, S. Casey and B. P. Lee, *Soft Matter* **9**, 3825 (2013).
10. J. L. Dalsin, B. P. Lee, L. Vollenweider, S. Silvary, J. L. Murphy, F. Xu, A. Spitz and A. Lyman, U.S. Patent No. 8,119,742 (21 February 2012).
11. B. P. Lee, J. L. Dalsin and P. B. Messersmith, *Biomacromol.* **3**, 1038 (2002).
12. J. L. Murphy, L. Vollenweider, F. Xu and B. P. Lee, *Biomacromol.* **11**, 2976 (2010).
13. M. Brodie, L. Vollenweider, J. L. Murphy, F. Xu, A. Lyman, W. D. Lew and B. P. Lee, *Biomed. Mater.* **6**, 015014 (2011).

Mater. Res. Soc. Symp. Proc. Vol. 1569 © 2013 Materials Research Society
DOI: 10.1557/opl.2013.797

# Inducible nitric oxide releasing poly-(ethylene glycol)-fibrinogen adhesive hydrogels for tissue regeneration

Margaret Brunette, Hal Holmes, Michael G. Lancina, Weilue He, Bruce P. Lee, Megan C. Frost, Rupak M. Rajachar*

Department of Biomedical Engineering, Michigan Technological University, Houghton, MI 49931, U. S. A.

## ABSTRACT

Nitric oxide (NO) release can promote healthy tissue regeneration. A PEG-fibrinogen adhesive hydrogel that would allow for *inducible* NO release was created with mechanical properties that could be tailored to specific applications and tissue types. PEG (4-arm)-fibrinogen hydrogels of varying ratios were derivatized with S-nitroso-N-acetyl-D, L-penicillamine (SNAP)-thiolactone to create an active NO donor material. Controlled release from gels was established using light as the activating source, although temperature, pH, and external mechanical loading are also means to induce active NO release. Gels with varying ratios of fibrinogen to PEG were made, derivatized, and tested. Gels below a ratio of 1.5:1 (fibrinogen:PEG) did not gel, while at ratio of 1.5:1 gelation occurs and NO release can be induced. Interestingly, the release from 1.5:1 gels was significantly lower compared to 2:1 and 3:1 gel formulations. Rheometric data show that lower ratio gels are more elastic than viscous. Derivatized gels exhibited linear elastic moduli, behaving more like other more synthetic hydrogels. Swelling data indicates that as the ratio of fibrinogen to PEG increases the swelling ratio decreases, likely due to the hydrophobic nature of the NO donor. Cells remain viable on both derivatized and non-derivatized gels.

## INTRODUCTION

In tendinopathy, overload of tendon tissue is central to the pathologic process. Overload may result in incremental weakening and failure of tendon tissue [1]. Repetitive micro-trauma (loss of structure in underlying collagen matrix) can eventually reach a threshold that results in altered mechanical properties, ultimately producing abnormal loading, as well as localized fiber degeneration and pain in injured tissues [2].

Nitric oxide (NO) is a necessary factor in various wound-injury healing conditions including tendinopathy [3]. The beneficial effects of NO are bound to its controlled availability. Inhibition of NO production during tendon repair results in a significant reduction in stable reparative tissue volume. Although recent work has shown the efficacy of NO therapy (i.e. injected and topical NO-donor group delivery) on soft tissue healing and regeneration [3], inducible NO release systems remain limited. The overall goal of this work is to develop a novel inducible NO releasing adhesive-hydrogel. We hypothesize that these adhesive-hydrogels could be used as injectable delivery systems to bind to injured tissue and deliver controlled NO as a regenerative support to accelerate wound healing. The adhesive-hydrogel is composed of a base adhesive poly(ethylene glycol) (PEG)-fibrinogen with the fibrinogen unit derivatized into an inducible NO donor.

For this work we chose to use PEG-fibrinogen as our model adhesive-hydrogel material, although the approach is readily amenable to any material combination containing free primary amines. The fibrinogen/fibrin system is a highly compatible natural biopolymer that can be prepared from autologous sources and is already an established material used as a tissue sealant (e.g. Evicel-fibrin sealant), making it a strong candidate for this work. Fibrinogen has the unique ability to directly attract and stably harbor different cells types, which in turn can aid in degrading the PEG-fibrinogen over time and its replacement with extracellular matrix (ECM) [4]. Fibrinogen-based hydrogel materials have been shown to inherently facilitate cell adhesion and promote regeneration through enhanced angiogenesis and cell viability. Additionally, these materials can be cross-linked to control stability and degradation behavior, i.e. hydrolysis of PEG-linkage to fibrinogen or fibrinogen itself [5]. Our 4-arm PEG is end-capped with reactive *N*-hydroxysuccinimidyl (NHS) moieties that readily react with primary amine groups found on tissue surface, mimicing PEG-based adhesive found on the market (e.g., DuraSeal). We can further fine-tune the degradation rate, cross-linking density (affecting mechanical and adhesive properties), and curing rate by changing the PEG MW and number of arms. Using both components (fibrinogen and PEG) we can control degradability, mechanical and adhesive properties, and curing rate.

In this work we constructed a PEG-fibrinogen NO releasing adhesive hydrogel as a potential tissue adhesive and regenerative support based on a 4-arm PEG-NHS Succinimidyl Glutarate-fibrogen-monomer hydrogel. The materials properties, release behavior, and biocompatibility were determined.

## EXPERIMENT

### Hydrogel preparation and NO release

4-arm PEG Succinimidyl Glutarate (PEG-NHS, JenKem USA) and fibrinogen (Sigma Aldrich) were mixed with 10 mM Tris buffer, pH 7.40 in separate vials. PEG and fibrinogen solutions were then mixed in a beaker with varying ratios of fibrinogen to PEG (3:1, 2:1, 1.5:1, and 1:1) at a final concentration equal to 100 mg/mL for all samples. After gelation, samples were derivatized by soaking in t-butyl nitrite to covalently link the NO-donor (nitrosothiol, S-nitroso-N-acetyl-D-penicillamine (SNAP)) for 30 minutes and allowed to sit overnight (Figure. 1A, schematic) [6, 7]. Light activated NO release was determined using a nitric oxide analyzer (Sievers 280i Nitric Oxide Analyzer; Boulder, CO, USA). Samples were exposed to light (Intensity (I) = 116 $W/m^2$) at five-minute intervals.

**Rheometry**

Derivatized and non-derivatized gels were prepared and tested using a Bohlin C-VOR 200 NF rheometer. All samples were cut with a biopsy punch (dia.=15mm). Oscillatory frequency (0.1-10 Hz at 0.1 strain) and strain sweep measurements were made with a gap width of 0.88 the height of the gel. Frequency sweep data is shown in Figure 2.

**Swelling Behavior**

Gels were cut into 10 mm disks, placed in phosphate buffered saline (PBS), and allowed to swell for 24 hours. Samples were weighed after 24 hours to determine the swollen weight ($W_s$). The samples were then freeze-dried overnight to determine the dry weight ($W_d$). The swelling ratio was determined using the following equation (Figure 3):

$$\text{Swelling Ratio} = W_s/W_d \qquad (1).$$

**Biocompatibility- Cell Viability**

Fibroblasts (L929, ATCC) were cultured directly on un-derivatized and derivatized PEG-fibrinogen adhesive hydrogels (thin discs; thickness~2mm, diameter=10mm) at a cell concentration of 2 x $10^4$ cells/$cm^2$. Cells were allowed to adhere for one day. Light-exposed hydrogels (Intensity=116 $W/m^2$) were stimulated for 60 seconds at 24 hours. Cell viability was measured after 2 days using a live cell stain (calcein-AM) and fluorescence imaging (Olympus BX51).

**DISCUSSION**

**Hydrogel preparation and NO release**

Data establishes controlled NO release in response light exposure from derivatized PEG-fibrinogen hydrogels (Figure 1B). A ratio of 1:1 (fibrinogen: PEG) does not allow for gelation, while a ratio of 1.5:1 gels and exhibits NO release. Interestingly, as shown in Figure 2B, the release from 1.5:1 gels was significantly lower compared to 2:1 and 3:1 gel formulations. Fibrinogen contains free amine groups that serve two functions: 1) macromolecular cross-linking when reacted with PEG-NHS and 2) functionalization with the NO-donor. As such, a higher fibrinogen concentration was necessary for network formation and elevated level of NO release.

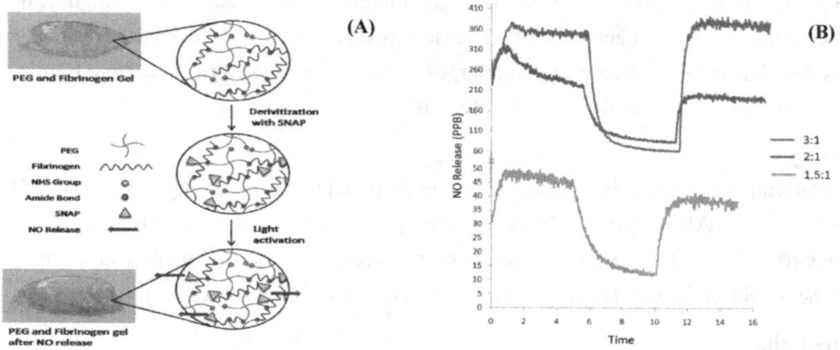


**Figure 1. (A)** PEG-Fibrinogen hydrogel synthesis and derivatization. **(B)** NO release from derivatized hydrogels was controllable- on/off activation with light source.

**Rheometry**

Frequency sweeps for PEG-fibrinogen gels exhibited behavior of a cross-linked network where G' was independent of frequency and values for G' were higher than G" (Figure 2). Prior to derivatization, G' values increased with PEG content. The low ratio (1.5:1) gels exhibited the highest measures of G'. The 4-armed PEG provides a branching point for network formation which contributed to a more elastic behavior. After derivatization, gels with higher fibrin content demonstrated a jump in both G' and G" values. SNAP is relatively hydrophobic and attaches to the amine groups on fibrinogen. Hydrophobic-hydrophobic interaction of SNAP functional group likely increases physical cross-linking density of these gels. There is 4-5 fold and 6-8 fold increase in G' and G", respectively, for 2:1 and 3:1 gels after derivatization with no change for 1.5:1 gels. SNAP is not only more hydrophobic but also replaces the hydrophilic $NH_2$ group.

**Swelling Behavior**

Both 2:1 and 3:1 non-derivatized gels displayed reduced water content compared to 1.5:1 gels, with a similar trend seen between derivatized gels. The hydrophobic nature of SNAP likely reduced the ability of derivatized gels to swell, seen more predominantly in 3:1 and 1.5:1 gels (Figure 3). Given that SNAP modifies fibrinogen, it is reasonable that the greatest change in swelling between derivatized and non-derivatized gels occurred for the 1.5:1 gels.

**Biocompatibility- Cell Viability**

Derivatized and non-derivatized gels were determined suitable to sustain viable cells in culture (> 95-percent cell survival). Interestingly, qualitative morphological assessment suggests that derivatized gels promote greater cell spreading, indicative of a more adhesive phenotype and/or stiffer substrate character. Further measures of cell morphology and correlation to surface stiffness measures may provide more insight into these observations and will be part of future studies of cellular function associated with these gels with and without NO release (Figure 4).

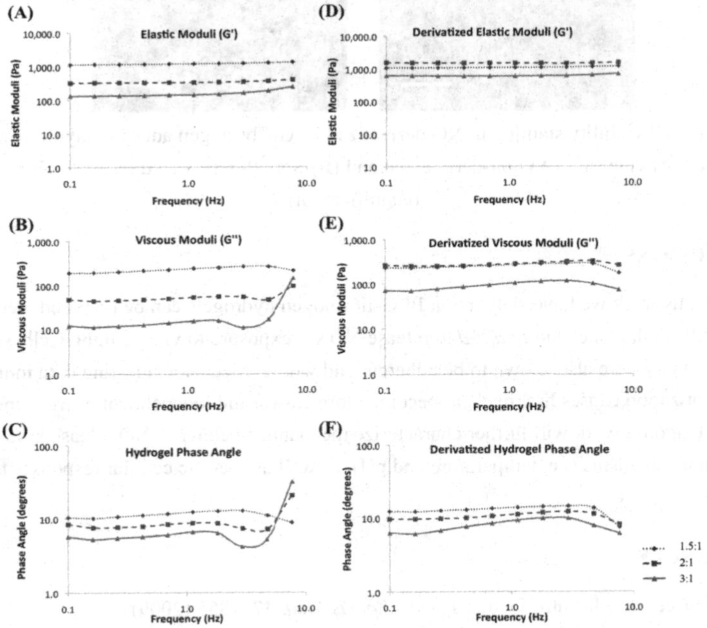


**Figure 2.** Rheometric characterization of PEG-fibrinogen derivatized hydrogels.

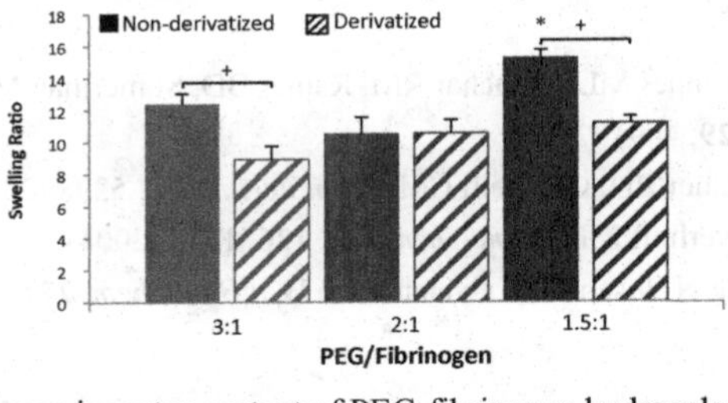


**Figure 3.** Percent change in water content of PEG-fibrinogen hydrogels after SNAP-derivatization. Significant differences were seen between non-derivatized 1.5:1 gels and all

others (*), and derivatized and non-derivatized gels (+). An ANOVA with standard t-test was used, p-values <0.05 considered significant. Error bars indicate ± SEM.

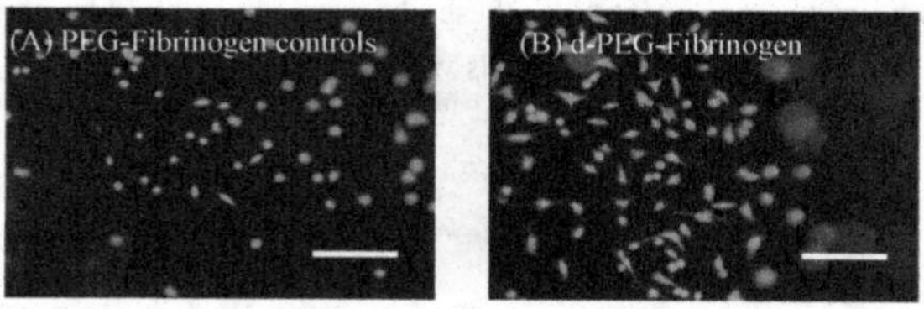


**Figure 4.** Cell viability staining in NO derivatized PEG-fibrinogen adhesive-hydrogels showed no loss of viability, **(A)** non-derivatized and **(B)** SNAP-derivatized (scale = 25um, 20x magnification).

## CONCLUSIONS

In this work we have shown that PEG-fibrinogen hydrogels can be prepared, derivatized with SNAP-thiolactone, and *induced* to release NO via exposure to visible light. Cells seeded on these hydrogels were also shown to be adherent and viable. Mechanical testing data indicates that derivatization causes hydrogels to become more elastic and independent of frequency changes. Ongoing work will further characterize mechanical behavior, NO release kinetics and activation mechanisms (i.e. temperature and pH), as well as specific cellular responses to these hydrogels.

## REFERENCES

1. JD Rees, N Maffulli, J Cook J, *Am J Sports Med.* **37**, 1855 (2009).
2. Clayton RA, Court-Brown CM. *Injury.* **39**, 1338 (2008).
3. Isenberg JS, Ridnour LA, Espey MG, Wink DA, Roberts DD. *Microsurgery.* **25**, 442 (2005).
4. Osathanon T, Linnes ML, Rajachar RM, Ratner BD, Somerman MJ, Giachelli CM. *Biomaterials.* **29**, 4091 (2008)
5. Linnes MP, Ratner BD, Giachelli CM. *Biomaterials.* **28**, 5298 (2007).
6. Frost MC, Meyerhoff ME. *J Am Chem Soc.* **126**, 1348 (2004).
7. Singh RJ, Hogg N, Joseph J, Kalyanaraman B. *J Biol Chem.* **271**, 18596 (1996).

Mater. Res. Soc. Symp. Proc. Vol. 1569 © 2013 Materials Research Society
DOI: 10.1557/opl.2013.1059

# Characterizing the mechanical properties of tropoelastin protein scaffolds

Audrey C. Ford[1,2], Hans Machula[2], Robert S. Kellar[1,2], Brent A. Nelson[1]
[1]Department of Mechanical Engineering and [2]Department of Biology, Northern Arizona University, Flagstaff, AZ, 86011

## ABSTRACT

This paper reports on mechanical characterization of electrospun tissue scaffolds formed from varying blends of collagen and human tropoelastin. The electrospun tropoelastin-based scaffolds have an open, porous structure conducive to cell attachment and have been shown to exhibit strong biocompatibility, but the mechanical character is not well known. Mechanical properties were tested for scaffolds consisting of 100% tropoelastin and 1:1 tropoelastin-collagen blends. The results showed that the materials exhibited a three order of magnitude change in the initial elastic modulus when tested dry vs. hydrated, with moduli of 21 MPa and 0.011 MPa respectively. Noncrosslinked and crosslinked tropoelastin scaffolds exhibited the same initial stiffness from 0 to 50% strain, and the noncrosslinked scaffolds exhibited no stiffness at strains >~50%. The elastic modulus of a 1:1 tropoelastin-collagen blend was 50% higher than that of a pure tropoelastin scaffold. Finally, the 1:1 tropoelastin-collagen blend was five times stiffer from 0 to 50% strain when strained at five times the ASTM standard rate. By systematically varying protein composition and crosslinking, the results demonstrate how protein scaffolds might be manipulated as customized biomaterials, ensuring mechanical robustness and potentially improving biocompatibility through minimization of compliance mismatch with the surrounding tissue environment. Moreover, the demonstration of strain-rate dependent mechanical behavior has implications for mechanical design of tropoelastin-based tissue scaffolds.

## INTRODUCTION

Biomaterials are materials that interface with living tissues, applications of which range from contact lenses to vascular grafts and hernia patches [1]. Because biomaterials are surrounded by living tissues, the interface between foreign material and the living tissue is an essential consideration for evaluating the utility of a biomaterial [2]. The biocompatibility of a material is measured by a range of techniques, including changes in mechanical properties or surface conditions and chemical or histological identification of inflammatory indicators [3]. A candidate biomaterial must also have the required strength or elasticity for its given application [4], but mechanical characterization can be more complex than just evaluating load and extension behavior. There is evidence that the mechanical behavior of a material can play a role in its biocompatibility, with materials that better match the compliance and mechanical properties of the surrounding tissue eliciting a more favorable response [5].

Electrospinning is a method that has been applied to biomaterial production to produce micro- and nano-architectures more similar to the open micro-architecture of native extracellular matrix and promote cell growth [4,6]. In the electrospinning process, a protein solution is sprayed across a gap at high voltage, allowing customization of the porosity, fiber diameter, and potentially the mechanics of the synthesized material [7,8,9]. Electrospinning has been used on a variety of synthetic polymers and natural proteins for proposed use in blood vessel and wound-dressing applications [10,11,12,13].

Tropoelastin is the soluble precursor to elastin, an extracellular matrix protein which confers elasticity to tissues and organs [14]. Elastin has shown favorable biocompatibility exhibited both *in vitro* by cellular ingrowth and *in vivo* by minimal immune response when implanted [15]. Elastin is compliant and one of the most durable proteins, able to endure billions of loading and unloading cycles without failure in blood vessels [16].

Electrospinning tropoelastin scaffolds provides a spectrum of variables that can be adjusted to affect the properties of the resultant material [13,17]. In addition to being able to alter the architecture, porosity, and degree of crosslinking of the material [18], electrospun scaffolds can be produced from blends of multiple proteins or synthetic polymers in solution together to better replicate in vivo extracellular matrix [7,8,19]. Little work has been done to characterize the effect of continuous variation of the protein ratios on the mechanics of the material. Similarly, little work has been done to characterize the effect of varying degrees of crosslinking on the material. Mechanical testing of electrospun human tropoelastin materials has shown a percent elongation at failure of 75 ±5% and a shift in the elastic modulus between 40-50% strain from 0.15 ±0.03MPa to 0.99 ±0.17Mpa [7]. However, no studies have considered the effect of strain rate on the material or looked at the properties of the noncrosslinked state.

Electrospun tropoelastin scaffolds offer significant potential as candidate biomaterials, having shown favorable cell attachment as well as similar mechanical character to living tissues. Determining the mechanical behavior of tropoelastin scaffolds synthesized in various forms will aid in understanding the potential applications of tropoelastin-based biomaterials. This paper describes the mechanical characterization of electrospun tropoelastin at various strain rates, hydration conditions, degrees of crosslinking, and degrees of blending with collagen.

## EXPERIMENTAL

### Electrospinning

Recombinant human tropoelastin was provided by Dr. Burt Ensley (Protein Genomics, Inc., Sedona, AZ). PureCol®, Type 1 bovine collagen was obtained from Advanced Biomatrix. An HV 350 R adjustable power supply was used as the voltage source. A 1 $in^2$ aluminum target was secured to a plastic stand and connected to the ground electrode. For each scaffold, 100mg of protein, either tropoelastin or a 1:1 collagen:tropoelastin blend, was dissolved into 1mL of hexafluoroisopropanol, loaded into a 1.0cc syringe with a blunt-tip 18 gauge needle, and secured to a syringe pump set to 2.5 mL/hour. The distance from the needle tip to the grounded aluminum target was approximately 8cm with a voltage of 18-22kV. Crosslinking was performed at room temperature for 24 hours with glutaraldehyde vapor. Figure 1 shows a scanning electron microscope image of a synthesized electrospun scaffold. Crosslinking was verified through fiber growth in electron microscope images of the scaffolds and was further evidenced by the lack of solubility of the scaffolds in water when crosslinked. Fiber diameter and porosity were evaluated from scanning electron microscope images of the synthesized scaffolds using custom image processing scripts written in Matlab (Mathworks), with porosity determined from the void fraction, as described previously [20].

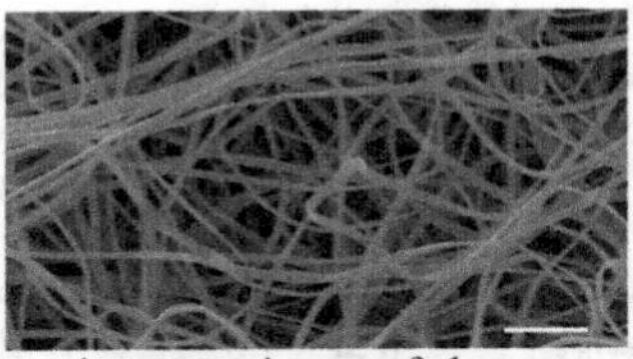

**Figure 1**. Scanning electron microscope image of electrospun tropoelastin scaffold. The scale bar in the figure represents 10 microns.

**Mechanical Characterization**

The mechanical properties of electrospun tropoelastin were characterized using ASTM standard D882-10 on a home-built servomotor uniaxial tensile testing system. Torsional spring clips gripped the material, with the lower clip mounted to a rigid base and the upper clip attached to the lever arm of the servomotor. The internal compliance of the testing system and grip attachments was calibrated and subtracted from all material testing data in order to isolate the extension of the material being tested. Force and displacement measurements were acquired using a data acquisition board and software interface.

Mechanical tests were performed on two different materials: 100% tropoelastin, and a 1:1 blend of tropoelastin and collagen. For each condition, tests were performed both for crosslinked and noncrosslinked samples. Testing samples had a width of 10mm, thickness of under 0.25mm, and an initial gage length of 1mm, with dimensions measured by a micrometer. Stress was calculated using the measured thickness of the scaffold materials. Each material was tested at two strain rates and repeated two times at room temperature. The dry testing was done at 0.0083mm/s and 0.0017mm/s and the hydrated testing was done at 0.1667mm/s and 0.0333mm/s. The slower of these strain rates is prescribed by the ASTM standard and the faster rate is five times that rate. The maximum range of motion of the system was 3.5mm. Hydrated testing was done by submerging the entire system in phosphate buffered saline (PBS) for 60 seconds, and then remaining in PBS during completion of the test.

Force and distance data were converted to engineering stress and strain. Stress and strain were plotted and a linear regression was used to determine the elastic modulus. Because the mechanical characteristics of the materials were nonlinear, elastic moduli are reported for both the low strain (<50-100%) and high strain (>100%) regions of the data. The system compliance and slack were removed numerically. Modulus results are reported for both runs, while for clarity the figures show only a single representative graph due to similarity between the runs.

**RESULTS**

Synthesized scaffolds had a dry thickness ranging from 0.3-0.8mm. Sample porosities were close to 50% and fiber diameters ranged from 350nm to 3μm, with an average of 1μm, similar to previous reports of similar scaffolds [8]. The effect of hydration on the mechanical properties of the scaffolds is shown in Figure 2. The low strain elastic modulus measurements of 1:1 tropoelastin-collagen blends were 28 and 15 MPa when dry, and decreased to 0.0091 and .012 MPa when hydrated, a change of over three orders of magnitude. The high strain elastic modulus measurements of the blend were .067 and .063 MPa. Since any implantable scaffold application would be in a hydrated environment, all further testing was performed hydrated.

Swelling of the scaffolds upon hydration was 1-3% as evidenced by initial slack during testing. The right image in Figure 2 shows the effect of different protein blends on the mechanics of the scaffolds. The low strain elastic modulus of the pure tropoelastin was 0.0063 and .0046 MPa and the high strain elastic modulus was 0.015 and .0097 MPa, suggesting a progressive shift from the elasticity of tropoelastin to the stiffer collagen.

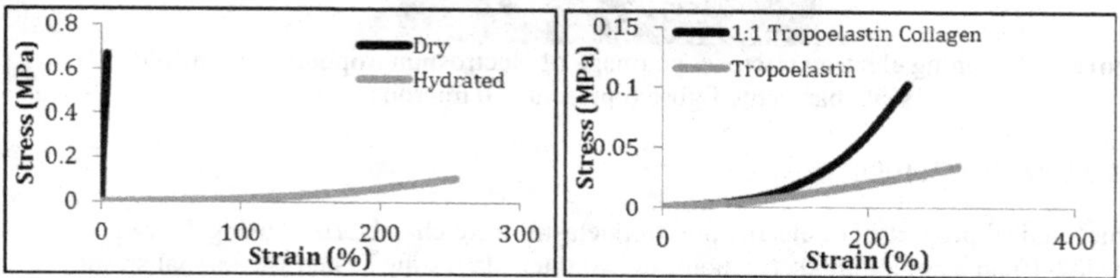


**Figure 2**. Mechanical testing of synthesized scaffolds. Left: effect of hydration on crosslinked 1:1 tropoelastin-collagen. Right: comparison of mechanical properties of crosslinked tropoelastin and 1:1 tropoelastin-collagen blends.

The left image of Figure 3 shows the stress strain characteristics for crosslinked and non-crosslinked tropoelastin scaffolds. The low strain elastic modulus of noncrosslinked pure tropoelastin was 0.0042 and .0067 MPa, which was very similar to the crosslinked material. However, above 50% strain the noncrosslinked scaffold deviated from the crosslinked scaffold, exhibiting virtually no stiffness. The difference between crosslinked and noncrosslinked materials became more pronounced at strains above 100%. Noncrosslinked tropoelastin-collagen blends were not tested due to instability of noncrosslinked collagen when hydrated.

To test if the material exhibited viscoelasticity, the 1:1 tropoelastin-collagen blend was tested at two different strain rates, shown in the right image of Figure 3. At five times the standard strain rate, the low strain elastic modulus was 0.057 and .037 MPa, nearly five times the magnitude of the low-strain modulus of the sample at the standard strain rate. At strains beyond 50%, the modulus was 0.13 and 0.15 MPa, again higher at the faster strain rate. Results of high strain rate testing of tropoelastin scaffolds were inconclusive and are under further investigation.

## DISCUSSION

Most synthetic polymer biomaterials do not significantly change their mechanical character when they are hydrated, for example the elastic modulus of expanded Polytetrafluroethylene (ePTFE) was measured dry and hydrated with this system, and the results were similar to reported values, and exhibited no change upon hydration. Comparatively, hydration of the 1:1 tropoelastin-collagen blend caused a reduction in elastic modulus of three orders of magnitude. That such a drastic change occurs in the mechanical characteristics of electrospun elastin and collagen scaffolds upon hydration underscores the importance of performing mechanical testing of such scaffolds in their end-use environment. Moreover, mechanical properties of such tissue scaffolds when dry only have significance for the durability of the material during handling, and should not be assumed to be representative of the mechanical characteristics when implanted.

Scaffolds made from a blend of proteins showed a progressive change in mechanics, with the incorporation of collagen resulting in an increased scaffold stiffness as compared to pure tropoelastin. This suggests that there is a continuous variation between the mechanics of either of the two pure protein materials that can be leveraged by adjusting the ratio of the blend. As different tissues in the body have unique protein compositions, the protein blend of the material could be altered to meet the mechanical and biological needs.

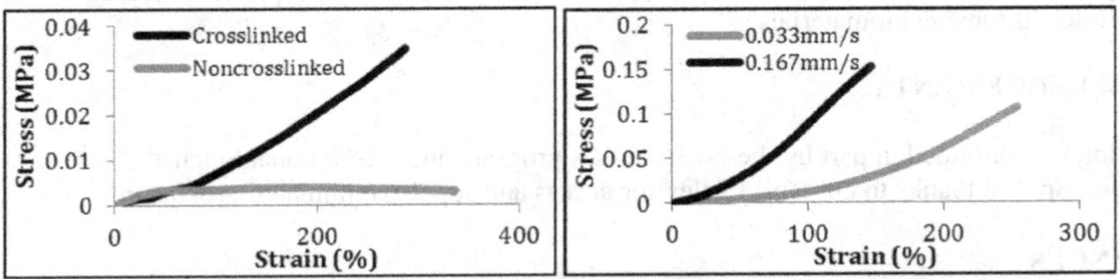


**Figure 3.** Left: Noncrosslinked tropoelastin scaffolds exhibit no stiffness at higher strains, but similar stiffness to crosslinked scaffolds at lower strains. Right: Higher strain rates resulted in an increased stiffness for 1:1 tropoelastin-collagen blend.

While noncrosslinked collagen readily dissolves when hydrated, noncrosslinked tropoelastin maintained the same low strain elastic modulus as crosslinked tropoelastin. This suggests that tropoelastin may have the flexibility to be used as a resorbable material, with the resorption half-life potentially controlled by the degree of crosslinking. Additionally, while the results presented here only showed data for fully crosslinked and noncrosslinked materials, partial crosslinking or different crosslinking techniques may also allow the synthesis of scaffolds that maintain the low stiffness elasticity at high strain exhibited by the noncrosslinked tropoelastin scaffold.

The majority of existing data on electrospun tropoelastin and collagen scaffolds have focused on single strain rates. However, the increased stiffness exhibited when the scaffolds were tested at five times the ASTM standard strain rate indicates that a more thorough understanding is needed of the viscoelastic behavior of these scaffolds. Whether in cardiac, dermal, or other applications, the tissue scaffolds may be subject to both static and dynamic loads, as they will be attached to the tissues of a dynamic, living, organism. As such, the strain-rate dependence of the mechanical characteristics of electrospun tropoelastin and collagen must be considered when designing scaffolds.

## CONCLUSION

Hydration, protein composition, crosslinking, and strain rate are all variables that can affect the mechanics of the protein scaffolds. The modulus of the 1:1 blend of tropoelastin-collagen was three orders of magnitude lower when hydrated than when dry. Blending collagen with tropoelastin made the material stiffer than a pure tropoelastin scaffold. Noncrosslinked material was less stiff than crosslinked material, but in the case of tropoelastin, maintained a similar modulus at low strains before losing stiffness at higher strains. Finally, the 1:1 tropoelastin-collagen blends demonstrated viscoelastic behavior with increased stiffness at higher strain rates. By manipulating the composition and crosslinking, protein scaffolds made from tropoelastin and collagen offer the potential for being adjusted for specific mechanical requirements. This will allow for electrospun biomaterials to be engineered to mechanically match the compliance of the tissue in their given application. Further characterization is needed to understand the broader range and limits of the mechanics of these materials, including evaluating different protein blends to more accurately capture the native protein composition in tissues, evaluating variable levels and methods of crosslinking, performing fatigue testing, and performing further dynamic characterization to better understand the mechanical characteristics of such scaffolds over a wider range of strain rates. These results show promise for the use of electrospun tropoelastin-based scaffolds as tunable biomaterials.

## ACKNOWLEDGEMENTS

Funding was provided in part by the NAU HURA program and the National Science Foundation. Special thanks to Dr. Burt Ensley for access and supply of human tropoelastin.

## REFERENCES

1. P. Chen, J. McKittrick, and M.A. Meyers, Prog. in Mat. Sci. 57, 1492-1704 (2012).
2. S.V. Bhat, *Biomaterials,* 2nd ed. (Alpha Science International Ltd. Harrow, 2007) p. 1.
3. A.F. von Recum, *Handbook of Biomaterials Evaluation,* 2nd ed. (Edwards Brothers, Ann Arbor, 1999) p. 1.
4. C.P. Barnes, S.A. Sell, E.D. Boland, D.G. Simpson, and G.L. Bowlin. Adv. Drug Del. Rev. 59, 1413-1433 (2007).
5. S.J. Hollister, R.D. Maddox, and J.M. Taboas. Biomat. 23, 4095-4103 (2002).
6. J.J. Norman, and T.A. Desai, An. of Biomed. Engin. 34, 89-101 (2005).
7. K.A. McKenna, M.T. Hinds, R.C. Sarao, P.C. Wu, C.L. Maslen, R.W. Glanville, D. B. Gregory, and K.W. Gregory. Acta Biomat. 8, 225-233 (2012).
8. J. Rnjak-Kovacina, S.G. Wise, Z. Li, P.K.M. Maitz, C.J. Young, Y. Wang, and A.S. Weiss, Acta Biomat. 8, 3714-3722 (2012).
9. M. Li, M.J. Mondrinos, M.R. Gandhi, F.K. Ko, A.S. Weiss, P.I.Lelkes, Biomat. 26, 5999-6008 (2005).
10. S.J. Lee, J.J. Yoo, G.L. Lim, A. Atala, and J. Stitzel. J. Biomed. Mat. Res. A. 10, 999-1008 (2007).
11. X. Zhang, V. Thomas, Y.K. Vohra, J. of Mat. Sci. 21, 541-549 (2009).
12. S.G. Wise, M.J. Byrom, A. Waterhouse, P.G. Bannon, M.K.C. Ng, and A.S. Weiss, Acta Biomat. 7.1, 295-303 (2011).
13. J.A. Matthews, G.E. Wnek, D.G. Simpson, and G.L. Bowlin, Biomacro. 3, 232-238 (2002).
14. W.F. Daamen, J.H. Veerkamp, J.C.M. van Hest, and T.H. van Kuppevelt, Biomat. 28, 4378-4398 (2007).
15. K.A. Woodhouse, P. Klement, V. Chen, M.B. Gorbet, F.W. Keeley, R. Stahl, et al. Biomat. 25, 4543–4553 (2004).
16. C.M. Bellingham, M.A. Lillie, J.M. Gosline, G.M. Wright, B.C. Starcher, A.J. Bailey, K.A. Woodhouse, and F.W. Keeley, Biopol. 70.4, 445-55 (2003).
17. Y.F.Goh, I. Shakir, R. Hussain, J. of Mat. Sci. 48, 3027-3054 (2013).
18. F. Sabrana, M. Lorusso, C. Canale, B. Bochicchio, M. Vassalli, J. of Biomech. 44, 2118-2122 (2011).
19. L. Buttafoco, N.G. Kolkman, P. Engbers-Buijtenhuijs, A.A. Poot, P.J. Kijkstra, I.Vermes, and J. Feijen, Biomat. 27, 724-734 (2006).
20. S. Soliman, S. Pagliari, A. Rinaldi, G. Forte, R. Fiaccavento, F. Pagliari, O. Frenzese, M. Minieri, P. Di Nardo, S. Licoccia, and E. Traversa, Acta Biomat. 6, 1227-1237 (2010)

**Mater. Res. Soc. Symp. Proc. Vol. 1569 © 2013 Materials Research Society**
**DOI: 10.1557/opl.2013.764**

# Designing Nanostructured Hybrid Inorganic-biological Materials via the Self-assembly

Evan Koufos and Meenakshi Dutt
*Chemical and Biochemical Engineering, Rutgers- The State University of New Jersey, Piscataway, New Jersey 08854, USA*

## ABSTRACT

Our objective is to design nanostructured hybrid inorganic-biological materials using the self-assembly of functionalized nanotubes and lipid molecules. In this presentation, we summarize the multiple control parameters which direct the equilibrium morphology of a specific class of nanostructured biomaterials. Individual lipid molecules are composed of a hydrophilic head group and two hydrophobic tails. A bare nanotube encompasses an ABA architecture, with a hydrophobic shaft (B) and two hydrophilic ends (A). We introduce hydrophilic hairs at one end of the tube to enable selective transport through the channel. The dimensions of the nanotube are set to minimize its hydrophobic mismatch with the lipid bilayer. We use a Molecular Dynamics-based mesoscopic simulation technique called Dissipative Particle Dynamics which simultaneously resolves the structure and dynamics of the nanoscopic building blocks and the hybrid aggregate. The amphiphilic lipids and functionalized nanotubes self-assemble into a stable hybrid vesicle or a bicelle in the presence of a hydrophilic solvent. We demonstrate that the morphology of the hybrid structures is directed by factors such as the temperature, the molecular rigidity of the lipid molecules, and the concentration of the nanotubes. We present material characterization of the equilibrium morphology of the various hybrid nanostructures. A combination of the material characterization and the morphologies of the hybrid aggregates can be used to predict the structure and properties of other hybrid materials.

## INTRODUCTION

Our goal is to design hybrid nanostructured biomaterials for use in controlled release applications such as drug delivery, sensing and imaging. These applications require a material platform that can store active compounds and release them upon demand. Our earlier investigations on hybrid nanostructured biomaterials composed of functionalized nanotubes and amphiphilic lipid molecules demonstrated the formation of hybrid lipid bilayers [1-5.] We demonstrated that the morphology of the hybrid aggregate depended upon the rigidity of the aggregate which was controlled by the concentration of the nanotubes [2] and the molecular chain stiffness of the lipid hydrophobic tails [4.] The aggregate morphology can also be tuned via the composition of the soft material which supports the rigid nanoscopic components. In this paper, we explore the effect of the molecular architecture of the lipid molecules which constitute the bilayer on the mechanical properties of the bilayer.

Via the Dissipative Particle Dynamics (DPD) approach [1-6] we will investigate the role of the architecture of amphiphilic lipids on the properties of lipid bilayer membranes. We focus on four lipid architectures with differing molecular geometries arising due to the head group shape and the hydrocarbon tail length. We investigate the properties of the bilayer membranes arising from each of the lipid architectures by characterizing the membrane thickness and the interfacial tension as a function of the inter-lipid spacing or area per lipid These investigations can potentially guide the design and creation of effective controlled-release vehicles for lab-on-chip devices for applications in the areas of bionanotherapeutics, sensing and catalysis.

## THEORY

Dissipative Particle Dynamics (DPD) [1-6] is a mesoscopic particle-based simulation method that captures the molecular details of the components through soft-sphere coarse-grained (CG) models, and reproduces the hydrodynamic behavior of the system over extended time scales. The DPD method is highly effective in resolving the morphology and dynamics of complex fluids over extended spatio-temporal scales due to the soft-repulsive interaction forces. The DPD technique involves integrating the dynamics of soft spheres via Newton's equation of motion through similar numerical integrators used in other MD-based simulation methods [7-8.] The soft spheres interact via a soft-repulsive force ($\mathbf{F}_{c,ij} = a_{ij}(1 - {r_{ij}}/{r_c})\hat{\mathbf{r}}_{ij}$, for $r_{ij} < r_c$ and $\mathbf{F}_{c,ij} = 0$, for $r_{ij} \geq r_c$), a dissipative force ($\mathbf{F}_{d,ij} = -\gamma\omega^d(r_{ij})(\hat{\mathbf{r}}_{ij} \bullet \mathbf{v}_{ij})\hat{\mathbf{r}}_{ij}$) and a random force

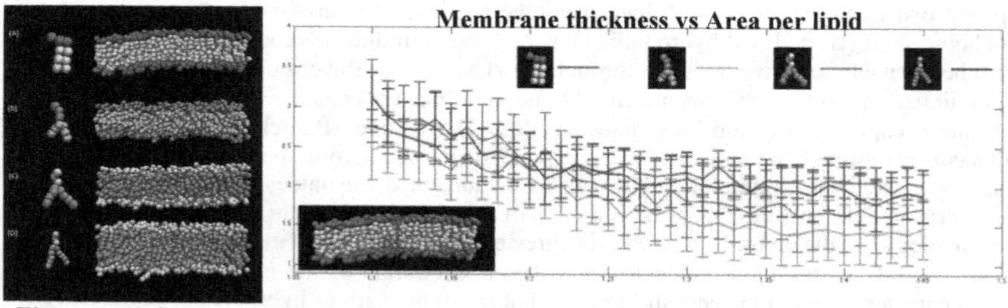


**Figure 1.** (*left*) Images of the lipid molecules and the corresponding tensionless lipid bilayer membranes for lipid architectures (a) A, (b) B, (c) C and (d) D. (right) Measurements of the membrane thickness as a function of the area per lipid for the four lipid architectures.

($\mathbf{F}_{r,ij} = -\sigma\omega^r(r_{ij})\theta_{ij}\hat{\mathbf{r}}_{ij}$), where $\omega^d(r) = [w^r(r)]^2 = (1-r)^2$ (for $r < 1$), $\omega^d(r) = [w^r(r)]^2 = 0$ (for $r \geq 1$) and $\sigma^2 = 2\gamma k_B T$. $a_{ij}$ is the maximum repulsion between spheres i and j, $\mathbf{v}_{ij} = \mathbf{v}_i - \mathbf{v}_j$ is the relative velocity of the two spheres, $\mathbf{r}_{ij} = \mathbf{r}_i - \mathbf{r}_j$, $r_{ij} = |\mathbf{r}_i - \mathbf{r}_j|$, $\hat{\mathbf{r}}_{ij} = \mathbf{r}_{ij}/r_{ij}$, $r = r_{ij}/r_c$, $\theta_{ij}(t)$ is a randomly fluctuating variable from Gaussian statistics, $\omega^d$ and $\omega^r$ are the separation dependent weight functions which become zero at distances greater than or equal to the cutoff $r_c$. The maximum repulsion $a_{ij}$ can be related to the Flory interaction parameter $\chi$ for a given bead number density $\rho = 3$ by the following relation $\chi = (0.286 \pm 0.002)(a_{ij} - a_{ii})$ [6.] Each force is pairwise additive and conserves linear and angular momentum [9-10.] The random and dissipative forces are constrained by certain relations that ensure that the statistical mechanics of the system conforms to the canonical ensemble [7-8.] Conservation of momentum allows for a thermostat to be put in place, which stabilizes the simulation and enables longer time scales with the proper hydrodynamics to be run [10-13.] The total force acting on a soft sphere i arises due to its interactions with neighboring soft spheres j ($j \neq i$), which are within a cutoff distance of $r_c$, is given by $\mathbf{F}_i = \sum_{j \neq i} \mathbf{F}_{c,ij} + \mathbf{F}_{d,ij} + \mathbf{F}_{r,ij}$.

The lipid molecules are represented by bead-spring models with individual lipids composed of a hydrophilic head group and two hydrophobic tails. The bonds between two consecutive beads in a chain are represented by the harmonic spring potential $E_{bond} = K_{bond}((r-b)/r_c)^2$, where $K_{bond}$ is the bond constant and $b$ is the equilibrium bond length. In addition to the bond potential between two consecutively bonded beads, the conservative soft repulsive force acts between 1-2 and 1-3 beads along a chain. The three-body stiffness potential along the lipid tails has the form $E_{angle} = K_{angle}(1+\cos\theta)$ where $\theta$ is the angle formed by three adjacent beads along the lipid tails. This stiffness term increases the stability and bending rigidity of the bilayers [4.]

The soft repulsive pair potential parameters for the lipid molecule beads were selected in reference to experimental systems. Measurements of the average interfacial areas per lipid and the diffusion coefficients provide the physical spatiotemporal scales for our particle-based DPD model [1-5.] The soft repulsive interaction parameters between the head (h), tail (t) beads of the amphiphilic lipid head (h), tail (t) and the solvent (s) beads are assigned the following values: $a_{ss} = 25$, $a_{hh} = 25$, $a_{tt} = 25$, $a_{ht} = 100$, $a_{hs} = 25$ and $a_{ts} = 100$.

In DPD simulations, it is convenient to use dimensionless reduced units [6,14] so that the mass $m$, cutoff radius $r_c$, and the Boltzmann constant $k_B$ are equal to 1 [15.] We run our simulations at a reduced temperature $T$ of 1 which corresponds to the temperature at which the DPPC bilayer is in the fluid state. We take $r_c$ as the characteristic length scale and $k_BT$ as the characteristic energy scale in our simulations. The characteristic time scale is given by $\tau = \sqrt{mr_c^2 / k_BT}$ and the integrating time step is $0.02\tau$ [11.] In our simulations, the bead number density was set at $\rho = 3/r_c^3$ [6.]

To develop a correspondence between the reduced units and physical units, we relate the experimental measurements of the area per lipid $a_p$ for a tensionless membrane and the diffusion coefficient of a lipid molecule in a membrane in the fluid state [16-18.] Table 1 lists the physical correspondence of the reduced units for the four lipid models. We note that the lipid model A is based upon the DPPC molecule.

| | model A | model B | model C | model D |
|---|---|---|---|---|
| **Experimental $a_p$ (nm$^2$)** | ~0.6 | ~0.61 | ~0.63 | ~0.65 |
| $r_c$ ***(nm)*** | 0.71 | 0.69 | ~0.69 | 0.71 |
| $\tau$ ***(ps)*** | 0.12 | 0.052 | 0.35 | 1.23 |

**Table 1.** Physical correspondence of the reduced units of the various lipid models.

**Lipid Models**

We explore the effect of the lipid molecule architecture in the bilayer properties. We study four models of double-tail lipids with variations in the molecular packing parameter [19] and the hydrocarbon chain length. The molecular packing parameter [19] for each lipid architecture depends upon its area per lipid, hydrocarbon chain volume and the chain length. Lipid model A has a cylindrical shape with a packing parameter around 1, and is comprised of three head group beads (which are hydrophilic in nature) and two hydrocarbon tails with three hydrophobic beads each (see Figure 1 (a)). Lipid models B, C and D have molecular packing parameters in the range of 3 with successively longer hydrocarbon chain lengths (see Figure 1

(b) – (d).) The head group architectures for models B, C and D are identical and are comprised of three hydrophilic beads. Each hydrocarbon tail is comprised of three, four and five hydrophobic beads respectively, for lipid models B, C and D. In summary, lipid models A and B differ in their head architecture; lipid models A, C and D differ in their head and tail architectures, and lipid models B, lipid C, and lipid D differ in their hydrocarbon tail length.

Figure 1 (a) – (d) depicts the tensionless lipid bilayer membrane corresponding to each lipid model for a cubic simulation box with dimension 20 $r_c$ and with 24000 beads. The area per lipid for a tensionless lipid bilayers for model of lipid A contains 664 lipids at a lipid area of $A/r_c^2$=1.2053. Similarly, the tensionless bilayer for (1) model B contains 582 lipids at an area

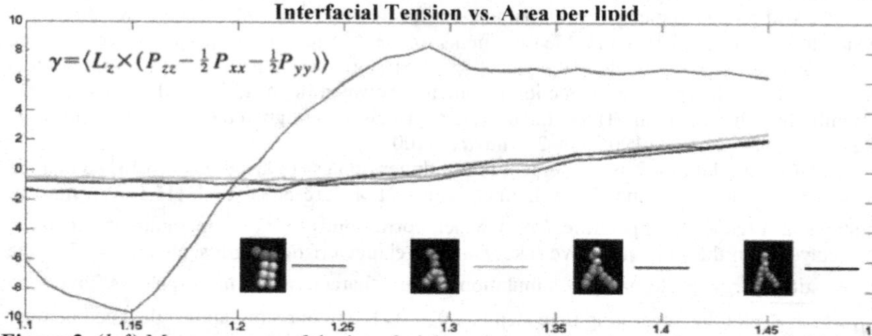


**Figure 2.** (*left*) Measurements of the interfacial tension as a function of the area per lipid for the four lipid architectures.

per lipid of $A/r_c^2$=1.2876; (2) model C contains 560 lipids at an area per lipid of $A/r_c^2$=1.3399, and model D contains 584 lipids at an area per lipid of $A/r_c^2$=1.2805.

**RESULTS**

We measured the membrane thickness and the interfacial tension for the lipid bilayer membranes resulting from each model. The average thickness of each tensionless lipid model was calculated using a program written in MatLab. The program read the trajectory file from the simulation and divided the bilayer membrane into square patches and extracted the x, y, and z coordinates for each patch. Those coordinates were utilized to determine the diametrically opposing x and y coordinates of the lipids in the top and bottom monolayers, for a specific patch. The thickness of the membrane was determined by taking difference in the z coordinates corresponding to each matched pair of coordinates in the patch. The average thickness of the bilayer was calculated using thickness measurements for all patches. These measurements were done for the area per lipid area ranging from 1.1 to 1.45, for each lipid model. Figure 1 (right) shows the bilayer thickness measurements for the different lipid architectures and the areas per lipid.

We expect that the combination of a large head group and the cylindrical shape of lipid model A to favor tight hydrocarbon chain packing in the hydrophobic region of the membrane and not exhibit *interdigitation* of the lipid tails. Interdigitation is known to occur in bilayers composed of the inverted wedge-shaped lipids where the hydrocarbon tails of lipids in opposing monolayers interlock with each other leading to a decreased bilayer thickness [19]. Our measurements of the membrane thickness as shown in Figure 1 supports our hypothesis by demonstrating less variability in the membrane thickness with increasing area per lipid. The inverted wedge-shapes of lipid models B, C and D demonstrate increasing splaying and interdigitation of the hydrocarbon tails with the chain length. We expect the interdigitation of the splayed hydrocarbon tails to increase with the inter-lipid spacing which is supported by the decrease in membrane thickness with the area per lipid, as shown in Figure 1 (*right.*) We also find the membrane thickness to increase with the chain length of the lipid molecule.

We also use measurements of the interfacial tension as a function of the area per lipid to characterize the mechanical properties of the bilayers resulting from the different lipid models. We utilize the principle pressure components $P_{xx}$, $P_{yy}$, and $P_{zz}$ to compute the interfacial tension using the following relation $\sigma r^2 / k_B T = (L_z \times (P_{zz} - \frac{1}{2}P_{xx} - \frac{1}{2}P_{yy}))$ [3-1] where $L_z$ is the dimension of the cubic simulation box [13.] The interfacial tension of a bilayer in solution encompasses enthalpic contributions from the solvent-head group and the head group- tail group interactions. An increase in the inter-lipid spacing or area per lipid is accompanied by more voids in the hydrophobic regions in the membrane. The response of the bilayer to the increasing voids in the hydrophobic region of the membrane is determined by the shape of the lipids. We find the interfacial tension of the bilayers formed from inverted wedge-shaped lipids to demonstrate lower sensitivity to the increased inter-lipid spacing or area per lipid. The lipid molecules are able to increase the splaying of their hydrocarbon tails and interdigitation with the area per lipid, thereby filling in the voids in the hydrophobic region without significant changes in the inter-head group and head group-solvent enthalpic interactions. We attribute this response to the smaller head groups of the lipids and the splayed hydrocarbon tails. In contrast, the bilayer membrane composed of cylindrical-shaped lipids with straight hydrocarbon tails and large head groups see a sharp increase in the inter-head group, head group-solvent and tail group-solvent enthalpic interactions with the area per lipid.

We have investigated the response of a lipid bilayer membrane to changes in the inter-lipid spacing via the area per lipid to understand the role of the lipid molecular architecture. We have characterized the response of the lipid bilayer membrane through the membrane thickness and the interfacial tension.

## CONCLUSIONS

In summary, we explore the properties of lipid bilayer membranes composed of fours distinct molecular architectures with different head and tail group size and shape. We characterize the properties of the bilayers by studying the response of the lipid bilayer membrane as a function of inter-lipid spacing through measurements of the membrane thickness and the interfacial tension. Our investigations indicate that bilayer membranes formed from cylindrical-shaped lipids will have a higher bending rigidity than the inverted wedge-shaped lipids. The results of investigations can be used to design next generation bionanoreactors for applications related drug delivery, sensing and imaging.

## ACKNOWLEDGEMENTS

The authors would like to acknowledge the use of high performance computing resources provided by the Rutgers Discovery Informatics Institute (Excalibur, an IBM Blue Gene/P system. URL:http://rdi2.rutgers.edu) and the Rutgers University Engineering Cluster.

**REFERENCES**

1. M. Dutt, M.J. Nayhouse, O. Kuksenok, S.R. Little and A.C. Balazs, Current Nanoscience 7, 699-715 (2011.)
2. M. Dutt, O. Kuksenok, M.J. Nayhouse, S.R. Little and A.C. Balazs, ACS Nano 5, 4769-4782 (2011.)
3. M. Dutt, O. Kuksenok, S.R. Little and A.C. Balazs, Nanoscale 3, 240-250 (2011.)
4. M. Dutt, O. Kuksenok, S.R. Little and A.C. Balazs. Designing Tunable Bio-nanostructured Materials via Self-assembly of Amphiphilic Lipids and Functionalized Nanotubes. *MRS Spring 2012 Conference Proceedings* (in press.)
5. M. Dutt, O. Kuksenok and A.C. Balazs, *submitted.*
6. R.D. Groot and P.B. Warren, J. Chem. Phys. 107, 4423-4435 (1997.)
7. M.P. Allen and D.J. Tildesley, Computer Simulations of Liquids (Clarendon Press, Oxford, 2001) p. 71.
8. D. Frenkel and B. Smit, Understanding Molecular Simulation, Second Edition: From Algorithms to Applications, 2nd ed. (Academic Press, Cornwall, 2001), p 63.
9. W. Jiang, J. Huang, Y. Wang, and M. Laradji, J. Chem. Phys. 126, 044901 (2007.)
10. J.C. Shillcock and R. Lipowsky, J. Chem. Phys. 117, 5048-5061 (2002.)
11. M. Kranenburg, M. Venturoli, and B. Smit, J. Phys. Chem. B. 107, 11491-11501 (2003.)
12. P. De Palma, P. Valentini, and M. Napolitano, Phys. Fluids. 18, 027103 (2006.)
13. T. Soddermann, B. Dunweg, and K. Kremer, Phys. Rev. E. 68, 046702 (2003.)
14. M. Kranenburg and B. Smit, J. Phys. Chem. B. 109, 6553-6563 (2003.)
15. S.J. Plimpton, J. Comp. Phys. 117, 1-19 (1995.)
16. L. Gao, J. Shillcock, and R. Lipowsky, J. Chem. Phys. 126, 015101 (2007.)
17. K.A. Smith, D. Jasnow, A.C. Balazs, J. Chem. Phys. 127, 084703 (2007.)
18. R. Goetz and R. Lipowsky, J. Chem. Phys. 108, 7397-7409 (1998.)
19. J. Israelachivili, Intermolecular and Surface Forces, Third Edition: Revised Third Edition. (Academic Press, Boston), p. 544-569.

**Mater. Res. Soc. Symp. Proc. Vol. 1569 © 2013 Materials Research Society**
**DOI: 10.1557/opl.2013.805**

# Influence of Coupling Agent on the Morphology of Multifunctional, Degradable Shape-Memory Polymers

Liang Fang, Wan Yan, Ulrich Nöchel, Michael Zierke, Marc Behl, Karl Kratz, Andreas Lendlein
Institute of Biomaterial Science and Berlin-Brandenburg Centre for Regenerative Therapies, Helmholtz-Zentrum Geesthacht, 14513 Teltow, Germany.

## ABSTRACT

Multifunctional polymer-based biomaterials, which combine degradability and shape-memory capability, are promising candidate materials for biomedical implants. An example is a degradable multiblock copolymer (PDC), composed of poly(*p*-dioxanone) (PPDO) as hard and poly(ε-caprolactone) (PCL) as switching segments. PDC exhibits a unique linear mass loss during hydrolytic degradation, which can be tailored by the PPDO to PCL weight ratio, as well as an excellent thermally induced dual-shape effect. PDC can be synthesized by co-condensation of two oligomeric macrodiols (PCL-diol and PPDO-diol) using aliphatic diisocyanates as coupling agent. Here, we investigated whether different morphologies could be obtained for PDCs synthesized from identical oligomeric macrodiols (PCL-diol with $M_n$ = 2000 g·mol$^{-1}$ and PPDO-diol with $M_n$ = 5300-5500 g·mol$^{-1}$) with 2, 2(4), 4-trimethyl-hexamethylene diisocyanate (TMDI) and 1, 6-hexamethylene diisocyanate (HDI), respectively. More specifically, atomic force microscopy (AFM) was utilized for an investigation of the surface morphologies in solution casted PDC thin films in the temperature range from 20 °C to 60 °C. The results obtained in differential scanning calorimetry (DSC) and AFM demonstrated that different morphologies were obtained when TMDI (PDC-TMDI) or HDI (PDC-HDI) were used as linker. PCL related crystals in PDC-HDI were more heterogeneous and less ordered than those in PDC-TMDI, while HDI resulted in a larger degree of crystallinity than TMDI. This research provides some new suggestions for choosing a suitable coupling agent to tailor the required morphologies and properties of SMPs with crystallizable switching segments.

## INTRODUCTION

Shape-memory polymers (SMPs) have attracted increasing interests recently because they can change from their programmed temporary shape to their permanent (original) shape upon exposure of an external stimulus such as heat [1, 2]. In addition to covalently crosslinked polymer networks with a suitable thermal glass or melting transition acting as switching domains [3, 4], thermoplastic SMPs have also been reported [5-10]. More recently, multifunctional SMPs such as thermoplastic copolyester urethanes, which combine degradability and the capability of a shape-memory effect, have been introduced as candidate implant materials, including the suture [5, 6]. Such copolyester urethanes can be synthesized by the prepolymer method via cocondensation of two oligomeric macrodiols coupled with an (aliphatic) diisocyanate as coupling agent. One example are multifunctional PDC copolyester urethanes, composed of poly(ε-caprolactone) (PCL) and poly(*p*-dioxanone) (PPDO) segments [5-8]. In PDC, PCL segments are forming the crystallizable switching domains, which stabilize the temporary deformed shape, while the crystallizable PPDO hard segments support the permanent shape [5-8].

Earlier studies showed that the coupling agent can affect the morphologies of polyurethane elastomers [11, 12]. For example, Hojabri et al. demonstrated that the polyurethane prepared using 1, 7-heptamethylene diisocyanate had less ordered crystal structures than that prepared using 1, 6-hexamethylene diisocyanate (HDI) [12]. Coupling agent, thus, is also expected to alter the morphologies of the multifunctional SMPs based on copolyester urethanes, which are synthesized using the similar route. More importantly, it has been widely accepted that the degree of crystallinity and morphologies of crystallizable block copolymers, such as spherulite size and lamellar structures, dominate the final practical properties [13-15]. For example, Lyoo et al. demonstrated that the biodegradation rate of a poly(butylene succinate) copolymer was mainly controlled by the crystal morphology [15]. Therefore, studying the effects of coupling agent on the morphology of multifunctional SMPs is helpful for the understanding of the morphology and designing SMPs.

In this work, we explored two PDCs prepared from PPDO and PCL macrodiols, which were coupled via 2, 2(4), 4-trimethyl-hexamethylene diisocyanate (TMDI) or HDI. The resulting PDCs were named as PDC-HDI and PDC-TMDI. Differential scanning calorimetry (DSC) was applied to provide a global view of the thermal properties of the two PDCs. Atomic force microscopy (AFM) was performed for the investigation of the surface morphologies of the two PDCs in solution casted thin films at various temperatures ranging from 20 °C to 60 °C.

## EXPERIMENTAL DETAILS

### Materials and Synthesis

Poly(ε-caprolactone)diol ($M_n$ = 2000 g·mol$^{-1}$, polydispersity (PD) = 1.7) was purchased from Solvay Caprolactones (Warrington, UK). Poly(*p*-dioxanone)diol ($M_n$ = 5300-5500 g·mol$^{-1}$, PD = 1.6-2.4) was prepared by ring-opening polymerization of *p*-dioxanone (Daiwa Kasei Industry Co., Japan) with octandiol (Fluka, Taufkirchen, Germany) as initiator according to the method described in ref. [8]. In order to synthesize PDC-TMDI, 250 g PCL-diol and 250 g PPDO-diol were reacted in 750 mL 1,3-dioxolane (Acros Organics, Geel, Belgium) with 44.5 mL TMDI (Aldrich, Taufkirchen, Germany) at 75 °C for 28 days. Afterwards, the reaction mixture was diluted with 1, 3-dioxolane and precipitated in liquid nitrogen (Line AG, Wiesbaden, Germany), and subsequently freeze dried. PDC-TMDI with $M_n$ of 54000 g·mol$^{-1}$ and a polydispersity index of PD = 2.9 was achieved with a yield of 97%. The synthesis of PDC-HDI with a $M_n$ of 74000 g·mol$^{-1}$ and a PD around 12.3 was described previously in ref. [7].

### Solution Casting

The two PDCs were dissolved in chloroform overnight at ambient temperature with a polymer concentration of 0.3 wt%. The solution casting was carried out in glass petri dishes (diameter = 80 mm, Duran Group, Wertheim, Germany) with subsequent evaporation of the solvent for 3 days. The resulting film thickness was 20±1 µm, measured using a thickness gauge (Hans Schmidt, Waldkraiburg, Germany).

**Characterization**

DSC experiments were conducted on a Netzsch DSC 204 Phoenix (Selb, Germany) at heating and cooling rates of 1 K·min$^{-1}$ in sealed aluminum pans. The polymer samples were first increased from room temperature to 150 °C before it was cooled down to -100 °C. Subsequently, the sample was reheated to 150 °C. The crystallization temperatures ($T_c$), melting temperatures ($T_m$) and the related melting enthalpies were determined from the second cooling and heating run. Furthermore, the degree of crystallinity ($x_c$) of PDC-HDI and PDC-TMDI was calculated from DSC curves using equation (1). Here, $\Delta H_m$ is the melting enthalpy, representing the area of the melting peak. The $\Delta H_m^{100}$ is the specific melting enthalpy for a 100% crystalline PCL or PPDO segment, which were chosen as 135 [16] and 103 J·g$^{-1}$ [17] respectively.

$$x_c = \frac{\Delta H_m}{\Delta H_m^{100}} \cdot 100\% \qquad (1)$$

The solution casted PDC films were investigated by AFM (Nanoscope V, Veeco Instruments Inc., New York, USA) in AC mode. The temperature was controlled using a Veeco Thermal Application Controller. Typical scan rate was 0.3 Hz. The silicon cantilever (Olympus OMCL AC200TS-R3), having a driving frequency of around 150 kHz and a spring constant of 9 N/m, were used. The tip has the radius of 7 nm. The tip back and side angles are 35 and 18°, respectively. Before scanning, the samples were kept at newly set temperatures for 10 min for equilibration.

**RESULTS**

Compared to HDI, the three methyl groups in TMDI could reduce the susceptibility of the diisocyanate to be attacked by the hydroxyl groups of the diols for the reason of spatial restriction. This behavior could result in a low reactivity of TMDI as a coupling agent for PDC compared to HDI. Therefore, synthesis of PDC-TMDI required 28 days, while the preparation of PDC-HDI was completed after 48 hours. This distinct reaction kinetic caused a significant difference in the PD of the two PDCs, i.e., PDC-HDI possessed a greatly larger PD of 12.3 than that in PDC-TMDI (2.9) (see experimental section). On the basis of this difference in the PD the influence of the coupling agent on the morphologies of the two PDCs are discussed in the following.

## DSC Results

Prior to investigate the morphologies of PDC-HDI and PDC-TMDI, the thermal properties were explored first using DSC, as shown in Figure 1. A slow cooling rate of 1 °C·min$^{-1}$ was used here to ensure that polymeric crystals at their thermal equilibrium were formed during the crystallization. In addition, the heating rate was also chosen as 1 °C·min$^{-1}$, in order to avoid the cold crystallization and melting recrystallization, which were found in the heating run at a high heating rate of 5 or 10 °C·min$^{-1}$ (data not shown).

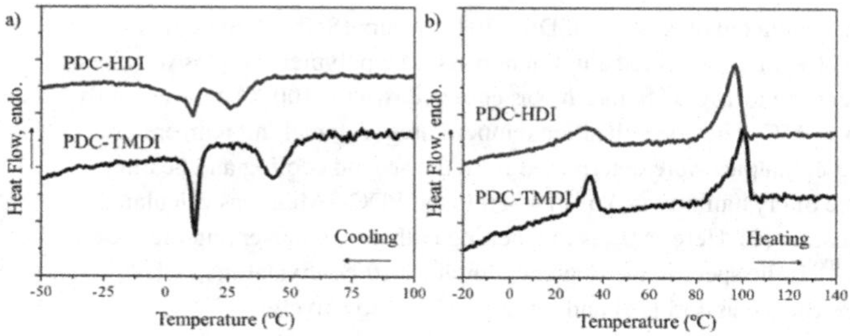


**Figure 1.** DSC thermograms of the second (a) cooling and (b) heating scans of two PDCs at the rate of 1 °C·min$^{-1}$. The curves were moved vertically for more clarity. The grey curves represent the data achieved fro PDC-HDI and the black curves for PDC-TMDI, respectively.

Figure 1a presents DSC thermograms obtained during the second cooling scans for the two PDCs. As shown in Figure 1a, two distinct crystallization temperatures ($T_c$) were evident for both PDC-HDI and PDC-TMDI. The crystallization peaks at low and high temperatures were attributed to PCL and PPDO domains, respectively. More importantly, the $T_c$ of PPDO domains in PDC-HDI was lower than that in PDC-TMDI, while the coupling agent did not vary the $T_c$ of PCL domains greatly (Table 1). The reason could be the larger polydispersity for PDC-HDI, which result in a higher number of short polymeric chains. These short chains could behave as plasticizer, deteriorating the ability of PPDO domains to crystallize. In other words, the short polymer chains in PDC-HDI could retard the organization of polymer chains into polymer crystals. The $T_c$ of PPDO domains in PDC-HDI, thus, was reduced. Subsequently, while further decreasing the temperature, also due to their high mobility, the PPDO segments in the short polymer chains in PDC-HDI can be packed into crystal lamellae more easily than the PPDO segments in the long polymer chains in PDC-HDI. These firstly packed PPDO crystals might facilitate the crystallization of remaining polymer chains, since they may behave as nucleation points. For the same reason, the firstly crystallized PCL segments might also increase the tendency of remaining PCL chains to crystallize. Therefore, the effects of molecular weight on the crystallization of PCL domains to some extent became negligible, resulting in similar $T_c$ of PCL domains in PDC-HDI and PDC-TMDI. In addition, it can also be seen from Figure 1a that the crystallization peak of PCL domains in PDC-HDI was substantially broader than that in

PDC-TMDI. The PCL chains in PDC-HDI with non-uniform distribution of the molecular weight preferred to pack into crystals having a broad range of size.

Figure 1b presents DSC thermograms obtained during the second heating scans for PDC-HDI and PDC-TMDI. As shown in Figure 1b, two separated melting peaks were observed in both PDC-HDI and PDC-TMDI, demonstrating that the PPDO and PCL crystal phases were immiscible in the two PDCs. It is necessary to mention that at slow heating and cooling rate of 1 °C·min$^{-1}$, the glass transition temperatures in PDC-HDI and PDC-TMDI were not obvious. While the heating rate was 10 °C·min$^{-1}$ (not shown), two separate glass transition temperatures can be observed ($T_{g, PCL}$ = -56 ~ -57 °C, $T_{g, PPDO}$ = -16 ~ -17°C), indicating that the amorphous phases in the two PDCs were also immiscible. In addition, the melting curve of PCL crystals in PDC-HDI was broader than that in PDC-TMDI. As mentioned above, slow cooling and heating rates of 1 °C·min$^{-1}$ were used during both the crystallization and melting. Therefore, instead of other kinetic reasons, the broad peak of PCL domains only indicated that more heterogeneous PCL crystal phases were produced in PDC-HDI. Furthermore, as shown in Table 1, the $T_m$ of both PCL and PPDO crystals in PDC-HDI were lower than those in PDC-TMDI, suggesting the formation of less perfect crystals in PDC-HDI [19]. As mentioned above, the firstly packed PCL and PPDO crystals from the short polymer chains did not grow into large crystal structures, leading to the formation of less ordered crystal structures in PDC-HDI. DSC curves can also be used to calculate the crystallinity degree based on the melting enthalpy. It can be seen from Table 1 that both PCL and PPDO domains in PDC-HDI possessed a higher $x_c$ than PDC-TMDI. It has been clear from the forgoing discussion that the small polymer chains in PDC-HDI easily packed into crystals, which might facilitate the crystallization of other polymer chains.

**Table 1.** Crystallization temperatures ($T_c$), melting temperatures ($T_m$) and degree of crystallinity ($x_c$) of PDC-HDI and PDC-TMDI obtained or calculated from the DSC.

| | $T_c$ PCL | $T_c$ PPDO | $T_m$ PCL | $T_m$ PPDO | $x_c$ PCL | $x_c$ PPDO |
|---|---|---|---|---|---|---|
| PDC-HDI | 11 °C | 26 °C | 34 °C | 97 °C | 12.2% | 36.7% |
| PDC-TMDI | 12 °C | 43 °C | 35 °C | 101 °C | 11.1% | 34.5% |

**AFM Images**

In terms of visualizing the surface morphologies, AFM in tapping mode was used to characterize the surface morphologies of the PDC thin films. Figure 2 illustrates the AFM phase images of PDC-HDI surface morphologies when the temperature was increased from 20 to 60 °C. Similarly, the reason for choosing 60 °C is that at this certain temperature, the PCL crystals melted while only PPDO crystals remained in the morphology, which offered an opportunity to distinguish the two crystals. Two different types of bright regions were found at the low temperature of 20 °C (Figure 2). The first type of the bright regions, for example the region as indicated by the circle, can be recognized as spherulites consisted of closely packed edge-on crystalline lamellae, grown from the central nuclei and branched outwards continuously. The second type of the bright regions can also be considered as crystal structures. These crystal structures, however, were essentially more heterogeneous than the first type. The examples could be the several regions as presented by the square. These crystal structures may be spherulites with flat-on crystalline lamellae. When the temperature was increased gradually from 20 to 60 °C, the variations in the shape and size of the large spherulites (Type 1) can be neglected. An essential variation, however, was found in the crystal structures with high heterogeneity (Type 2). This variation was negligible until the temperature reached 40 °C, i.e., the small crystal structures were no longer visible when the film was heated up to 40 °C. Since the $T_m$ of PCL domains in PDC-HDI was 33.5 °C (Table 1), the heterogeneous small crystal structures are attributed to PCL crystal phase. The surface morphologies of PDC-TMDI in a solution casted film were also characterized using AFM in an AC mode (Figure 2). Similarly, two immiscible types of crystal areas were observed. Both crystal structures can be recognized as spherulites, consisted of closely packed edge-on crystalline lamellae. While increasing the temperature up to 40 °C, the second type of crystal structures disappeared (examples are marked by the square), which was similar to PDC-HDI (Figure 2). Therefore, these vanished crystals were contributed by PCL domains, while the first type of spherulites was attributed to PPDO crystal phases.

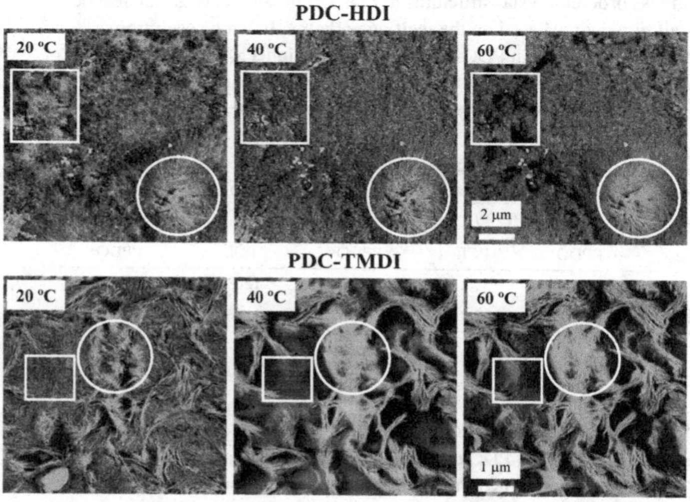


**Figure 2**. AFM phase images of a solution casted film of PDC-HDI and PDC-TMDI at different temperatures varied from 20 to 60 °C. The scanning area for PDC-HDI and PDC-TMDI was 10 and 5µm, respectively. The inserted circle and square marks indicate the locations of PPDO and PCL crystals, respectively.

The results from AFM also indicated that PCL crystals in PDC-HDI were more heterogeneous than those in PDC-TMDI, matching with the results from DSC. It is worthy mentioning that the crystal structures can be controlled by the film thickness [20]. Mareau et al. concluded that in ultrathin films, PCL preferred to crystallize with lamellae oriented flat on. Increasing the film thickness resulted in the formation of branched edge-on lamellae and a shift to the spherulitic morphology. Therefore, in this case of solution casted thick film, the spherulites were observed as expected.

## DISCUSSION

The results of DSC and AFM demonstrated that the coupling agent altered the morphologies of two PDCs. More specifically, PCL crystals in PDC-HDI were more heterogeneous and less ordered than those in PDC-TMDI. The HDI also resulted in a larger degree of crystallinity in PDC than TMDI. The reason was mainly related to the high PD in PDC-HDI, due to the different polymerization kinetic caused by the two coupling agents. In addition, the structure sequence of the main polymer chain may also affect the morphologies of the two PDCs. N'Goma et al. demonstrated that in PDC-TMDI, the PPDO and PCL segments showed a broad block sequence due to the low reactivity of TMDI. A low amount of PCL was found in the main PPDO segments, while a minor fraction of PPDO also incorporated into the main PCL segments [8]. The more random block sequence of the PCL and PPDO segments in PPDO-TMDI could retard the organization of PCL and PPDO segments into respective crystals. In addition, the heterogeneity of the composition of the macromolecules due to the more random block sequence could also reduce the growth of PCL and PPDO crystals in PDC-TMDI. These two behaviors resulted in a smaller degree of crystallinity in PDC-TMDI than PDC-HDI. Furthermore, the mobility of the two PCL and PPDO segments is also affected by the coupling agent itself. Compared to TMDI, which has three more methyl groups, HDI has a more symmetric chemical structure. The symmetric structure, being easy to rotate, has smaller spatial restriction for the two polymeric segments, which resulted in the formation of a much stronger diurethane structure than TMDI. Therefore PDC-TMDI was originally selected. For the same reason, in PDC-HDI, the covalently connected neighboring PPDO and PCL segments were easier to adjust their spatial conformation, facilitating the formation of crystals. A similar result was found in previous research. Kojio et al studied the effect of 1, 2-bis(isocyanate)ethoxyethane (TEGDI) on the structures and properties of conventional polyurethane elastomer [11]. Compared with HDI, the ether bond in the main chain of TEGDI also increased its flexibility, leading to a weak phase separation in the TEGDI-based polyurethane elastomer.

## CONCLUSION

The results of DSC and AFM indicated that the coupling agent (HDI and TMDI) affected the thermal properties and morphologies of PDC, which is a multifunctional SMP. PDC-HDI possessed more heterogeneous and less ordered PCL crystals than PDC-TMDI, which we attributed to the existence of a higher number of short chains in PDC-HDI, indicated by the larger polydispersity. For the same reason, HDI also facilitated the formation of more crystals in PDC, represented by the larger degree of crystallinity which was indicated by DSC results. This work provided some new suggestions for choosing or designing the suitable coupling agent in order to satisfy required morphologies and properties of multifunctional SMPs for specific applications.

## REFERENCES


1. C. Wischke, A. T. Neffe and A. Lendlein, *Adv. Polym. Sci.* **226**, 177 (2010).
2. K. K. Julich-Gruner, C. Löwenberg, A. T. Neffe, M. Behl and A. Lendlein, *Macromol. Chem. Phys.* **214**, 521 (2013).
3. K. Kratz, S. A. Madbouly, W. Wagermaier and A. Lendlein, *Adv. Mater.* **23**, 4058 (2011).
4. C. M. Yakacki, R. Shandas, R. D. Safranski, A. M. Ortega, K. Sassaman and K. Gall, *Adv. Funct. Mater.* **18**, 2428 (2008).
5. A. Kulkarni, J. Reiche, J. Hartmann, K. Kratz and A. Lendlein, *Eur. J. Pharm. Biopharm.* **68**, 46 (2008).
6. A. Lendlein and R. Langer, *Science* **296**, 1673 (2002).
7. K. Kratz, R. Habermann, T. Becker, K. Richau, and A. Lendlein, *Int. J. Artif. Organs* **34**, 225 (2011).
8. P. Y. N'Goma, W. Radke, F. Malz, H. J. Ziegler, M. Zierke, M. Behl and A. Lendlein, *Int. J. Artif. Organs* **34**, 110 (2011).
9. K. Kratz, U. Voigt, and A. Lendlein, *Adv. Funct. Mater.* **22**, 3057 (2012).
10. Y. Feng, M. Behl, S. Kelch and A. Lendlein, *Macromol. Biosci.* **9**, 45 (2009).
11. K. Kojio, T. Fukumaru, M. Furkawa, *Macromolecules* **37**, 3287 (2004).
12. L. Hojabri, X. Kong, S. S. Narine, *Biomacromelcules* **10**, 884 (2009).
13. G. Seretoudi, D. Bikiaris and C. Panayiotou, *Polymer* **43**, 5405 (2002).
14. M. Nagata, T. Machida, W. Sakai and N. Tsutsumi, *Macromolecules* **31**, 6450 (1998).
15. W. S. Lyoo, J. H. Kim, W. S. Yoon, B. C. Ji, J. H. Choi and J. Cho, *Polymer* **41**, 9055 (2000).
16. S. Jiang, X. Ji, L. An and B. Jiang, *Polymer* **42**, 3901 (2001).
17. N. Bhattarai, H. Y, Kim, D. I. Cha, D. R. Lee and D. I. Yoo, *Polymer* **43**, 5405 (2002). Eur. Polym. J., **39**, 1365 (2003).
18. L. Wang, X. Wang, W. Zhu, Z. Chen, J. Pan and K. Xu, *J. Appl. Polym. Sci.* **116**, 1116 (2010).
19. U. Nöchel, C. S. Reddy, N. K. Uttamchand, K. Kratz, M. Behl and A. Lendlein, *Eur. Polym. J.* (2013) DOI 10.1016/j.eurpolymj.2013.01.022.
20. V. H. Mareau and R. E. Prud'homme, *Macromolecules* **38**, 398 (2005).

# Interface of (Multifunctional) Biomaterials

**Mater. Res. Soc. Symp. Proc. Vol. 1569 © 2013 Materials Research Society**
**DOI: 10.1557/opl.2013.829**

# First Evidence of Singlet Oxygen Species Mechanism in Silicate Clay for Antimicrobial Behavior

Jiun-Chiou Wei[1], Yi-Ting Wang[1] and Jiang-Jen Lin[1*]
[1]Institute of Polymer Science and Engineering, National Taiwan University, Taipei 10617, Taiwan.

## ABSTRACT

Thin silicate nanoplatelets, derived from the exfoliation of natural Sodium montmorillonite ($Na^+$-MMT) clays, show an unexpected antimicrobial property. A physical trapping mechanism has been proposed because the clay nanoplatelets can indiscriminately inhibit the growth of a broad spectrum of bacteria, including drug-resistant species such as methicillin-resistance *S. aureus* (MRSA) and silver ion-resistant *E. coli*. The ability to generate singlet oxygen species was first observed for the clay platelets that showed a high-aspect-ratio geometric shape and the presence of surface ionic charges. By comparison, the pristine clay with a multilayered structure failed to generate any singlet oxygen species. The ability to emit singlet oxygen species provides direct evidence for the antimicrobial ability of clay through a non-chemical mechanism, which opens the potential for medical use.

## INTRODUCTION

Recent advances in our understanding of the nanostructures of inorganic materials has often led to the discovery of new functions such as quantum effect [1], conductivity [2], electricity [3], and biomedicinal effects [4]. Various nanosized materials, including $C_{60}$ derivatives [5] and ZnO [6], are able to generate reactive oxygen species and subsequently inactivate bacteria and viruses. Inorganic materials of particular interest are natural silicate minerals [7] such as MMT, fluorinated micas, hectorite, and saponite, which are conventionally considered to be inert to microorganisms [8,9]. Recent developments have revealed that the hydrated minerals of iron-rich clays could exhibit remarkable antibacterial activity. This bactericidal ability is derived from compositions such as the $Fe^{2+}$ species that are released from the clay under aqueous buffer pH conditions [10].

Finding new alternatives to inhibit these antibiotic-resistant bacteria is an important research task. Recently, we developed a method to manipulate the natural silicate clay of a multilayered structure into nanoscale silicate platelets (NSP) [11,12]. Examination of these exfoliated nanoplatelets revealed their unexpected antimicrobial property [13]. Previously, a series of tests on cytotoxicity and genotoxicity has demonstrated that these platelets have a low level of acute toxicity [14]. The unusual characteristics of the NSP nanoplatelets prompted us to investigate further their mechanism of antimicrobial activity against the drug-resistant bacterial strains. Here, we used a physicochemical method (electron paramagnetic resonance radical trapping) to detect the generation of reactive oxygen species (ROS) and propose a physical mechanism for their antimicrobial properties.

## EXPERIMENTAL DETAILS

$Na^+$-MMT, a natural smectite aluminosilicate, was obtained from Nanocor Co. (USA). The surfactants, cationic dodecyltrimethylammonium bromide (DDTMA) were obtained from

Aldrich Chemical Co. Two spin probes, 2,2,6,6-tetramehtyl-4-piperidone (4-oxo-TEMP) and 5,5-dimethyl-1-pyrroline N-oxide (DMPO), were obtained from Aldrich Chemical Co. Bacterial strains obtained from Super Laboratory Co. (Taiwan) were used as target antibiotic organisms: silver ion-resistant *E. coli* (J53pMG101) and methicillin-resistant *S. aureus*.

Antimicrobial Tests: Cells were cultivated overnight in LB broth: 100 μL of the suspension was inoculated in fresh medium to restart the cell cycle. A 0.1 mL aliquot of bacterial suspension containing ca. $10^8$ colony forming units (CFU)/mL was added to 9.9 mL of NSP solution. Bacterial suspensions were then adjusted to $1 \times 10^6$ CFU/mL. After agitation, portions of the culture were taken at periods of 0, 3, and 24 h, spread onto agar plates, and subsequently incubated at 37°C for 24 h. Finally, the number of colonies was counted to determine antibacterial activity [15].

EPR analyses were performed using a Bruker EMX spectrometer under the following conditions: temperature = 298 K, microwave frequency = 9.773 GHz, microwave power = 10.106 mW, modulation amplitude = 1.6 G at 50 KHz, and scan time = 42 s. Production of singlet oxygen was monitored using 4-oxo-TEMP as a spin-trapping reagent [16]. Experimental solution with NSP of 500 mg/L was prepared by mixing 1 M 4-oxo-TEMP (30 μL), 10 mg/mL NSP (50 μL), and phosphate buffer at pH 7.4 (920 μL). Rose Bengal was employed as a positive control for the generation of singlet oxygen. The spin-trapping reagent DMPO was used to trap hydroxyl radical. The slurry solution of the NSP sample at a concentration of 500 mg/L was prepared by mixing 100 μL $CuSO_4$ (10 mM), 100 μL DMPO (100 mM in phosphate buffer pH 7.4), 15 μL NSP (10 mg/mL), and 85 μL phosphate buffer at pH 7.4. The solution was introduced into a flat quartz cell and subjected to EPR measurement after light irradiation for 0–10 min. Hydrogen peroxide was employed for the generation of the hydroxyl radical as a positive control. UV-Vis spectroscopy measurements were performed with a Hitachi U-4100 spectrophotometer. Interfacial tensions were measured by the Wilhelmy method using a Kruss-K10 digital tension meter equipped with a spherical ring.

## DISCUSSION

A representative smectite clay, $Na^+$-MMT, was exfoliated into a random form of NSP according to procedures reported previously [11,12]. The 2-step process involved exfoliation of layered silicates by a polyamine-salt ionic exchange reaction and subsequent isolation of the silicate platelets by toluene/water extraction. In the process, the original clay stacks were totally randomized into individual platelets, which are dispersible in water after the organic polyamine agent is removed. Repeated extractions allowed isolation of the NSP material at > 94 wt% purity [11,12] based on the organic analysis. It is noteworthy that the exfoliated NSP have the same composition as pristine $Na^+$-MMT clay, but with totally exposed surface areas and surface-tethered ionic charges of ca. 18,000 ions per nanoplatelet. The silicate exhibited characteristic UV-Vis absorption, showing the presence of a new absorption band corresponding to 2 peaks at the maximum of 230 and 280 nm (Figure 1). By comparison, the pristine $Na^+$-MMT clay stacks had only a very weak or broad absorbance in the area around 200–300 nm [17,18]. The new UV absorption for the exfoliated nanoplatelets has not been previously reported for this range of particle size. The geometric shape of the silicate nanoplatelets and presence of surface charges may well contribute to the UV property, that is, the absence of the layered structure characteristic of natural clay.

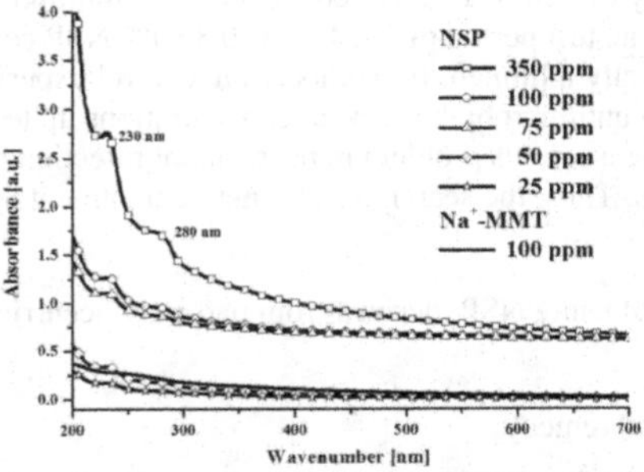


**Figure 1.** Absorption spectra of various concentrations of NSP and pristine $Na^+$-MMT suspension in DI-water.

**Antimicrobial Potency**

The randomized silicate nanoplatelets with polyvalent surface charges are believed to have strong a tendency for interaction with polar organic molecules and hence the microbial cell surface. To understand the non-covalent affinity of the NSP for polar organics, we selected 3 types of commonly used surfactants. The surface tension of NSP in association with cationic DDTMA, non-ionic Triton X-100, and anionic SDS was measured. The CMC (critical micelle concentration) of DDTMA at 1 wt% was found to be interrupted by the addition of NSP to the cationic surfactant at the weight ratios of NSP/DDTMA = 1/1, 2/1, 3/1, and 4/1. The surface tension increased to 42, 48, 68, and 70 mN/m, respectively. This finding implies that ionic exchange reactions or the formation of ionic bridges (i.e., $SiO^-/C_{12}H_{25}N(CH_3)_3^+$), through the replacement between $C_{12}H_{25}N(CH_3)_3^+\ Br^-$ and the exchangeable sodium, occurred on the NSP surface.

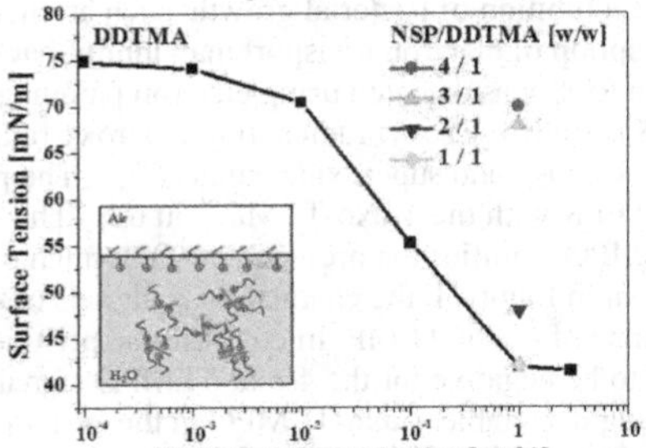


**Figure 2.** Surface tensions of NSP with cationic surfactants DDTMA in air/water.

By viewing the NSP/organic interacting behaviors, we expected NSP to interact strongly with the microbial polar surface. Here, we further tested the general effectiveness of NSP for other bacteria, including the drug-resistant strains silver ion-resistant *E. coli* and MRSA. It appeared that inhibition by NSP was effective indiscriminately for all of the tested bacteria in a 0.85%

saline solution. As shown in Figure 3, 2 of the drug-resistant bacteria were affected by the presence of NSP. The efficacy of NSP was dose dependent and inhibition of microorganism growth was observed during the test period of 3–24 h. At 0.3 wt% NSP concentration, both of the bacterial growths were totally inhibited. By comparison, control experiments using pristine $Na^+$-MMT failed to show any antimicrobial effects at concentrations up to 1.0 wt%. Drug-resistant bacteria have become a serious problem in nosocomial infection owing to the widespread uses of antibiotics. Thus, the search for alternative treatment agents remains a challenge. The effectiveness of using NSP, derived from naturally occurring silicate clays, may provide an alternative potential remedy.

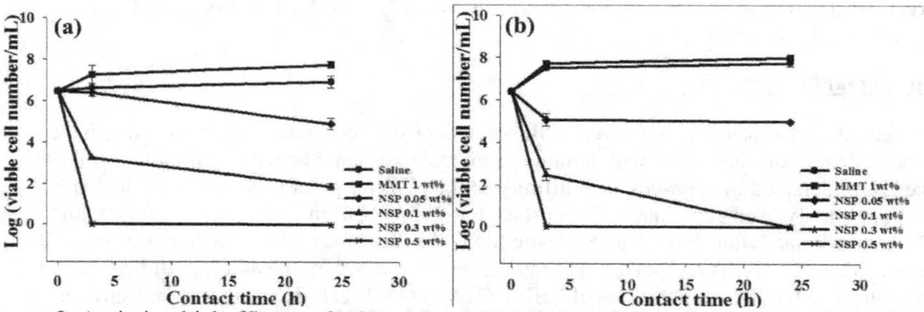


**Figure 3.** Antimicrobial efficacy of NSP against MRSA (a), and silver ion-resistant *E. coli* (b) in saline solution compared with control experiments using pristine MMT clay at 1.0 wt%.

**Detection of singlet oxygen species ($^1O_2$)**

Various mechanisms for the inhibition of bacterial growth, such as chelation of trace metals, generation of ROS, and interruption of electron transport may inhibit bacterial growth. The possibility for NSP to generate ROS was detected using electron paramagnetic resonance (EPR) spectroscope. In general, ROS include hydroxyl radical (OH), peroxy radicals ($HO_2$), hydrogen peroxide ($H_2O_2$), singlet oxygen ($^1O_2$), and superoxide anions ($O_2^-$). The presence of singlet oxygen may be detected using EPR with the 4-oxo-TEMP spin trap. The specific EPR characterization for 4-oxo-TEMPO signifies the presence of $^1O_2$, which instantly converts the 4-oxo-TEMP precursor. As shown in Figure 4, the characteristic signals of 4-oxo-TEMPO were observed for NSP in the presence of 4-oxo-TEMP. In contrast, the pristine $Na^+$-MMT before the exfoliation to NSP was found to be negative for the 4-oxo-TEMPO signal. The positive control experiment was performed using detectable 4-oxo-TEMPO in the system with the $^1O_2$ species from a photo-excited Rose Bengal.

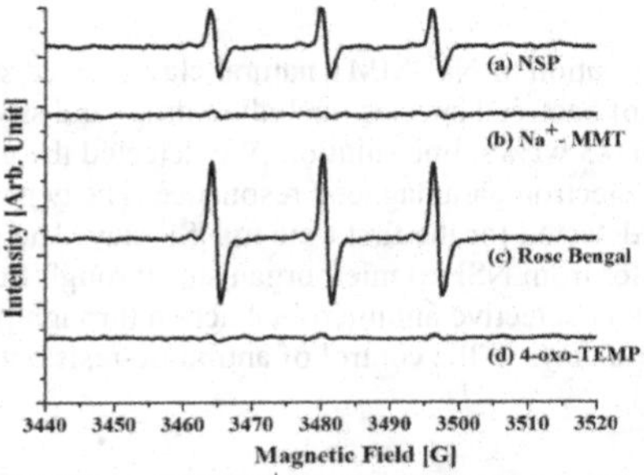


**Figure 4.** EPR spectra of (a) NSP and (b) $Na^+$-MMT (500 ppm) in the presence of 4-oxo-TEMP; (c) a positive control of $^1O_2$ generation by Rose Bengal (20 ppm) and (d) 4-oxo-TEMP, under UV-Vis for 20 min.

**Detection of a hydroxyl radical**

The compound DMPO was used as a spin-trapping agent for hydroxyl radical production. As shown in Figure 5, a DMPO-OH signal was not detected for NSP under UV irradiation. The negative control of the $FeSO_4$ iron catalyst in the absence of NSP failed to generate the DMPO-OH signals. The addition of hydrogen peroxide as the positive control did generate a hydroxyl radical signal. Therefore, we conclude that the hydroxyl radical species would not be generated by NSP in an aqueous solution. It is also posited that the generated ROS-like $^1O_2$ could appear for a short time and require intimate contact between NSP and bacteria cell surfaces to exhibit a remarkable antimicrobial activity. This NSP adherence with polar organics had been demonstrated in the abovementioned experiments showing NSP interacting with surfactants as well as cationic, nonionic, and anionic species. All of these observations show the affinity of NSP to organics. Therefore, the generation of short-lived ROS and the similar effectiveness for different bacteria is consistent with our previously proposed mechanism involving the physical trapping of the platelets on the bacterial cell surface.

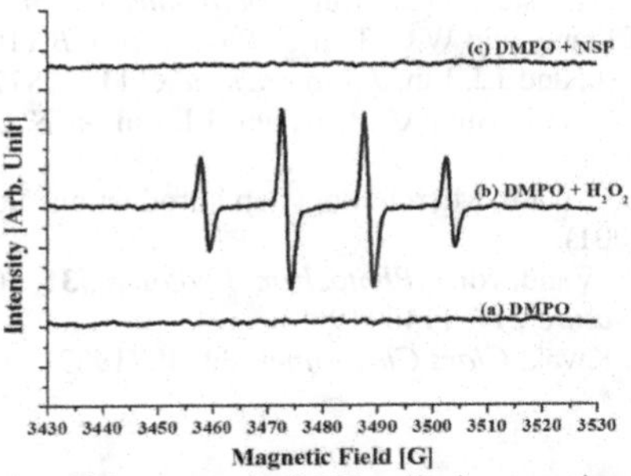


**Figure 5.** EPR spectra of (a) positive control of OH radical generation by DMPO in the presence of hydrogen peroxide, (b) DMPO, and (c) DMPO in the presence of NSP, all under the UV-Vis illumination for 20 min.

## CONCLUSIONS

NSP prepared from the exfoliation of $Na^+$-MMT natural clay showed similar antibacterial activity for a broad spectrum of bacterial species, including drug- and silver ion-resistant strains. NSP efficacy was observed in 0.3 wt% saline solution. We detected the generation of ROS by NSP under UV irradiation by electron paramagnetic resonance. The generation of $^1O_2$ species by photoexcitation in water was detected for the first time in NSP clay. The presence of $^1O_2$ may contribute to the energy transfer from NSP to microorganisms through free radical species. Further understanding of the non-selective antimicrobial action through physical interaction may suggest the potential for natural clays in the control of antibiotic-resistant bacteria.

## ACKNOWLEDGEMENTS

We acknowledge financial support from the National Science Council of Taiwan. We also thank Prof. Fu-Chuo Peng and Prof. Hong-Lin Su for the gifts of bacterial strains.

## REFERENCES

1. M. Aeschlimann, T. Brixner, A. Fischer, C. Kramer, P. Melchior, W. Pfeiffer, C. Schneider, C. Strüber, P. Tuchscherer, and D.V. Voronine, *Science* **333**, 1723 (2010).
2. A.P. Alivisatos, *Science* **271**, 933 (1996).
3. M.S. Kang, A. Sahu, D.J. Norris, and C.D. Frisbie, *Nano Lett.* **10**, 3727 (2010).
4. H.J. Yen, S.H. Hsu, and C.L. Tsai, Small **5**, 1553 (2009).
5. J. Lee, Y. Mackeyev, M. Cho, D. Li, J.H. Kim, L.J. Wilson, and P.J.J. Alvarez, *Environ. Sci. Technol.* **43**, 6604 (2009).
6. D. Yuvaraj, R. Kaushik, and K.N. Rao, *ACS Appl. Mater. Interfaces.* **2**, 1019 (2010).
7. S.S. Ray, and M. Okamoto, *Prog. Polym. Sci.* **28**, 1539 (2003).
8. H.V. Olphen, Clay colloid chemistry. 2nd ed. New York: John Wiley & Sons (1997).
9. B.K.J. Theng, The chemistry of clay-organic reactions. 2nd ed. New York: John Wiley & Sons (1974).
10. L.B. Williams, D.W. Metge, D.D. Eberl, R.W. Harvey, A.G. Turner, P. Prapaipong, and A.T. Poret-Peterson, *Environ. Sci. Technol.* **45**, 3768 (2011).
11. C.C. Chu, M.L. Chiang, C.M. Tsai, and J.J. Lin, *Macromolecules* **38**, 6240 (2005).
12. J.J. Lin, C.C. Chu, M.L. Chiang, and W.C. Tsai, *J. Phys. Chem. B* **110**, 18115 (2006).
13. J.C. Wei, Y.T. Yen, H.L. Su, and J.J. Lin, *J. Phys. Chem. C* **115**, 18770 (2011).
14. P.R. Li, J.C. Wei, Y.F. Chiu, H.L. Su, F.C. Peng, and J.J. Lin, *ACS Appl. Mater. Interfaces.* **2**, 1608 (2010).
15. G. Tortora, R.B. Funke, L.C. Case, Microbiology; an introduction. New York: Addison-Wesley Longman, Inc. (2001).
16. Y. Lion, E. Gandin, and A. Vandevorst, *Photochem. Photobiol.* **31**, 305 (1980).
17. A. Banin, and N. Lahav, *Nature* **217**, 1146 (1968).
18. L.L. Schramm, and J.C.T. Kwak, *Clays Clay Miner.* **30**, 40 (1982).

Mater. Res. Soc. Symp. Proc. Vol. 1569 © 2013 Materials Research Society
DOI: 10.1557/opl.2013.795

# Human Mesenchymal Stem Cell Response to 444 Ferritic Stainless Steel Networks

Antonia Symeonidou[1], Rose L Spear[1], Roger A Brooks[2] and Athina E Markaki[1]

[1]Department of Engineering, University of Cambridge, Trumpington Street, Cambridge, CB2 1PZ, UK
[2]Orthopaedic Research Unit, Addenbrooke's Hospital, Hills Road, Cambridge, CB2 2QQ, UK

## ABSTRACT

The aim of this work is to improve bone-implant bonding. This can, potentially, be achieved through the use of an implant coating composed of fibre networks. It is hypothesised that such an implant can achieve strong peri-prosthetic bone anchorage, when seeded with human mesenchymal stem cells (hMSCs). The materials employed were 444 and 316L stainless steel fibre networks of the same fibre volume fraction. The present work confirms that hMSCs are able to proliferate and differentiate towards the osteogenic lineage when seeded onto the fibre networks. Cellular viability, proliferation and metabolic activity were assessed and the results suggest higher proliferation rates when hMSC are seeded onto the 444 networks as compared to 316L. Cell distribution was found uniform across the seeded surfaces with 444 showing a somewhat higher infiltration depth.

## INTRODUCTION

Total joint replacement is one of the most important operations in the field of orthopaedics, involving replacement of the joint with an artificial prosthesis. Patients undergoing total joint replacement operations can experience complications during recovery due to failure of the implant to integrate with bone tissue. This failure relates to stability problems stemming from stress shielding and poor interfacial adhesion. In an effort to address this problem, different porous metals [1] are currently clinically used as coatings for the femoral and acetabular components. The use of bonded networks of ferromagnetic fibres has been proposed [2], as an anchoring technique for bone tissue in-growth. This anchoring could enhance implant fixation as in the course of time, near proximity bone cells will grow in, around and through the fibre network promoting osseointegration. By injecting hMSCs in the network and directing their differentiation towards the osteogenic lineage, bone will be formed from within the network and this will happen in parallel with osseointegration. When subjected to an external magnetic field, alignment of the fibres will impose mechanical strains on in-growing bone tissue. If the strains generated in the surrounding tissue lie in the beneficial range, bone cell growth will be promoted [3]. Cell responses to these networks have been assessed using human osteoblasts [4] and peripheral blood monocytes [5]. Metal fibre networks of this type [6] exhibit good mechanical properties even at high porosity levels as compared to typical open-cell metallic foams [7], whose properties are often very poor partly due unavoidable stress localization effects by severe inhomogeneities and gross defects and by the presence of embrittling constituents in the cell walls. In this study, the response of hMSCs is investigated, as they naturally populate the porous femoral implant surface following implantation. A comparison is made between 444 magnetic (ferritic) and 316L non-magnetic (austenitic) stainless-steel fibre networks of the same fibre volume fraction. Cellular viability, proliferation and metabolic activity are examined using the CyQuant® and AlamarBlue® Assays. Scanning electron microscopy (SEM) is used to investigate cellular distribution and penetration throughout the networks. Early osteoblastic differentiation is determined by measuring the alkaline phosphatase (ALP) activity.

## EXPERIMENTAL

### Substrates

The fibre networks were made of two types of fibres: AISI 444, a ferritic stainless steel, and AISI 316L, an austenitic stainless steel. 444 fibre networks were manufactured by shaving 60 µm fibres off a 100 µm thick coiled metal foil, which led to the rectangular cross-sectional shape of the fibres. 316L fibre networks were made from fibres produced by the bundle drawing process, which led to the hexagonal cross-sectional shape of the fibres. Both networks were bonded using solid state sintering at Bekaert S.A. in Belgium. The nominal porosity of both networks was 83%. Typical fibre distributions and cross-sectional shapes are illustrated in Figure 1, while their chemical composition [8] is given in Table I. The networks were supplied as a flat sheet (~ 1 mm thick) and cut into 9.5 mm diameter discs using a 9.84 mm die press. Samples were ultrasonically cleaned for 15 min sequentially in acetone, ethanol and ultrapure water, dried at 60 °C for 1 h and then sterilised by dry heat at 160 °C for 2 h.

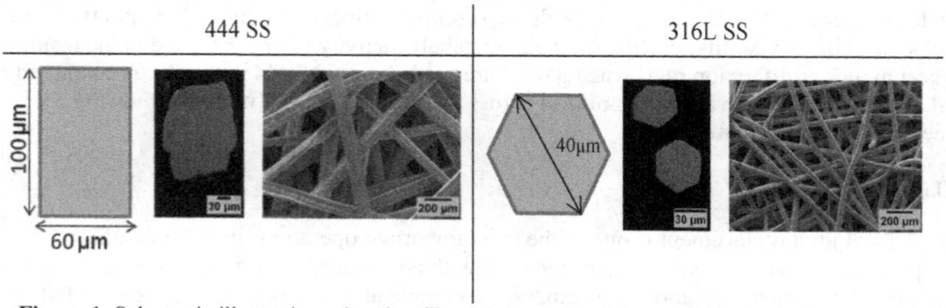


**Figure 1**: Schematic illustrations showing fibre cross-sectional shapes. Also shown are scanning electron micrographs of fibre cross-sections and top views of 444 and 316L networks [6].

**Table I**: Chemical composition of AISI 444 and 316L steels [8].

| AISI type | Composition, wt% | | | | | | | | | | |
|---|---|---|---|---|---|---|---|---|---|---|---|
| | C | Mn | Si | P | S | Cr | Ni | Mo | N | Ti + Nb | Fe |
| 444 | ≤0.025 | ≤1.00 | ≤1.00 | ≤0.040 | ≤0.015 | 17.00 - 20.00 | - | 1.80-2.50 | ≤0.03 | ≤ 0.45 | balance |
| 316L | ≤0.030 | ≤2.00 | ≤0.75 | ≤0.045 | ≤0.030 | 16.00 - 18.00 | 10.00 - 14.00 | 2.00-3.00 | ≤0.10 | - | balance |

**HMSCs Culture and Seeding**

hMSCs (PT-2501), obtained from Lonza, were selected for the cell culture studies. To assess cell proliferation, viability and differentiation, the seeding densities were adjusted according to the specific surface area of the networks [8] as follows: $5\times10^4$ cells per 444 and $1.3\times10^5$ cells per 316L fibre network. For cell distribution and network penetration, $5\times10^4$ cells were seeded onto both networks. The cell solution was diluted in 200 µL droplets of Lonza's growth medium (PT-3001) to allow complete covering of the samples. To prevent droplet dispersion, the networks were placed on hydrophobic polytetrafluoroethylene (PTFE) sheets. After 3h of incubation in 37 °C, 5% $CO_2$, the samples were each transferred to a well of a 24 well plate and osteogenesis was induced by adding 1ml of Lonza's osteogenic medium (PT-4120) per well. The results presented are of a single pilot experiment. In all the plots the standard deviation was calculated from three samples.

**Cell proliferation**

The CyQuant® Cell Assay was employed for the direct measurement of the number of cells present in the surface at each time point. The manufacturer's instructions were followed. Cell number was determined after 1, 3, 7, 15 and 21 days. At each time point the samples were washed twice with Dulbecco's Phosphate buffer Saline, free of calcium and magnesium cations, (DPBS, Sigma-Aldrich) and frozen at -80 °C until measurement. The samples ($n$ = 3) were lysed with 1 ml of CyQuant® lysis solution and CyQuant® dye was added to aliquots of the cell lysate. The DNA content was measured by means of a fluorometric assay using a BMG Labtech FLUOstar Optima plate reader (Offenburg, Germany) with excitation at 485 nm and emission at 520 nm. A standard curve, produced by using known numbers of cells, frozen at the day of seeding at -80 °C in eppendorf vials, was used to convert the measured fluorescence values into cell numbers.

**Cell viability**

Cell viability was evaluated by using the AlamarBlue® Assay (ADB Serotec) according to the instructions of the manufacturer. The AlamarBlue reagent uses resazurin, a non-fluorescent cell permeable compound that as a result of the metabolic activity is reduced to resofurin, which is highly fluorescent. At every time point (1, 3, 7, 15 and 21 days), the samples ($n$ = 3) were incubated for 4 h with osteogenic medium supplemented with 10% (v/v) AlamarBlue reagent and then 100 µL medium from each well were transferred in triplicate to a black 96-well microplate. Fluorescence signal (excitation 530 nm, emission 590 nm) was measured on the aforementioned plate reader.

The % reduction of AlamarBlue was calculated using the formula provided by the manufacturer:

$$\%\text{ reduction of AlamarBlue} = \frac{S_{AB}^{x} - S_{AB}^{control}}{S_{AB}^{100\%\ reduced} - S_{AB}^{control}}$$

where $S_{AB}^{x}$ is the AlamarBlue fluorescence signal of the sample at day $x$, $S_{AB}^{100\%\ reduced}$ is the signal of the 100% reduced form of AlamarBlue and $S_{AB}^{control}$ is the signal from the control. The

100% reduced form of AlamarBlue was produced by autoclaving osteogenic medium supplemented with 10% AlamarBlue dye (121 °C, 15 min).

## Cell differentiation

The early stages of hMSC differentiation towards osteoblasts were characterised by ALP activity, determined at days 1, 3, 7, 15 and 21. 50 μL of each sample's lysate was mixed with 50 μL of ALP Assay buffer containing 6,8-difluoro-4-methylumbelliferyl phosphate (DiFMUP) and samples ($n = 3$) were incubated (37 °C, 15 min) to release the fluorescent compound 6,8-difluoro-7-hydroxy- 4-methylcoumarin (DiFMU). ALP was measured by means of a fluorometric assay using the aforementioned reader (excitation 358 nm, emission 455 nm). The emission intensity units of the signal were converted to the quantity of DiFMU produced using a DiFMU standard curve.

## Cell distribution and infiltration

Scanning electron microscopy (SEM) was employed to investigate cell distribution and penetration into the fibre networks. Samples were washed twice with DPBS, fixed in 4% (v/v) glutaraldehyde in 0.1% 2-[4-(2-sulfoethyl)piperazin-1-yl]ethanesulfonic acid (PIPES) buffer (pH 7.4) (4 °C, 60 min) and washed thrice with DPBS, free of calcium and magnesium cations. Following fixation, top and cross-sectional views of the cell-seeded networks were prepared. Top views were rinsed in distilled water, dehydrated to 100% ethanol and then critical-point dried. Cross-sectional views were dehydrated in ascending concentrations of ethanol, frozen in liquid nitrogen, cleaved by impact with a cold razor blade and subsequently critical-point dried. For critical-point drying (Polaron E5000), the cell-seeded networks were post-fixed in 1% osmium tetroxide overnight, rinsed in distilled water, stained with 2% uranyl acetate in 0.05 M maleate buffer at pH 5.5 overnight. All samples were mounted on aluminium stubs, gold coated and observed under a Zeiss Evo MA 15 scanning electron microscope.

# DISCUSSION

## Cell proliferation and viability

Figure 2a shows the number of cells attached to the networks, obtained by the CyQuant® Assay. In terms of the initial seeding concentration, about a third of the seeded cells were measured on 316L and about half on 444 after 1 day of culture. While at the early stages of culture cell numbers are lower in 444 compared to 316L, since more cells were seeded on 316L according to the specific surface area, it can be seen that from day 3 cells are replicating faster in 444, reaching this way a higher cell number at days 15 and 21.

Cell viability, presented in Figure 2b, was determined by the AlamarBlue® Assay. The increase in metabolic activity in the early time points (until day 7) is an indication of healthy cells. The decline at day 15 is not related to cellular apoptosis, as a decrease in cell number was not observed in the cell proliferation data. Such decline suggests that the environment provided by the fibre networks, the osteogenic medium used and the production of extracellular matrix induced differentiation that is marked by the cessation of the proliferative capacity. Similar behaviour has been observed [9] at day 15, when hMSCs were cultured onto ceramic scaffolds cultured in osteogenic medium.

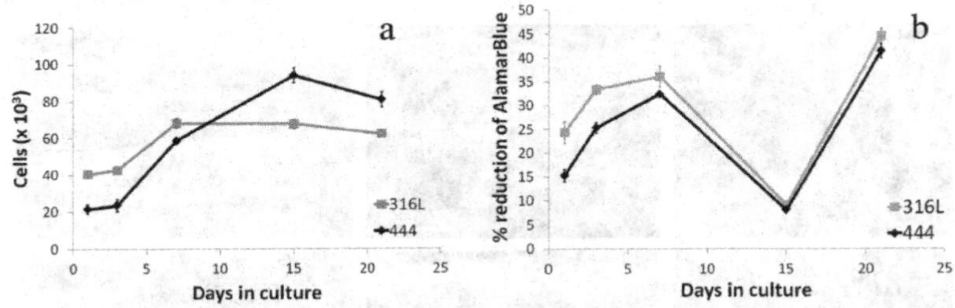


**Figure 2**: (a) Proliferation and (b) viability of hMSCs seeded onto 444 and 316L networks.

## Cell differentiation

The decline in metabolic activity attributed to osteogenesis was in agreement with the measured ALP activity of hMSCs seeded onto the networks (Figure 3a). SEM images of cells cultured onto 316L networks at day 7, such as that shown in Figure 3b, illustrate the formation of philopodia (white arrows) that grow overlapping, slowing down the proliferation due to production of nodules and eventually promoting differentiation. The highest ALP activity was observed at day 15 for the 316L networks, at which point mineralisation was already evident (Figure 3c). The fact that 3 times more cells were seeded on 316L compared to 444 resulted in the cells populating mostly the top layers of the network and therefore producing confluence and differentiation, which explains the higher ALP values for 316L.

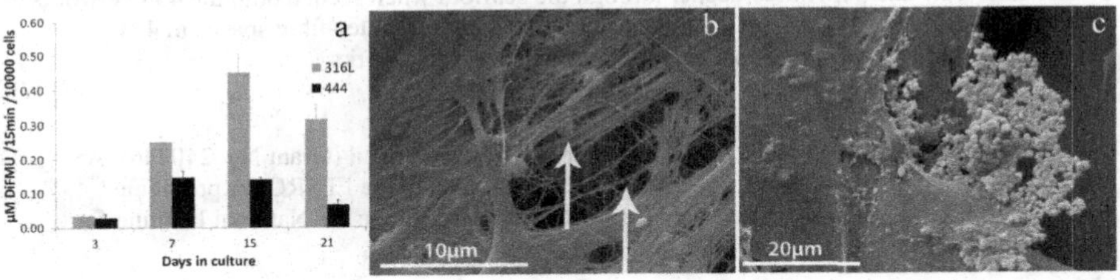


**Figure 3**: (a) ALP activity of hMSCs seeded onto the networks as function of time in culture. Scanning electron micrographs of 316L seeded networks after (b) 7 and (c) 15 days of culture.

## Cell distribution and infiltration

Figure 4 shows SEM images of top and cross-sectional views of hMSCs cultured on 444 and 316L networks, after 7 days of culture. Top views (seeded side) show uniformly distributed cells attaching onto the fibres and spreading along their length. At cross-over points, cells often formed bridges between adjacent fibres. Cross-sectional views suggest that cell infiltration at day 7 was higher in 444 networks compared to 316L. This may be attributed to the larger inter-fibre spaces in 444, which allowed the cells to be seeded deeper into the networks. Deeper cell infiltration may also account for the higher proliferation rates measured in 444 compared to 316L (Figure 2a).

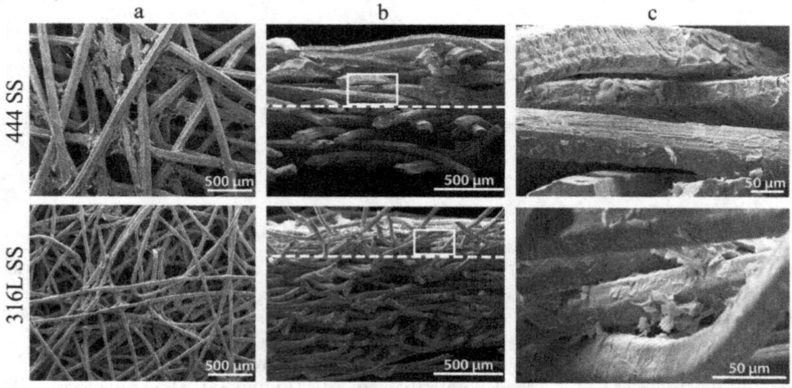


**Figure 4**: SEM images showing (a) top and (b) cross-sectional views of 444 and 316L networks, after 7 days in culture. Cell infiltration is indicated by the dashed lines. (c) High magnification images of the regions in the white squares in (b).

## CONCLUSIONS

The results demonstrate that 444 stainless steel fibre networks support hMSC proliferation, differentiation and mineralisation during short term culture. HMSCs had a higher proliferation rate and penetrated deeper through the scaffold when seeded onto the 444 networks as compared to 316L networks. This was attributed to the larger inter-fibre spaces in 444 compared to 316L, allowing deeper cell penetration into the networks.

## ACKNOWLEDGMENTS

This research was supported by the European Research Council (Grant No. 240446). AS would like to acknowledge the European Research Council and the EPSRC for providing financial support. Financial support for RAB has been provided via the National Institute for Health Research (NIHR).

## REFERENCES

1. BR Levine & DW Fabi, Materialwissenschaft & Werkstofftechnik. 41(12), 1001–1010 (2010)
2. AE Markaki, TW Clyne, Biomaterials. 25, 4805-15 (2004)
3. HM Frost, Bone Miner. 19(3), 257–271 (1992)
4. VN Malheiro, RL Spear, RA Brooks, AE Markaki, Biomaterials. 32(29), 6883-92 (2011)
5. RL Spear, RA Brooks, AE Markaki, J Biomed Mater Res Part A. 101(5), 1456-1463 (2013)
6. AE Markaki, TW Clyne, Acta Mater 51(5), 1351-1357 (2003)
7. AE Markaki, TW Clyne. Mat Sci Eng A. 323(1-2), 260-269 (2002)
8. VN Malheiro, JN Skepper, RA Brooks, AE Markaki,J Biomed Mater Res Part A 101(6), 1588-1598 (2013).
9. T Cordonnier, P Layrolle, J Gaillard, A Langonné, L Sensebé, P Rosset & J Sohier, J Mater Sc. Materials in Medicine. 21(3), 981–7 (2010)

Mater. Res. Soc. Symp. Proc. Vol. 1569 © 2013 Materials Research Society
DOI: 10.1557/opl.2013.962

# Quantification of Protein Adsorption on Polyglycerol-based Polymer Network Films

Duygu Ekinci[1], Adam L. Sisson[1], and Andreas Lendlein[1]
[1]Institute of Biomaterial Science and Berlin Brandenburg Center for Regenerative Therapies, Helmholtz-Zentrum Geesthacht, Kantstr. 55, 14513 Teltow, Germany.

## ABSTRACT

Neutral, hydrophilic, polymer-based architectures are widely investigated for a wide range of biomedical applications from drug-conjugates to delivery systems and scaffolds for regenerative therapies. In most cases, it is crucial that biomaterials provide a blank, inert background in order to hinder unspecific cell-material interactions so that protein mediated biological events leading to foreign body reactions are prevented. Hydrophilic polyglycerol-based polymer network films are a recently developed class of amorphous macroscopic materials, which offer great versatility in design and control of resultant properties. In this study, protein adsorption on polyglycerol-based polymer network films is investigated by using Micro BCA protein assay for three types of proteins having critical roles in the human body, and various copolymer networks with differing sidechains and crosslink densities.

## INTRODUCTION

In the course of biomaterial development, functionality is one of the main concerns in order to conceptualize how the material transduces its structural makeup to direct or control the response of the organisms in vivo [1]. It is critical to recognize that the bulk and surface properties of biomaterials used for medical implants have been shown to influence the dynamic interactions that take place at the tissue-implant interface [1]. In construction of biocompatible artificial implants, one of the main strategies to accommodate cell-material interactions is creating an inert surface not allowing the adsorption of proteins and adhesion of cells. It is generally acknowledged that surfaces that resist protein adsorption will also resist cell adhesion [1]. And above all, resistance of such nonfouling surfaces to protein adsorption would prevent a cascade of protein mediated biological events such as activation of the immune system, blood coagulation, thrombosis, extracellular matrix deposition and other interactions between the material and the surrounding environment.

It is generally accepted that hydrophilic surfaces possessing high wettability and overall neutral charge generally resist protein adsorption and thus protein mediated biological events. Hydrogels among the vast number of synthetic biomaterials stand as the best candidate to nonfouling surface materials. Poly(ethylene glycol) is one the polymers studied extensively for its antifouling properties, especially when coated on planar model surfaces as self assembled monolayers (SAMs) [2]. In addition, bulk structures based on poly(ethylene glycol) showed antifouling properties [3]. Polymers based on poly(2-hydroxyethyl methacrylate) [4], and poly(vinyl alcohol) [5] are also among some of the bulk structures studied for their protein repellent properties.

Notably, it has also been presented that hydrophobic surfaces can be functionalized successfully in order to improve protein resistance [6, 7]. Recently we have shown a versatile method for preparation of novel polyglycerol based network films for potential biomedical applications [8]. The outstanding property of this material comes primarily from its bulk structure which can be used as a coating for various surfaces. Herein, the polyglycerol-based network films are analyzed for their protein adsorption characteristics, especially by the variation in their compositions. Protein adsorption studies are conducted with three different plasma proteins (albumin, fibronectin and gamma globulin) on the material; the quantification is performed by the Micro BCA protein assay.

## EXPERIMENTAL DETAILS

The Micro BCA protein assay with bovine serum albumin standard ampules (2 mg/mL) and bovine gamma globulin standard ampules (2 mg /mL) were purchased from Thermo Fisher Scientific (Schwerte, Germany). Bovine serum fibronectin (1 mg) were purchased from Sigma Aldrich Chemie Gmbh (Munich, Germany). Polyglycerol-based polymer network films are synthesized as previously described [8], according to Figure 1. Disks of 5 mm in diameter are cut from the crude films and extracted with ethanol and subsequently with millipore water in order to remove impurities. Following extraction, disks are further washed and equilibrated with millipore water. Compositions of the polymer film samples used for the test are shown in Table 1 below.

**Table 1** Chemical composition of polyglycerol-based networks

| | | Monomer | | Crosslinker[a] | |
|---|---|---|---|---|---|
| Monomer | Notation | Wt% | Mol% | Wt% | Mol% |
| None | $G_{100}$ | 0 | 0 | 100 | 100 |
| MGE | $M_{80}G_{20}$ | 20 | 43 | 80 | 57 |
| MGE | $M_{60}G_{40}$ | 40 | 66 | 60 | 34 |
| MGE | $M_{40}G_{60}$ | 60 | 82 | 40 | 18 |
| EGE | $E_{80}G_{20}$ | 20 | 39 | 80 | 61 |
| EGE | $E_{60}G_{40}$ | 40 | 63 | 60 | 37 |
| EGE | $E_{40}G_{60}$ | 60 | 79 | 40 | 21 |
| IGE | $I_{80}G_{20}$ | 20 | 36 | 80 | 64 |
| IGE | $I_{60}G_{40}$ | 40 | 60 | 60 | 40 |
| IGE | $I_{40}G_{60}$ | 60 | 77 | 40 | 23 |
| BGE | $B_{80}G_{20}$ | 20 | 33 | 80 | 67 |
| BGE | $B_{60}G_{40}$ | 40 | 57 | 60 | 43 |
| BGE | $B_{40}G_{60}$ | 60 | 75 | 40 | 25 |

[a] GGE as the crosslinker, initiator concentration was kept at 2.0 wt% with respect to the initial mixture amount.

**Micro BCA Protein Assay**

Disks of 5 mm in diameter were incubated in 1 mL of different protein solutions (0,5 mg/mL) at 37 °C for 4 hours. Subsequent to incubation in protein solutions, films were rinsed gently with millipore water and then washed firmly in 1 mL of SDS solution (1 wt/v) for 1 hour at room temperature. Both steps, incubation of polymer films in protein solutions and washing in SDS solution were performed by using uncoated glass vial containers. 0.150 ml of the SDS solution (i.e. solution containing the eluted proteins from the film surface by SDS) was pipetted into 96-well plates and 0.150 ml of the Micro BCA working reagent was added into each well.

The well plate was covered and incubated at 37 °C for 2 hours. Subsequent to cooling the plate to room temperature, light absorbance of the protein solutions in each well was measured by a Spectra Max 340PC 384 spectrophotometer at 562 nm. The values obtained were analyzed based on the standard curve plotted for each protein solution. Depending on the light absorbance, protein concentrations for each solution were calculated. In order to improve the quality of the test conducted, 8 repetitions were performed for each series.

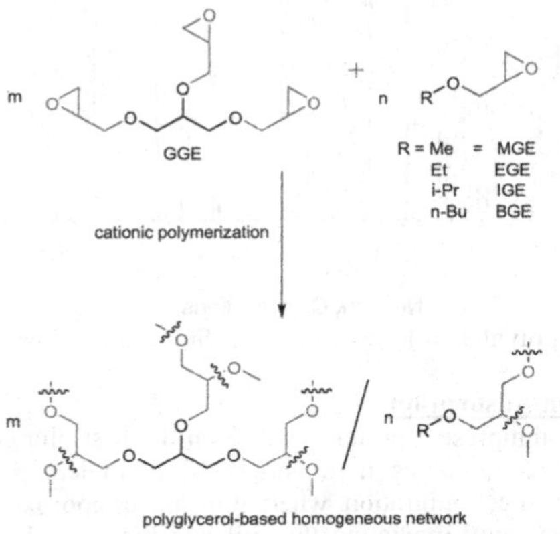


**Figure 1.** General scheme of the synthesis of polyglycerol-based polymer network films

## DISCUSSION

### Bovine Serum Albumin Adsorption

In humans, albumin is the most abundant plasma protein accounting for 55-60% of the measured serum protein and the total body albumin pool measures approximately 3.5 – 5.0 g/kg body weight for healthy adults [9]. Albumin, as a soft globular protein with its conformational adaptability, is capable of formation of bound-compounds in blood; regulating onconic pressure and acid-base balance in plasma buffer and moreover it has significant antioxidant potential [9]. Most importantly, it has effects on blood coagulation, exerting a heparin-like action and showing anticoagulant effects [9]. Figure 2 shows the adsorption profile of bovine serum albumin (BSA) on polyglycerol-based network films composed of different hydrophobic chain length alkyl glycidyl ether monomers and for their different compositions used in the network. Albumin adsorption on, p(HEMA) and HEMA copolymers, which are among the best candidates for antifouling surface materials, give values from 0.05 to 2,0 μg/cm$^2$ [10].

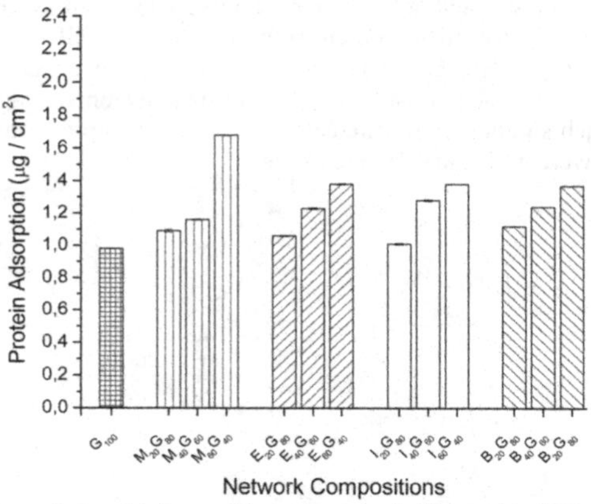


**Figure 2.** BSA adsorption polyglycerol-based network films with different compositions

**Bovine Serum Fibronectin Adsorption**

Fibronectin, as an omnipresent protein even from the first pluricellular eukaryotes, is an adhesive glycoprotein, which involves in a variety of biological processes, particularly in mediating cell attachment and cell migration, where it measures approximately 280 – 380 mg/l in healthy adults [11, 12]. Not only mediating the cell attachment and cell migration, it is also responsible for key functions such as cell differentiation and cell-cell and cell-substrate adhesion through integrin receptors [12]. Figure 3 shows the bovine serum fibronectin (FBS) adsorption on polyglycerol-based network films composed of different hydrophobic chain length alkyl glycidyl ether monomers and for their different compositions used in the network.

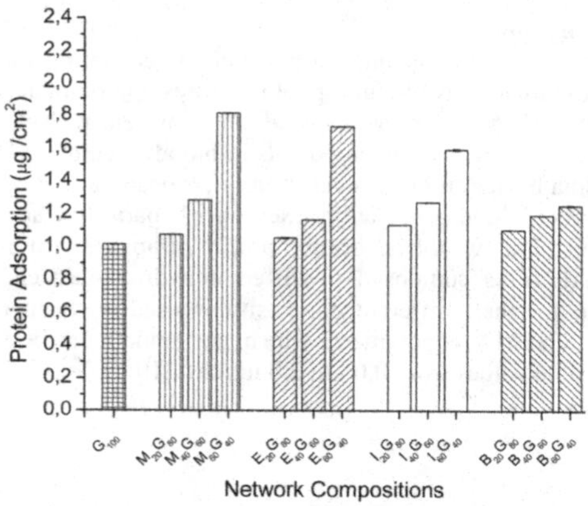


**Figure 3.** FBS adsorption polyglycerol-based network films with different compositions

**Bovine Gamma Globulin Adsorption**

The group of blood plasma proteins composing the γ-globulins, which are frequently named as immunoglobulins, play a key role in immune system [13]. The normal gamma globulin concentration of adult human serum has been reported between the range 0,7 – 1,7 g/100 mL [14]. Along with the antibodies, gamma globulins defend/prevent the body from infections and diseases. In figure 4, bovine γ-globulin (BGG) adsorption on polyglycerol-based network films composed of different hydrophobic chain length alkyl glycidyl ether monomers and for their different compositions used in the network is shown.

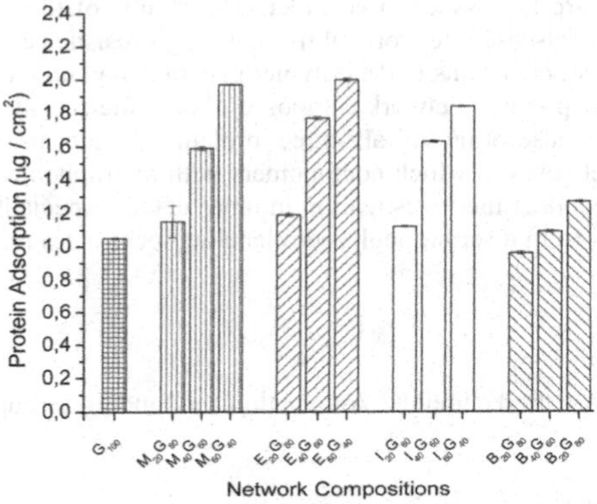


**Figure 4.** BGG adsorption polyglycerol-based network films with different compositions

It is clear that each composition studied has very low protein adsorption under 2 µg/cm$^2$ in all cases, comparable to those on p(HEMA) and HEMA copolymer materials [10, 15, 16]. A typical hydrophobic surface such as poly(ether imide)s may be expected to adsorb proteins around an order of magnitude higher [6]. There is not a remarkable difference in protein adhesion among the three proteins investigated and it can be summarized that these polyglycerol-based networks are generally protein resistant.

There is however, obviously a trend whereby protein adhesion is increased in copolymer networks containing higher ratios of monoglycidyl ether units. In previous investigations it was shown that network properties such as elasticity (Youngs modulus) and degree of swelling were proportional to the relative amount of feed monoglycidyl ether comonomer, being dependent on the crosslink density of the homogeneous networks [8]. Likewise, free volume in the networks is expected to increase with a higher ratio of monoglycidyl ether comonomers, a phenomenon reflected in varying glass transition temperatures of the networks also investigated previously. These network properties that are dependent on crosslink density may also influence polymer chain hydration leading to surface changes that influence protein adhesion [17]. Sidechain hydrophobicity of the monoglycidyl ether components may also influence protein adhesion, but the attained results show little deviation with varied side chains.

**CONCLUSIONS**

Conditioning of biomaterials by bodily fluids upon implantation brings the complications with itself due to the well-recognized phenomena called "nonspecific protein adsorption". Since blood encompasses a variety of proteins, interaction of the implant materials with those proteins must be studied in great detail. Nonfouling materials, preventing nonspecific protein adsorption on their surfaces are a class of materials befitting in numerous biomedical applications where interaction of the material with the body should be minimized. In this study, we have investigated adsorption profiles of three different proteins on the recently developed class of polyglycerol-based network films. Variation of the mechanical properties of these polymer network films depending on the comonomer feed in the structure is of great importance, making these materials promising candidates for various biomedical applications where elasticity and stiffness of the material used are key issues to consider. The results of the protein adsorption studies revealed that polyglycerol-based network films show high resistance to BSA, FSB and BGG adsorption. Glycidyl ether side chains in the polymer networks had minor influences on the protein adsorption profiles. The polymer network composed of only the crosslinker itself showed the highest resistance to the adsorption of all three proteins due to its relatively higher hydrophilic nature than the networks in which comonomers with hydrophobic alkyl side chains are incorporated. In order to conduct the investigation in more detail, especially for competitive adsorption of specific proteins from a serum pool, radio labeling techniques might be performed for the future studies.

## ACKNOWLEDGEMENTS

The authors would like to thank the Helmholtz Association for funding through grant No. SO-036.

## REFERENCES


1. B. Ratner, eds. *Biomaterials Science: An Introduction to Materials in Medicine*, (Elsevier, 2004) pp. 225-246.
2. I. Banerjee, R.C. Pangule, and R.S. Kane, *Advanced Materials*, **23**, 690 (2011).
3. G. Kang, Y. Cao, H. Zhao, and Q. Yuan, *Journal of Membrane Science*, **318**, 227 (2008).
4. C. Zhao, L. Li, Q. Wang, Q. Yu, and J. Zheng, *Langmuir*, **27**, 4906 (2011).
5. Q. An, F. Li, Y. Li, and H. Chen, *Journal of Membrane Science*, **367**, 158 (2011).
6. M. Lange, et al., *Macromolecular Rapid* Communications, **33**, 1487 (2012).
7. C. Siegers, M. Biesalski, and R. Haag, *Chemistry – A European Journal*, **10**, 2831 (2004).
8. D. Ekinci, A.L. Sisson, and A. Lendlein, *Journal of Materials Chemistry*, **22**, 21100 (2012).
9. J.P. Nicholson, M.R. Wolmarans, and G.R. Park, *British Journal of Anaesthesia*, **85**, 599 (2000).
10. E.J. Castillo, J.L. Koenig, J.M Anderson, and J. Lo, *Biomaterials*, **5**, 319 (1984).
11. K.M. Yamada, *Journal of Biological Chemistry*, **266**, 12809 (1991).
12. S.R. Sousa, M.M. Brás, P.M. Ferreira and M.A. Barbosa, *Langmuir*, **23**, 7046 (2007).
13. T.B. Tomasi, *Blood*, **25**, 382 (1965).
14. B. Balikov, and D.F. Solometo, *Clinica Chimica Acta*, **2**, 61 (1957).
15. S. Abraham, S. Brahim, K. Ishihara, and A.G. Elie., *Biomaterials*, **26**, 4767 (2005).
16. E.J. Castillo, J.L. Koenig, and J.M. Anderson, *Biomaterials*, **7**, 89 (1986).
17. S. Chen, L. Li, C. Zhao, and J. Zheng, *Polymer*, **51**, 5283 (2010).

Mater. Res. Soc. Symp. Proc. Vol. 1569 © 2013 Materials Research Society
DOI: 10.1557/opl.2013.832

# Bacterial attachment on poly[acrylonitrile-*co*-(2-methyl-2-propene-1-sulfonic acid)] surfaces

Petra Landsberger[1], Viola Boenke[1], Anna A. Gorbushina[1], Karsten Rodenacker[2], Benjamin F. Pierce[3], Karl Kratz[3], Andreas Lendlein[3]

[1]Free University of Berlin and Federal Institute for Materials Research and Testing (BAM), Unter den Eichen 87, 12205 Berlin, Germany
[2]Helmholtz Zentrum München, German Research Center for Environmental Health, Institute of Biomathematics and Biometry, Ingolstädter Landstr. 1, 85764 Neuherberg, Germany
[3]Institute of Biomaterial Science and Berlin-Brandenburg Centre for Regenerative Therapies, Helmholtz-Zentrum Geesthacht, Kantstr. 55, 14513 Teltow, Germany

## ABSTRACT

The influence of material properties on bacterial attachment to surfaces needs to be understood when applying polymer-based biomaterials. Positively charged materials can kill adhered bacteria when the charge density is sufficiently high [1] but such materials initially increase the adherence of some bacteria such as *Escherichia coli* [2]. On the other hand, negatively charged materials have been shown to inhibit initial bacterial adhesion [3], but this effect has only been demonstrated in relatively few biomaterial classes and needs to be evaluated using additional systems. Gradients in surface charge can impact bacterial adhesion and this was tested in our experimental setup.

Moreover, the evaluation of bacterial adhesion to biomaterials is required to assess their potential for biological applications. Here, we studied the bacterial adhesion of *E. coli* and *Bacillus subtilis* on the surfaces of acrylonitrile-based copolymer samples with different amounts of 2-methyl-2-propene-1-sulfonic acid sodium salt (NaMAS) comonomer. The content related to NaMAS based repeating units $n_{NaMAS}$ varied in the range from 0.9 to 1.5 mol%.

We found a reduced colonized area of *E. coli* for NaMAS containing copolymers in comparison to pure PAN materials, whereby the bacterial colonization was similar for copolymers with different $n_{NaMAS}$ amounts. A different adhesion behavior was obtained for the second tested organism *B. subtilis*, where the implementation of negative charges into PAN did not change the overall adhesion pattern. Furthermore, it was observed that *B. subtilis* adhesion was significantly increased on copolymer samples that exhibited a more irregular surface roughness.

## INTRODUCTION

Bacterial attachment and subsequent biofilm formation on surfaces give rise for difficulties in the use of materials in biomedical applications. Biofilms are surface associated bacterial communities. On the surface of biomedical devices biofilms mean a risk for bacterial infections and they may further serve as a reservoir for plasmids carrying antibiotic-resistance genes.

Many attempts were made to prevent biofilm formation on biomedical materials. One starting point is the modification of existing materials by incorporation of antimicrobial substances, such as silver salts, chlorhexidine and antibacterial peptides, or antibiotics to kill bacteria on the surface of the material. On the other hand the prevention of biofilm formation might be achieved by the development of materials that are inherently resistant to biofilm formation. The chemical composition of the material itself, but also its physical properties such as surface charges, hydrophilicity and surface roughness can influence the attachment of bacteria. Materials with charged surfaces may inhibit initial bacterial adhesion. In this study we used negatively charged copolymers (P(AN-*co*NaMAS)) prepared from acrylonitrile (AN) and sodium-2-methyl-2-propene-1-sulfonate (NaMAS). The standard production methods for commercially available AN-based copolymers are radical polymerizations in *N,N'*-dimethylformamide (DMF) or *N,N'*-dimethylsulfoxide (DMSO) as solvent. This makes the resulting material potentially toxic for cells as these solvent residues cannot be completely removed. Therefore, a water born suspension polymerization was applied for the synthesis of AN-based copolymers in which only water, the necessary monomers, and the water soluble initiators were present. The resulting materials were characterized by several physico-chemical methods to explore the impact of the processing on the material properties [4].

For the evaluation of the influence of material surfaces to bacterial adhesion we have developed an approach to study and quantify bacterial attachment to polymeric materials under standardized conditions.

## EXPERIMENTAL DETAILS

### Materials:

The water born synthesis and the characterization of a series of P(AN-*co*-NaMAS) was reported recently [4]. Four different polymers with a content related to the NaMAS based repeating units $n_{NaMAS}$ of 0.0 mol% (PAN), 0.9 mol% (P(AN-*co*-NaMAS)09), 1.2 mol% (P(AN-*co*-NaMAS)12) and 1.5 mol% (P(AN-*co*-NaMAS)15) as determined by $^1$H-NMR spectroscopy were obtained with number averaged molecular weights ranging from 34000 to 51000 g·mol$^{-1}$ and a polydispersity around 5 (table I).

**Table I**: Composition, molecular weight, thermal properties and surface characteristics of P(AN-*co*-NaMAS) copolymers

| Sample ID | $n_{NaMAS}$ [a] [mol%] | $M_n$ [b] [g·mol$^{-1}$] | $T_g$ [c] [°C] | $\theta_{Adv}$ [d] [°] | $R_q$ [e] [µm] |
|---|---|---|---|---|---|
| PAN | 0.0 | 34000 | 97 | 57.7± 5.3 | 0.110 ± 0.014 |
| P(AN-*co*-NaMAS)09 | 0.9 | 51000 | 99 | 61.2 ± 3.7 | 0.084 ± 0.007 |
| P(AN-*co*-NaMAS)12 | 1.2 | 36000 | 105 | 63.6 ± 4.2 | 0.063 ± 0.042 |
| P(AN-*co*-NaMAS)15 | 1.5 | 46000 | 105 | 57.4 ± 5.3 | 0.071 ± 0.014 |

**a)** Content related to the NaMAS based repeating units in P(AN-*co*-NaMAS) determined by $^1$H-NMR spectroscopy and the last two-digit number was given accordingly; **b)** Number averaged molecular weight as obtained from gel permeation chromatography (GPC); **c)** Glass transition temperature ($T_g$) determined by DSC measurements; **d)** Advancing water contact angle achieved via contact angle measurements using captive bubble method; **e)** Surface roughness analysed by optical profilometry.

Test specimen with a diameter of 13 mm and a thickness of 1 mm were produced via a sintering method at temperatures in the range from 125 °C to 140 °C under a load of 2000 kg utilizing a custom made tool. The disc shaped samples were sterilized with ethylene oxide (45 °C, exposure time: 180 min) prior to biological tests or surface characterization.

As shown in table I, all copolymers exhibited similar thermal properties with a glass transition in the range of 100 to 104 °C, almost identical surface characteristics with advancing contact angles around 58 to 63° and a surface roughness of $R_q = 0.06 – 0.08$ µm as determined by optical profilometry [4].

**Bacterial attachment to material surfaces:**

As material properties cannot be possibly reflected and assessed by just one indicator organism, two model organisms known for their medical relevance and their biofilm forming capacity were chosen for the bacterial adhesion experiments. These selected bacterial strains with different Gram-characteristics are *E. coli* (K-12 W3110) and *Bacillus subtilis* (DSM 10). They were routinely grown on Luria-Bertani (LB) or 3-(N-morpholino)propane sulfonic acid (MOPS) minimal medium (modified) [6] agar plates or in broth with 130 rpm (both at 37 °C). MOPS minimal medium was used in biofilm experiments for standardizing the cultivation conditions. Bacterial adhesion was tested on ethylene oxide sterilized acrylonitrile-based copolymer test specimen submerged in glass flasks with diluted (OD 0.03) bacterial cell suspensions from overnight cultures. After an initial sedimentation/growth phase of 1 h without shaking the polymer samples were incubated at 37 °C with 60 rpm shaking for 4 h or 18 h. After inoculation the test samples were removed and incubated in phosphate buffered saline (PBS) for three minutes. Non-adhering bacteria were removed by multiple washing in PBS (room temperature, 8 times). Fixation of attached bacteria was done with 3 % para-formaldehyde (PFA) overnight at 4 °C. PFA was removed and the material samples were washed six times with PBS before storage in 50 % EtOH at 4 °C. The samples were dried for 10 minutes at 46 °C before dehydration in an ethanol series of 50 %, 80 % and 96 % (all steps of 3 min duration). After air drying, samples were examined using a Nikon N*i*-U Microscope with NIS-elements imaging software (Nikon, Japan). Bacterial attachment on all materials was tested in three or more replications for every material and both reference organisms. Surface colonization by attached bacteria was documented by incident light microscopy. A minimum of 30 pictures was taken for every tested material sample. Surface colonization was assessed on digitalized images and the area covered by bacteria on the surface quantified using ImageJ 1.47 software [7]. Statistical calculations were done with an R-script [8].

## DISCUSSION

Bacterial attachment on the surfaces of biomedical materials is undesired. In our studies we analysed medical grade P(AN-*co*NaMAS)s in respect to bacterial adhesion. To assess the attachment of different bacterial species to the negatively charged P(AN-*co*NaMAS) test specimen and the uncharged pure PAN samples as reference the ratio of the surface covered by bacteria to the total sample surface was analysed.

The results of our experiments (table II) indicate that reference organisms *E. coli* and *B. subtilis* demonstrate a different adhesion behavior to the materials tested.

**Table II**: Mean area of P(AN-*co*-NaMAS) copolymers covered by bacteria

| Sample ID | Mean area[a] [%] covered by | |
|---|---|---|
| | *E. coli* | *B. subtilis* |
| PAN | 5.62 ± 0.32 | 2.44 ± 0.29 |
| P(AN-*co*-NaMAS)09 | 1.82 ± 0.15 | 2.60 ± 0.62 |
| P(AN-*co*-NaMAS)12 | 1.12 ± 0.16 | 4.33 ± 0.50 |
| P(AN-*co*-NaMAS)15 | 1.27 ± 0.15 | 1.58 ± 0.26 |

a) Calculation of the ratio mean and standard deviation of the area proportion covered by bacteria based on the analysis of 30 up to 120 digitalized pictures of colonized surfaces.

Material surfaces of acrylonitrile-based copolymers show a reduced adhesion of *E. coli* cells in comparison to the reference (figure 1).

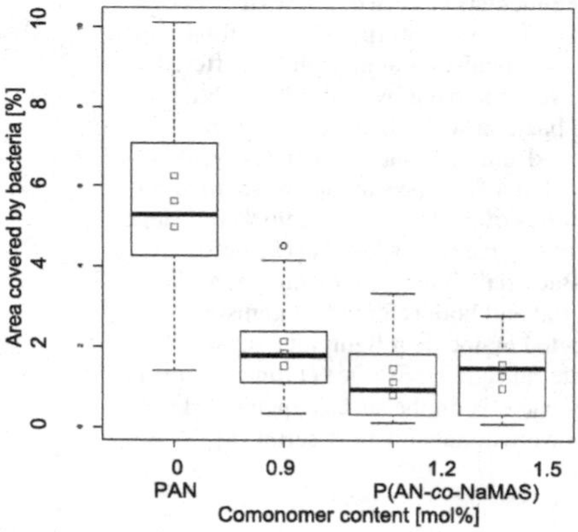


**Figure 1:** Colonization of *E. coli* on P(AN-*co*-NaMAS) copolymers.
Median (black bar), mean value and doubled standard-deviation (red squares), 25 % and 75 % percentiles (small edges of the rectangle), minimum and maximum values (thin black bars) and outlier (circle) for the calculated area covered by bacteria are displayed.

In comparison to the pure PAN materials the incorporation of 2-methy-2-propene-1-sulfonic acid sodium salt (NaMAS) seems to cause a reduction in the attachment of *E.* coli whereby no statistically significant differences in bacterial colonization were recorded between copolymers with varying $n_{NaMAS}$ amounts.

A different adhesion pattern has been observed for the second reference organism *B. subtilis* (figure 2). Bacterial adhesion to the copolymers with 0.9 and 1.5 % $n_{NaMAS}$ did not differ from the cell attachment to the uncharged PAN.

On the contrary, spread of *B. subtilis* cells significantly increased on copolymer samples with 1.2 % $n_{NaMAS}$. These materials exhibited a significantly higher standard deviation of the surface roughness values (*Rq*) (table I). Therefore, we attributed the observed differences in the average surface area covered by the Gram-positive *B. subtilis* cells seem to correlate to the variability of surface roughness.

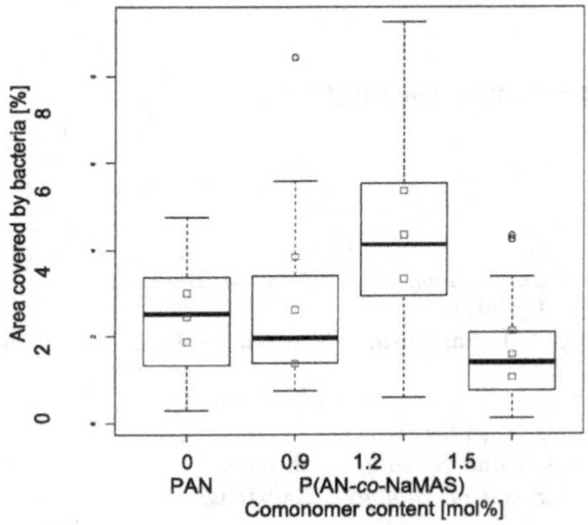


**Figure 2:** Colonization of *B. subtilis* on P(AN-*co*-NaMAS) copolymers.
Boxplot legend see Figure 1.

A different attachment behavior was observed for the two bacterial strains on the tested biomaterial samples. In comparison with the uncharged pure PAN reference material, an influence of negatively charged P(AN-*co*-NaMAS) copolymers on initial bacterial adhesion was only observed in tests with *E. coli* strain. The area colonized by *E. coli*, however, did not significantly change with increasing $n_{NaMAS}$ content.

Varying levels of attachment were observed for the two tested bacterial species. This finding is not surprising because Gram-positive and Gram-negative bacteria possess different surface properties and macromolecular surfaces and therefore use diverse attachment mechanisms (e.g. surface proteins, flagella and exopolysaccharides).

*B. subtilis* attachment seems to be influenced by the interaction of two parameters: electrical charge and surface roughness. The influence of negative charge of the copolymers on bacterial attachment could not be revealed clearly. Further experiments shall clarify the influence of single parameters (e. g. surface roughness, wettability, electrical charge) on bacterial adhesion. It would be especially interesting to analyze the influence of the surface roughness on bacterial adhesion.

## CONCLUSIONS

The chosen experimental setup with two different reference microorganisms enables a preliminary estimation of the tendency of materials to be colonized by microbial cells. Our first results suggest that negatively charged P(AN-*co*NaMAS) may inhibit initial bacterial adhesion. This further shows that tests with only one bacterial strain are not sufficient for a general assessment of material susceptibility to microbial colonization. Further studies with additional bacteria strains should be carried out to evaluate the first results. A set of reference microbial strains for the standardized investigation of diverse (bio)materials should be generated. This would enable the estimation of the influence of physical and chemical surface parameters on bacterial colonization. Results of such microbiologically standardized tests could support the development of new materials for biomedical applications.

## ACKNOWLEDGMENTS

This work was supported by the Bundesministerium für Bildung und Forschung (BMBF) within project 0315696A "Poly4bio BB".

## REFERENCES

1. H. Murata, R. R. Koepsel, K. Matyjaszewski, A.J. Russell, *Biomaterials* **28**, 4870 (2007).
2. G. Harkes, J. Feijen, J. Dankert, *Biomaterials* **12**, 853 (1991).
3. N. N. Lawrence, J. M. Wells-Kinsbury, M. M. Ihrig, T. E. Fangman, F. Namavar, C. L. Cheung, *Langmuir* **28**, 4301 (2012).
4. N. Scharnagl, B. Hiebl, K. Trescher, M. Zierke, M. Behl, K. Kratz, F. Jung, A. Lendlein, *Clinical Hemorheology and Microcirculation* **52**, 295–311 (2012).
5. K. Hayashi, N. Morooka, Y. Yamamoto, K. Fujita, K. Isono, S. Choi, E. Ohtsubo, T. Baba, B. L. Wanner, H. Mori, T. Horiuchi, *Molecular Systems Biology* **2**:2006.0007 (2006).
6. F. C. Neidhardt, P. L. Bloch, D. F. Smith, *Journal of Bacteriology* **119** (3), 736-747 (1974).
7. W.S. Rasband. ImageJ, U.S. National Institutes of Health, Bethesda, Maryland,USA, imagej.nih.gov/ij/ (1997—2012).
8. R Core Team, R: A language and environment for statistical computing. R Foundation for Statistical Computing, Vienna, Austria. URL http://www.R-project.org/ (2013).

Mater. Res. Soc. Symp. Proc. Vol. 1569 © 2013 Materials Research Society
DOI: 10.1557/opl.2013.804

# Comparing Biocompatibility of Nanocrystalline Titanium and Titanium-Oxide with Microcrystalline Titanium

Raheleh Miralami[1], Laura Koepsell[1], Thyagaseely Premaraj[2], Bongok Kim[3], Geoffrey M. Thiele[4], J. Graham Sharp[5], Kevin L. Garvin[1], Fereydoon Namavar*[1]
[1]Department of Orthopaedic Surgery and Rehabilitation, UNMC, Omaha, NE
[2]Department of Growth & Development -, University of Nebraska-Lincoln, Lincoln, NE
[3]Department of Adult Restorative Dentistry, University of Nebraska - Lincoln, Lincoln, NE
[4]Department of Internal Medicine – Rheumatology, UNMC, Omaha, NE
[5]Department of Genetics, Cell Biology & Anatomy, UNMC, Omaha, NE
*Corresponding Author fnamavar@unmc.edu

## ABSTRACT

Titanium (Ti) is the material of choice for orthopaedic applications because it is biocompatible and encourages osteoblast ingrowth. It was shown that the biocompatibility of Ti metal is due to the presence of a thin native sub-stoichiometric titanium oxide layer which enhances the adsorption of mediating proteins on the surface [1]. The present studies were devised to evaluate the adhesion, survival, and growth of cells on the surface of new engineered nano-crystal films of titanium and titanium oxides and compare them with orthopaedic-grade titanium with microcrystals. The engineered nano-crystal films with hydrophilic properties are produced by employing an ion beam assisted deposition (IBAD) technique. IBAD combines physical vapor deposition with concurrent ion beam bombardment in a high vacuum environment to produce films (with 3 to 70 nm grain size) with superior properties. These films are "stitched" to the artificial orthopaedic implant materials with characteristics that affect the wettability and mechanical properties of the coatings.

To characterize the biocompatibility of these nano-engineered surfaces, we have studied osteoblast function including cell adhesion, growth, and differentiation on different nanostructured samples. Cell responses to surfaces were examined using SAOS-2 osteoblast-like cells. We also studied a correlation between the surface nanostructures and the cell growth by characterizing the SAOS-2 cells with immunofluorescence and measuring the amount alizarin red concentration produced after 7 and 14 days. The number of adherent cells was determined by means of nuclei quantification on the nanocrystalline Ti, $TiO_2$, and microcrystalline Ti and analysis was performed with Image J. Our experimental results indicated that nanocrystalline $TiO_2$ is superior to both nano and microcrystalline Ti in supporting growth, adhesion, and proliferation. Improving the quality of surface oxide, i.e. fabricating stoichiometric oxides as well as nanoengineering the surface topology, is crucial for increasing the biocompatibility of Ti implant materials.

## INTRODUCTION

Osseointegration of orthopaedic implants is dependent upon different parameters of the implant surface such as surface properties and physicochemical properties including surface energy, charge, wettability, chemistry, and topography. Surface topography can impact cellular behavior including adhesion, migration, morphology, and orientation in addition to focal adhesion, the development of cytoskeleton, and differentiation [2]. Surface modifications on implant surfaces can be manipulated by etching [3], plasma deposition [4], sintering powders [5], machining/micromachining [6], and ion-beam assisted deposition [7].

When an implant is placed in the body, a multi-step process occurs. First, serum/plasma proteins adsorb on the implant surface and cells attach to the protein layer by integrins which recognize presented extracellular ligands and mediate the initial interactions of cells and the implant material [8]. After the cells adhere, there is a rearrangement of cytoskeleton proteins, formation of tight focal adhesion contacts, activation of focal adhesion kinase, and the induction of several intracellular signal transduction pathways, which leads to proliferation and differentiation of the cells [8, 9]. Jager et al. and Stevens et al. have shown that roughened titanium surfaces can increase the focal contacts for cellular adhesion and they are able to guide membrane receptor organization and cytoskeletal assembly [10, 11]. In vitro experiments have shown that rough implant surfaces promote the adsorption of fibronectin and albumin [12, 13]. In vivo osteointegration has also been improved on roughened surfaces compared to smoother surfaces, which may suggest that the surface modulates the bone response including osteoblast differentiation, ECM deposition, and calcification [14, 15]. Hence, if there is an improvement in the symbiosis of the implant and osteointegration, it should accelerate the healing time, increase the implant longevity, and reduce the necessity of revision surgery [8].

In the present work, osteoblast function including cell adhesion, growth, and differentiation on different nanosamples were investigated. Cell responses to surfaces were examined using SAOS-2 osteoblast-like cells. We studied a correlation between the surface topography and the cell growth by characterizing the SAOS-2 cells with immunofluorescence and measuring the alizarin red concentration produced after 7 and 14 days.

## MATERIALS AND METHODS

**Sample Preparation** The scientific community has been actively pursuing the study of IBAD for specific applications such as tribological coatings, anticorrosion coatings, optical coatings, superconducting buffer layers, and coatings for temperature sensitive substrates such as polymer. IBAD (Figure 1) combines evaporation with concurrent ion beam bombardment in a high vacuum environment. Energetic ions (with a depth penetration of typically less than 20 nm) were employed to produce engineered nanocrystals "stitched" to a substrate (utilizing billions and billions of directed and parallel ionic hammers). Ion bombardment is also the crucial factor for controlling other film properties such as surface morphology, density, stress level, crystallinity, grain size, grain orientation, and chemical composition [7].

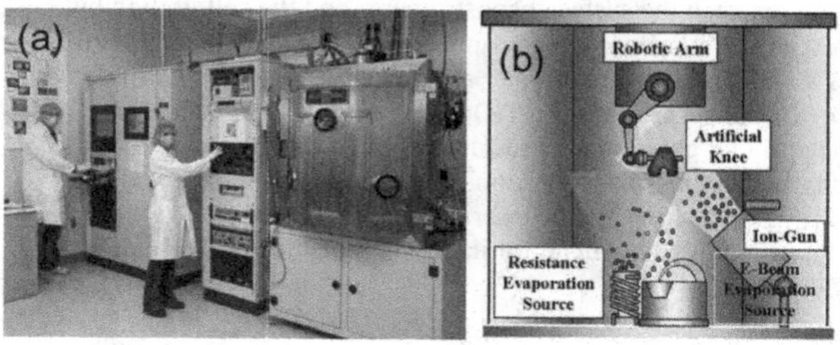


**Figure 1.** (a) IBAD system and (b) schematic. The process combines physical vapor deposition with concurrent ion beam bombardment to produce a wide range of nanocrystalline and coating [7].

The nanocrystalline $TiO_2$ and Ti samples were prepared by ion beam assisted deposition at the Nanotechnology Laboratory of the University of Nebraska Medical Center (Figure 1). The IBAD system is composed of a Veeco 12 cm RF ion gun that supplies ions at energies up to 1500 eV with a total current density of 500 mA, which provides a broad uniform ion beam of oxygen, nitrogen, and argon. IBAD experiments were carried out in an ultrahigh vacuum environment at a base pressure of $10^{-8}$ Torr. Using this ion beam technique, we can easily create a gradual transition between the substrate material and the deposited film with less built-in stress than other techniques. These properties result in films with a much more durable adhesion to the substrate, even at room temperature. Ion bombardment also aids the production of stress-free films, eliminating stress-induced problems such as buckling, microcracking, or peeling. IBAD samples were cut into 1 $cm^2$ sections and placed in a 50:50 mixture of acetone: methanol. Samples were sonicated for 1 hour at room temperature, rinsed in ethanol, and dried under $N_2$ air. Samples were wrapped in foil and autoclaved before use.

**Cell Culture and Cell Seeding**

A human osteoblast-like cell line (SAOS-2) was used in this research. SAOS-2 cells were purchased from ATCC. Cells were grown in McCoy's 5A medium (ATCC) supplemented with 15% fetal bovine serum (FBS) and 1% gentamycin (Invitrogen) in a humidified 5% CO2 atmosphere at 37° C. Cells were seeded at a cell density of 10,000, 25,000, 50,000, and 75,000 cells/ml/$cm^2$ for cell adhesion and cell growth.

For differentiation experiments, cells were incubated in the standard McCoy's medium (supplemented with FBS and antibiotics) for 4 days. On day 4, 0.3 mM ascorbic acid and 10 mM glycerol phosphate were added to the medium (differentiating medium). The medium was changed every 3rd day and the cells were incubated for 7 and 14 days (mineralization).

**Cell Adhesion and Fluorescence Imaging**

The number of adherent cells was determined by means of nuclei quantification on the $TiO_2$-nano, Ti-nano, and Ti-micro substrates. Therefore, after 48 hours of incubation, SAOS-2 cells were washed gently with PBS. Then the cells were fixed with 4% formaldehyde for 10 minutes and stained with 300 nM DAPI (6-diamidino-2-phenylindole) for 5 minutes and rinsed. From each sample, 10 digital images were randomly taken with a Nikon camera using a planar 10x/0.25 objective. DAPI-stained nuclei were automatically counted after single-channel image segmentation and binary masking. The analysis was performed with Image J (Rasband, W.S., Image J, U. S. NIH, Bethesda, Maryland, USA, http://imagej.nih.gov/ij/, 1997-2012).

The morphology of focal adhesions and actin of SAOS-2 cells cultured on the samples were characterized by microscopy after immunofluorescent staining of actin (5:200, Invitrogen, CA, USA Alexa Fluor 546 Phalloidin) and DAPI (300 nM, Invitrogen, CA, USA 6-diamidino-2-phenylindole). Briefly, samples were gently rinsed in PBS and fixed with 4% formaldehyde for 10 minutes at room temperature. After another rinse in 1% PBS-bovine serum albumin (BSA; Sigma), cells were permeabilized with 0.1% Triton-X 100 solution in PBS. Samples were washed with PBS 3 times for 5 minutes at room temperature. Actin stain was added to each sample at a dilution of 5:200 for 30 minutes and rinsed with PBS 3 times for 5 minutes at room temperature. DAPI was added to each sample (300 nM) for 5 minutes and washed with PBS 3 times for 5 minutes at room temperature. Samples were mounted to slides and cover-slipped until further examination.

**Alizarin red staining and quantification**

Alizarin red staining was used in order to determine the calcium deposition of SAOS-2 osteosarcoma cells incubated on the nanosamples at day 7 and 14 using the Millipore Ostegenesis Assay Kit (ECM815) according to the manufacturer. Briefly, media was removed from the samples and rinsed with PBS. The nanosamples were fixed in 10% formaldehyde and stained with alizarin red stain for 20 minutes. The nanosamples were placed in 10% acetic acid for 30 minutes to remove the monolayer, placed in a microcentrifuge tube, and heated to 85°C for 10 minutes and centrifuged. The supernatant was neutralized with 150 μl of 10% ammonium hydroxide and the absorbance was read at 405 nm.

**Statistical Analysis** All data presented were derived from three independent experiments (n=3) and within one experiment three separate samples were analyzed. The results are presented as mean ± SD. Statistical significant differences between three different substrates for alizarin red experiments were evaluated using ANOVA with post-hoc Bonferroni's multiple comparison tests.

## RESULTS

Morphology of IBAD nanocrystalline Ti and $TiO_2$ determined by atomic force microscopy (AFM) and is shown in Figure 2.

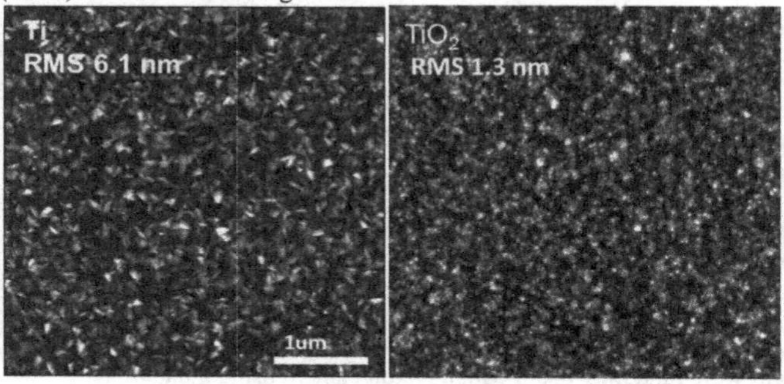


**Figure 2.** AFM images (5μm scan size) of ion beam deposited nano crystalline Ti and $TiO_2$ deposited at room temperature (a) Ti with a RMS of 6.1 nm (b) $TiO_2$ with a RMS of 1.3 nm.

**Cell Adhesion** SAOS-2 cells were monitored for adhesion to the nanocrystalline $TiO_2$, Ti, and medical grade of Ti substrates at 48 hours. The number of adherent cells was determined by nuclei quantification with DAPI in Figure 3. Figure 3a shows a higher number of nuclei on nanostructured surfaces compared to biomedical grade Ti, which indicates more adhesion and growth on nano surfaces. However, by only observing DAPI stained cells, it is impossible to know if cells are healthy and prolific on the surface or not; besides DAPI staining, we also monitored actin fiber shapes (Figure 3b and 3c) that show a significant difference in cell shape on nano crystalline $TiO_2$ and Ti as compared to micro crystalline Ti.

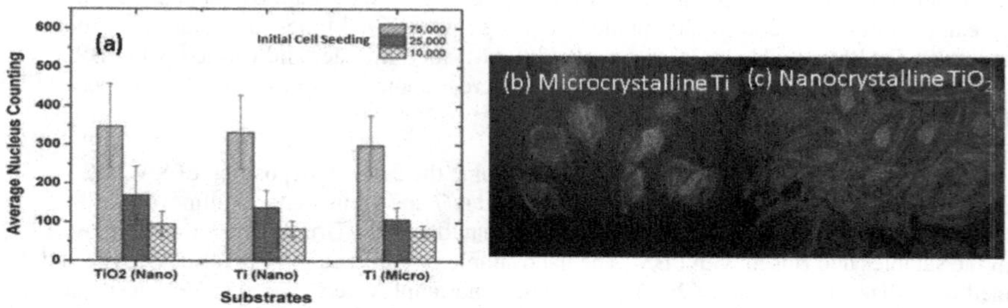

**Figure 3.** (a) Comparing the number of nuclei of SAOS-2 cells on different substrates using DAPI at 48 hours, (b) and (c) merged actin and DAPI stained cells on microcrystalline Ti and nanocrystalline $TiO_2$, respectively.

**Cell Morphology**

In order to evaluate the morphology of cells adherent to the various substrates, the attached cells were labeled with actin and DAPI. Pronounced, large focal adhesions or thin, round shapes with actin staining indicate larger surfaces available for adhesion and thus stronger adhesion contacts between cell and material [16]. From Figure 2, the overall cell morphology is flattened, cells are well spread on the substrate, and the used timing (48 hours of incubation) allowed the cell to go through one cell cycle, which is also an indicator of the suitability of the substrate.

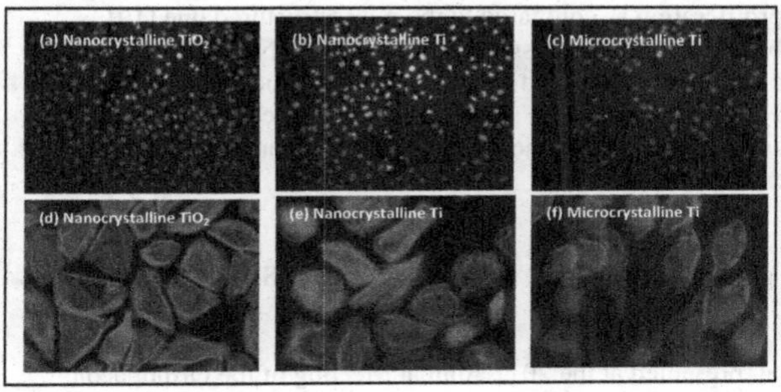


**Figure 4.** Comparing cell adhesion on nanocrystalline $TiO_2$, Ti, and biomedical grade of Ti by fluorescence images microscopy. 50000 cells were incubated for 48 hours. (a), (b) and (c) are DAPI and (d), (e) and (f) are actin stain experiments.

**Alizarin Red Quantification**

In order to determine the calcium deposition of SAOS-2 on osteosarcoma cells incubated on different substrates on the 7th day and 14th day of the culture, alizarin red staining was performed. Alizarin Red-S Staining (ARS) is a dye that binds selective calcium salts and is ideally used for calcium mineral histochemistry [17]. Results of a typical experiment performed in quadruplicate are shown. Values are given as means ± SD.

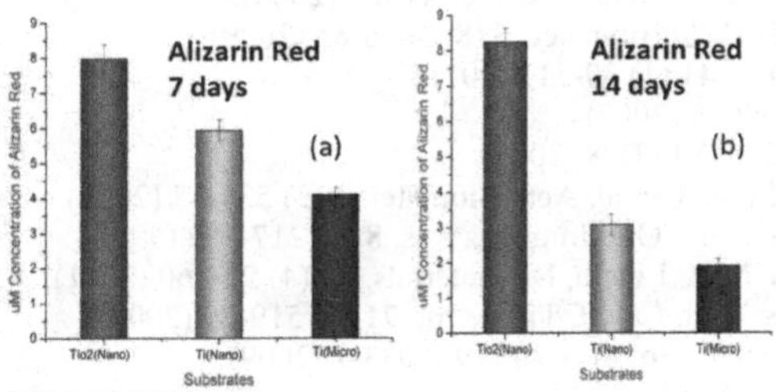


**Figure 5.** Comparing calcium deposition on different substrates using Alizarin Red Assay which indicates that more calcium deposited on IBAD nanocrystalline $TiO_2$ and Ti as compared to biomedical Ti.

## CONCLUSIONS

Fluorescence images presented in Figure 4 show that cell body shape is dependent on the surface of the substrate. If the cell appears round, it is possible that the cells recognize the surface as "flat" and unstructured. However, the nanoroughness of the nano-substrate surfaces appears comparable in size and structure, causing the cells to attach only slightly and to search for suitable places for optimal anchoring [18]. Figure 5 shows the observed matrix mineralization

(Calcium) in SAOS-2 cells. Osteoblastic cells deposit calcium in order to support bone construction. The cells cultured on the nanosurfaces clearly show bright orange-red staining which implies a higher degree of differentiation of the cells. Therefore, enhanced bone formation ability can be expected from the developed nanostructured surfaces. Our statistical analysis indicates that calcium deposition on nano-crystalline $TiO_2$ is significantly different from biomedical grade of Ti after 7 days. These results indicate that the surface nanostructures affect cell interactions at the surface and alter cell behavior when compared to conventional (microstructures) size topography.

Nanostructured surfaces possess unique properties that alter cell adhesion by direct (cell–surface interactions) and indirect (affecting protein–surface interactions) mechanisms [18]. Indeed, in recent work we have shown that the nano-engineered cubic zirconia is superior in supporting growth, adhesion, and proliferation as compared to the microstructured one [19]. Since cell attachment is mediated by adhesive proteins such as fibronectin (FN), we performed a comparative analysis of adsorption energies of FN fragment using quantum mechanical calculations and Monte Carlo (MC) simulation, both on smooth and nanostructured surfaces. We have found that an FN fragment adsorbs significantly more strongly on the nanostructured surface than on the smooth surface [20].

**Acknowledgment:**
This work was supported by DOE grant (DE-SC0005318), Material Science Smart Coating.

**References**

1. F. Namavar, R. Sabirianov, et al, presented at the 2012 Annual Meeting of the Orthopaedic Research Society, San Francisco, Feb 2012 (unpublished).
2. M. Lord, M. Foss, F. Besenbacher, Nano Today 5, 66-78 (2010).
3. T.J. Webster, C. Ergun, et al, J. Biomed. Mater. Res. A **67** (3), 975-80 (2003).
4. M. Svehla, P. Morberg, W. Bruce, B. Zicat, et al, J. Arthroplasty **17** (3), 304-11 (2002).
5. P.R. Klokkevold, P. Johnson, S. Dadgostari, et al, Clin. Oral. Implants. Res. **12**, (4), 350 (2001).
6. D.M. Brunette, B. Chehroudi, J. Biomech. Eng. **121** (1), 49-57(1999).
7. F. Namavar, C.L. Cheung, R.F. Sabirianov, et al, NanoLetters **8** (4) 988 (2008).
8. M. Kalbacova, B. Rezek, V. Baresova, et al, Acta Biomater. **5** (8) 3076-85 (2009).
9. J.B. Kim, P. Leucht, C.A. Luppen, et al, Bone **41** (1) 39-51 (2007).
10. M. Jager, et al, J. Biomed. Biotechnol. 8, 69036 (2007).
11. M.M. Stevens, J.H. George, Science **310** (5751) 1135-8 (2005).
12. L. Le Guehennec, M.A. Lopez-Heredia, B. Enkel, et al, Acta Biomater. **4** (3) 535-43 (2008).
13. P. Francois, P. Vaudaux, M. Taborelli, et al, Clin. Oral Implants Res. **8** (3) 217-25 (1997).
14. Y. Wu, J.P. Zitelli, K.S. TenHuisen, X. Yu, M.R. Libera, Biomaterials. **32** (4) 951-60 (2011).
15. B.D. Boyan, L.F. Boneweld, E.P. Paschalis, et al, Calcif. Tissue Int. **71** (6) 519-29 (2002).
16. O. Babchanko, A. Kromka, et al, Physica. Status Solidi A **206** (9) 2033-7 (2009).
17. H.T. Sasmazel, Int. J. Biol. Macromol. 49, 838–846 (2011).
18. G. Mendonca, D.B. Mendonca, F.J. Aragao, et al, Biomaterials **29** (28) 3822-3835 (2008).
19. F. Namavar, R.F. Sabirianov, et al in *Engineered Nanostructured Coatings for Enhanced Protein Adsorption and Cell Growth* (Mater. Res. Soc. Symp. Proc. **1418**, Boston, MA 2011).
20. R.F. Sabirianov, A. Rubinstein, F. Namavar, Phys. Chem. Chem. Phys. **13** (14) 6597 (2011).

**Mater. Res. Soc. Symp. Proc. Vol. 1569 © 2013 Materials Research Society**
**DOI: 10.1557/opl.2013.907**

# QUATERNARY AMMONIUM ANTIMICROBIAL POLYMERS

Abraham J. Domb[1], Nurit Beyth[2], Shady Farah[1]

[1] Institute for Drug Research, The Hebrew University of Jerusalem, Jerusalem, Israel.

[2] Department of Prosthodontics, Hebrew University - Hadassah School of Dental Medicine, Israel.

## ABSTRACT

Polymers possessing antimicrobial activity have been used for self sterilization surfaces as well as agents for treating contaminated water. Cationic polymers based on quaternary ammonium or guanidine groups have shown high inherent antimicrobial activity where the activity is related to the disruption of the microorganism cell wall. A range of antimicrobial nanoparticles possessing active quaternary ammonium groups with one of the alkyl is a an octyl chain have been synthesized. These nanoparticles were incorporated in dental restoration compositions to form self sterile composites. Quaternary ammonium polyethyleneimine nanoparticles with N-octyl dimethyl residues, demonstrated high antibacterial effect.

## INTRODUCTION

Antimicrobial agents of low molecular weight are widely used as antimicrobial drugs, food preservatives and for water and soil decontamination. However, they suffer from residual toxicity [1]. Polymeric antimicrobial agents are nonvolatile, chemically stable, have long-term antimicrobial activity, can be bound to the surface of interest and hardly permeate through biological membranes such as the skin [2]. Distinctively, polycationic antimicrobials have high surface density of active groups which might result in increased antimicrobial activity [3]. Quaternary ammonium compounds have a broad spectrum of antimicrobial activity against both Gram-positive and Gram-negative bacteria. Most of these polymers bearing quaternary ammonium groups appear to act by interacting non-specifically with the cell wall and disrupting negatively charged bacterial cell membrane [4]. The most common polyamines used for medical applications include: poly(L-lysine) (PLL), poly(N-alkyl-4-vinylpyridinium) salts (PVP), polyamidoamine (PAMAM) dendrimers, polyethyleneimine (PEI), and poly(2-dimethylaminoethyl methacrylate) (PDEAEM). These polycations are either homopolymers or random copolymers having linear, branched or dendritic architectures. A number of reviews and books issues discuss the physiochemical, biochemical and therapeutic aspects of these polymers [5].

Linear and crosslinked quaternary ammonium polyethyleneimine (QA-PEI) inhibits bacterial growth [6,7]. Lin et al. proposed a novel, non-release strategy for creating bactericidal surfaces which involves covalent coating with long hydrophobic polycationic chains based on PEI which was found effective against a variety of Gram-positive and Gram-negative bacteria [7]. Its antibacterial activity was found to be dependent on molecular weight of PEI indicating that N-alkylated PEI of low molecular weights, i.e. 2- and 0.8-kDa, had a weak bactericidal activity due to short chain length which probably do not penetrate bacterial cell.

Due to the positive charge of the amino groups in physiological medium, poly(L-lysine) (PLL) demonstrated a wide spectrum of antimicrobial activities against yeast, fungi, and Gram-positive or Gram-negative bacteria species [8]. Hiraki et al. examined inhibition effect of poly(ε-lysine) on bacillus strains and found that inhibitory concentrations for the spore germination of *B. coagulans*, *B. stearothermophilus* and *B. subtilis* were 12.5 μg/ml, 2.5 μg/ml and 12.5 μg/ml, respectively [9]. A correlation was found between antibacterial activity of poly(ε-lysine) based polymers and their molecular weight. Chain length of at least 9 L-lysine residues was found to be optimum for inhibition of microbial growth [8].

Pyridinium-type polymers possess antibacterial activity due to their positive charge and hydrophobic nature following N-alkylation of the monomer units [10]. The antimicrobial activity of insoluble poly(4-vinylpyridine-co-divinylbenzene) beads and block and random copolymer structures based on polystyrene and p-(4-VP) quaternerized with octyl iodide were reported [11]. Quaternerized polystyrene-*block*-poly(4-vinyl pyridine) demonstrated higher potency in the reduction of *Pseudomonas aeruginosa* and *Staphylococcus aureus* growth than quaternerized polystyrene-*random*-poly(4-vinyl pyridine). This behavior is attributed to the different concentration of 4-VP on the surface of the tested specimens prepared from the two copolymers. Imazato et al. described the potential use of pyridinium type polymers as antibacterial additives in restorative composite resins [12]. They showed that incorporation of an antibacterial monomer such as 12-methacryloyloxydodecyl pyridinium bromide (MDPB) in a dentin primer and bonding-resin provides the composite antibacterial properties [12]. Bonding-resin containing MDPB inhibited bacterial growth on its surface by means of the action of an immobilized bactericide after the resin is polymerized, without adversely affecting its bonding characteristics [13,14].

The antimicrobial activities of the chitosan based polysaccharides have been reported [15]. Quaternary ammonium chitosan derivatives can be prepared via reductive amination by introducing alkyl groups into the amino groups of chitosan via Schiff base intermediates followed by quaternization with methyl iodide to produce water-soluble cationic polyelectrolytes [15] or by N-alkylation with an appropriate alkyl halide [16].

In the present work we report on the synthesis and antimicrobial activity of polycationic nanoparticles based on crosslinked polyethyleneimine and poly(4-vinyl pyridine) bearing quaternary ammonium groups. These polymers were effective in inhibition of the bacterial growth for 6 months at 1% w/w of polymer particles in dental restoration compositions.

## EXPERIMENT

The synthesis of quaternary ammonium PEI nanoparticle was previously described [19]. In brief, PEI (10 g, 0.23 mol monomer units) dissolved in 100 ml ethanol was reacted with diiodopentane at a 1:0.04 mol ratio (monomer units of PEI/diiodopentane) under reflux for 24 h.

N-alkylation was conducted as follows: octyl halide was added at a 1:1 mol ratio (monomer units PEI/octyl halide). Alkylation was carried out under reflux for 24 h followed by neutralization with sodium hydroxide (1.25 equimolar, 0.065 mol) for an additional 24 h under the same conditions. N-methylation was conducted as follows: 43 ml of methyl iodide (0.68 mol) were added. Methylation was continued at 42 □C for 48 h followed by neutralization with sodium bicarbonate (0.23 mol, 19 g) for an additional 24 h. The supernatant obtained was decanted and precipitated in 300 ml of double distilled water (DDW), washed with hexane and DDWand then freeze-dried. The purification step was repeated using additional amounts of hexane and DDW. The average yield was 80% (mol/mol).
PVP particles were prepared by suspension polymerization of 4-vinylpyridine (4-VP) with divinyl benzene (DVB) as crosslinker. The polymerization reaction of 4-VP (1.08 ml, 9.9 mmol) and DVB was carried out at either 1:0.01, 1:0.1, 1:0.15, 1:0.2, 1:0.25, and 1:0.3 mole ratio (4-VP/DVB) in a three-necked round bottom flask equipped with a nitrogen inlet and reflux condenser. 4-VP and DVB were dissolved in 0.5 ml of N-methyl pyrrolidone. The mixture was added to the polymerization flask containing 100 ml of DDW, 10 mg of zoisobutyronitrile (AIBN and polyvinyl alcohol (0.8 g) as a dispersing agent. Reaction was carried out at 80°C under nitrogen atmosphere. White suspension was obtained within 17 hours. The resulting crosslinked particles were collected by centrifugation followed by washing with ethanol and DDW to remove unreacted 4-VP, DVB, N-methyl pyrrolidone and polyvinyl alcohol, respectively. The product was vacuum dried over $P_2O_5$ over night. Average yield: 80% (w/w). Quaternization of the pyridine rings previously polymerized nanoparticles of p-4-VP (0.2 g, 1.9 mmol of the pyridine groups) dispersed in 30 ml absolute ethanol were reacted with 0.5 ml of bromooctane or with 0.4 ml bromohexane (2.85 mmol, 1.5 equimolar). The reaction was carried out at reflux conditions with vigorous mixing for 48 hours. The obtained product was collected by centrifugation followed by washing with DDW and ethanol to remove unreacted bromooctane. The brown crude was vacuum-dried over $P_2O_5$ over night. Average yield: 95% (w/w).

QA nanoparticles analysis, IR spectra were recorded on a Perkin Elmer System 2000 FT-IR. Particle size was determined by dynamic light scattering using High Performance Particle Sizer (ALV-NIBS/HPPS, Langen, Germany). The measurements were done in DDW. Hydrophobicity of synthesized materials, before alkylation and for the final QA-PEI particles, was calculated by elemental microanalysis (C/N) using a Perkin-Elmer 2400/II CHN analyzer. Zeta-potential was measured on a Zetasizer 2000 (Malvern, UK) in DDW, in triplicate. Thermal analysis was determined on a Mettler TA 4000-DSC differential scanning calorimeter, at a heating rate of 10°C/min (typical sample weight was 10 mg) and on a Stuart Scientific Melting point SMP1 heater.

Bacterial strains and growth conditions: Clinically isolated *Escherichia coli*, *Pseudomonas aeruginosa, Enterococcus faecalis, Staphylococcus epidermidis* (the listed bacteria were clinically isolated at the Maurice and Gabriela Goldschleger School of Dental Medicine at Tel-Aviv University, Israel), and *Staphylococcus aureus* ATCC 8325-4 were used in this study. The bacteria were cultured aerobically overnight in 5 ml of brain–heart infusion (BHI) broth (Difco, Detroit, MI), at 37°C. The tested materials were prepared by adding the synthesized PEI nanoparticles to the dental restorative resin composite Filtek Flow (47% Zirconia/silica, average particle size 0.01–6.0 m; BIS-GMA, TEGDMA (3 M ESPE Dental St Paul, MN, USA)). The nanoparticle powder was added at 0, 1 or 2% w/w to the resin composite and homogeneously mixed in a dark room for 30 s with a spatula in a microtiter plate and were light cured according to the manufacturer's instructions. Through the aging process, the plates were kept at 37°C, and each well was filled with 250 µl PBS replaced every 48 hr. At the end of the aging process plates were dried. Direct contact test (DCT): The microtiter plate was positioned vertically and a 10 µl volume of bacterial suspension was placed on the surface of each tested material. The plate was then incubated vertically for 1 hr at 37°C, the suspension liquid evaporated and direct contact between all bacteria and the tested surfaces as ensured. The plate was then positioned horizontally and 220µl of BHI broth were added to each well. Surviving bacteria growth was measured in each well every 20 minutes for 14 hr using a temperature-controlled microplate spectrophotometer (VERSAmax, Molecular Devices Corporation, CA, USA), at 37°C with 5 sec mixing before each reading. Controls included equal bacterial suspension placed on the sidewall of eight wells and calibration experiment was performed simultaneously to the experimental setup in each microtiter plate. Data analysis: Absorbance measurements were plotted providing bacterial growth curves for each well, and the linear portion of the logarithmic growth phase was statistically analyzed. Data were expressed as the slope (ß) and the constant (α) of the linear function $ßx+\alpha=y$ derived from the ascending portion of the bacterial growth curve which correlates with the bacterial growth rate and initial bacterial number, respectively. Results were analyzed using ANOVA followed by a Tukey ($p<0.05$).

**RESULTS and DISCUSSION**

Incorporation of the quaternary ammonium polyamine particles with size of 160 ± 10nm (figure 1) within the matrix of restorative composite resins provides antibacterial properties. Quaternary ammonium polyamine based particles are attractive candidates as antibacterial additives because their target is the negatively charged microbial membrane and are driven by the membrane potential. The use of non biodegradable polyamine particles is especially suitable for the above application, because they do not degrade and do not release toxic quaternary ammonium residues to the medium.

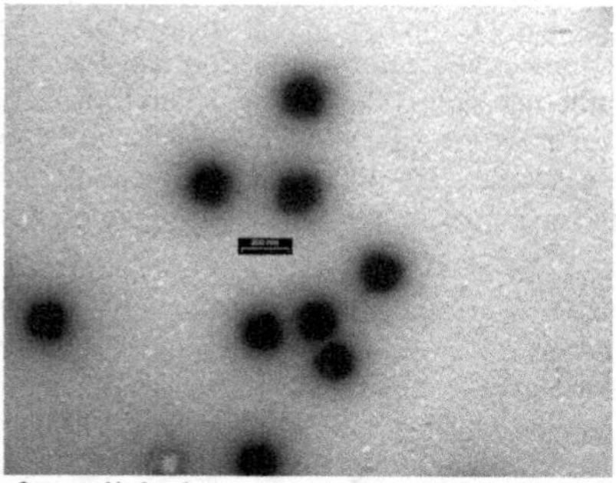

**Figure 1:** SEM image of crosslinked quaternary ammonium particles (QA-PEI) at constant acceleration voltage of 5KV.

The availability of a large number of amino group functionalities along the cationic polyamine based particles allows numerous and versatile chemical modifications for altered physiochemical properties. Such modifications including attachment of hydrophobic residues to enhance cell interaction, attachment of hydrophilic stretches (i.e. polyethylene glycol (PEG) or poly(vinyl pyrrolidone)) to increase the resistance towards extracellular fluids, to improve polycation's biocompatibility or to enhance antibacterial properties due to increased bacterial interaction with the bactericide surface, attachment of fluorescence markers for distribution studies, the both N-alkylation and N-methylation provide polyamine with hydrophobic nature and cationic charge, respectively, which result in antibacterial properties, etc, figure 2.

Crosslinked polyamine based particles were obtained by reacting polyethyleneimine in ethyl alcohol with 1,5-dihalidopentane as crosslinking agent, which causes intramolecular crosslinking, and by suspension polymerization of 4-vinyl pyridine with divinyl benzene. The crosslinking concept, where crosslinking agent is covalently bonded to the amino groups of the polyamine chain, prevents separation of the various polymeric chains that form the particle and dissolving of the particles within the medium. Usage of nanoparticles is advantageous as active antibacterial agents, since their surface area is exceedingly outsized relative to their size and scope. Thus, nanoparticles may provide high biological activity although only a small dosage of these particles is added to the composite materials. Polyethyleneimine was crosslinked at three crosslinking degrees, i.e. 1:0.01, 1:0.04 and 1:0.2 (monomer units of PEI/dihalidopentane) mole ratios. The obtained crosslinked polyethyleneimine based nanoparticles were allowed to react with the corresponding alkyl halide. Pyridinium type particles crosslinked with divinyl benzene at 1:0.01 to 1:0.3 (4-vinyl pyridine/divinyl benzene) mole ratios were prepared via suspension polymerization followed by N-alkylation leading to quaternerized particles. Polyamine based particles were N-alkylated with linear alkyl halides of different chain lengths varying from butyl to hexadecyl. Previously N-alkylated polyethyleneimine based nanoparticles were neutralized with sodium bicarbonate followed by N-methylation with methyl iodide to obtain quaternary ammonium polyethyleneimine (QA-PEI) nanoparticles. The majority of the polyamine particles were freely dispersed in water and alcohols, DMF and DMSO with regard to their chemical structure including polyamine backbone, alkyl chain length and alkylation degree.

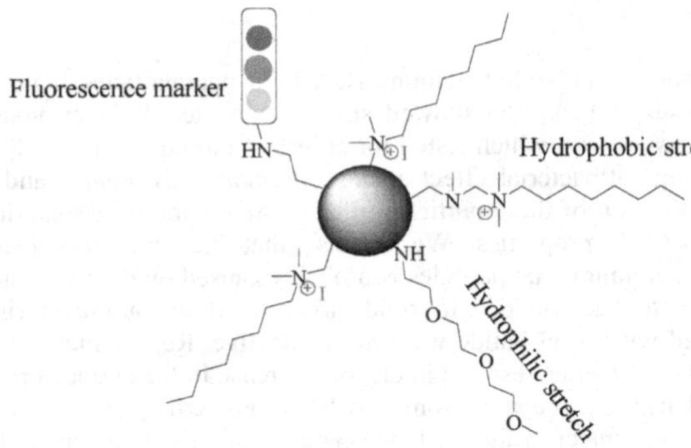


**Figure 2.** Schematic illustration of the polyethyleneimine based particles.

The synthetic polyamine based nanoparticles were precipitated and purified by extensive wash with hexane and double deionized water (DDW) to obtain pure particles uncontaminated with unreacted alkyl halides and salt as evidenced by elemental analysis. The % Iodine content of the QA-PEI varied between 35 % and 43 % was dependent mainly on the chain length of the alkylating agent, degree of alkylation and methylation. % Bromine content of the pyridinium-type particles ranging from 8 % and 25 % was dependent mainly on the crosslinking degree of the starting poly(4-vinyl pyridine) particles. Comparison of carbon/nitrogen (C/N) content of alkylated polyethyleneimine based particles with the corresponding unmodified polyethyleneimine indicated that more than 70 % of the alkyl is being substituted. Degree of quaternization, i.e. the percentage of pyridinium groups to the pyridine rings, was calculated from elemental analysis of bromine. Pyridinium-type particles with low crosslinking degree (i.e. 1:10 (4-vinyl pyridine/divinyl benzene) mole ratio) resulted in 70 % to 75 % of the quaternization rate, while higher crosslinking rates (i.e. 1:30 (4-vinyl pyridine/divinyl benzene) mole ratio) reduced the percentage of the pyridinium rings to 40 %. Although crosslinking rates were up to 1:30 (4-vinyl pyridine/divinyl benzene) mole ratio, this steep decrease in the quaternization degree is attributed to the reduced access of the pyridine sites to the alkyl halide due to rigid structure of the pyridine aromatic rings. In addition, hexyl bromide resulted in higher quaternization degree of poly(4-vinylpyridine) particles than octyl halide, suggesting that alkyl chain length plays key role in the quaternization yield. The mean particle size of QA-PEI particles was found to be ranging from 30 nm and 600 nm with regard to their chemical properties and structure, while mean particle size of pyridinium-type particles was varied from 400 nm to 1 micron. This steep difference in the mean particle size between small and large QA-PEI nanoparticles is mostly attributed to both alkyl chain length and degree of alkylation. In a detail, moderate degree of alkylation (i.e. 1:0.25 (monomer units of PEI/alkyl halide) mole ratio) resulted in low mean particle size of the QA-PEI nanoparticles due to probably moderate hydrophobic nature of the obtained particles, while high degree of alkylation (i.e. 1:1 (monomer units of PEI/alkyl halide) mole ratio) resulted in large nanoparticles. Zeta potential of the QA-PEI particles was found to be ranging from +60 mV and +106 mV, while zeta potential of pyridinium-type particles was varied from +43 mV to +48 mV.

**Antimicrobial activity:**

Quaternary ammonium polyethyleneimine (QA-PEI) nanoparticles being embedded in restorative composite resin at 1 % w/w showed strong antibacterial effect against cariogenic bacteria, i.e. *Streptococcus mutans*, which lasted for at least 6 months, figure 3. Pyridinium type polymers showed a strong antibacterial effect against *Streptococcus mutans* and *Enterococcus faecalis*, however incorporation of these particles in restorative composite materials completely diminished their antibacterial properties. We surmise that the surprising inactivity of the composites containing pyridinium type particles is probably caused by dis-exposure of the active alkyl pyridinium groups to bacteria due to rigid structure of the aromatic rings. QA-PEI nanoparticles N-alkylated with octyl halide were most effective. Replacement of octyl residues with longer or shorter alkyl moieties resulted in drastic decrease in the antibacterial activity. The most effective nanoparticles were prepared from crosslinked polyethyleneimine N-alkylated with octyl chains followed by quaternization of the amino groups with methyl iodide. The antibacterial studies were proceeded using three different restorative composite resins (bonding, hybrid and flowable) and showed similar inhibition effect with octyl alkylated QA-PEI the most active compound [15]. To emphasize the importance of grafted octyl content on the antibacterial activity, a series of QA-PEI nanoparticles alkylated with octyl halide at various degrees were prepared and tested for their antibacterial effect. The antibacterial efficiencies of these synthetic polycationic particles were found to be depended on the content of the substituted octyl. QA-PEI nanoparticles containing moderate octyl content (of at least 1:0.25 (monomer units of PEI/alkylating agent) mole ratio) resulted in complete inhibition of the bacterial growth at low concentrations, whereas low octyl content (i.e. 1:0.0625 (monomer units of PEI/alkylating agent) mole ratio) resulted in reduced antibacterial effect. After numerous antibacterial experiments, it was decided to focus on the most active derivative and further investigate the parameters affecting antibacterial activity. The factors affecting antimicrobial activity are summarized below:

Crosslinking degree: three QA-PEI derivatives crosslinked at various degrees were prepared and tested for their antibacterial activity being incorporated in dental composite resin at 1 % w/w. The starting polyamine was polyethyleneimine (750-kDa) crosslinked at 1:0.01, 1:0.04 and 1:0.2 (monomer units of PEI/dihalidopentane) mole ratios. As expected, high degree of crosslinking resulted in a reduced yield of octyl substitution (6.04 carbon/nitrogen), while lower degree of crosslinking (1:0.01 and 1:0.04 (monomer units of PEI/ dihalidopentane) mole ratios) resulted in increase of the carbon/nitrogen content (6.53 and 6.85, respectively). The antibacterial efficiencies of QA-PEI prepared from low degree of crosslinking resulted only in a slight inhibition of the bacterial growth, whereas QA-PEI nanoparticles prepared from high degrees of crosslinking inhibited more effectively bacterial growth, but less successfully than moderate crosslinked QA-PEI. The most effective compound embedded within the matrix of restorative composite resin was octyl alkylated QA-PEI crosslinked at 1:0.04 (monomer units of PEI/ dihalidopentane) mole ratio. The reason for the reduced activity of the low crosslinked compound can be attributed to the insufficient crosslinking degree of the nanoparticles which might result in separation of the various polymeric chains that form the particle. On the contrary, crosslinking at 1:0.2 (monomer units of PEI/dihalidopentane) mole ratio resulted in more compact particles in comparison with low degree of crosslinking which might be responsible for the reduced access of the hydrophobic chains to the bacterial membrane that might be critical for the effectiveness of the compound.

N-methylation effect: unlike QA-PEI based nanoparticles, non-methylated octyl-PEI based nanoparticles showed reduced antibacterial activity with bacterial recovery reduced to 34 % compared to the negative control, in which restorative composite resins were not treated with QA-PEI particles. Thus, N-methylation step is essential to the antibacterial activity of the particles. This behavior can be explained by the fact that quaternary methylation converts remained secondary and tertiary amines to quaternary amino groups. Therefore, use of foregoing alkylation and methylation methodology elevates antibacterial efficiency of the octyl alkylated QA-PEI being incorporated within the matrix of the clinically used dental composite materials.

Molecular weight of starting polyethyleneimine: QA-PEI nanoparticles prepared from crosslinked polyethyleneimine of various molecular weights (25- and 750- kDa) N-alkylated with octyl halide followed by quaternization with methyl iodide, were embedded in dental composite resin at 1% w/w and tested for their antibacterial activity. High antibacterial effect was obtained with QA-PEIs having average molecular weights of 25- and 750- kDa. The results show that QA-PEIs prepared from high molecular weight polyethyleneimine are efficient in inhibition of bacterial growth probably due to better access of the hydrophobic polymeric flexible chains to the bacterial surface.

Effect of particle size: Dental composite resin embedded with 1 % w/w QA-PEI microparticles was tested for its antibacterial effect in comparison with resin containing QA-PEI nanoparticles. QA-PEI particles up to 3.4 microns were found to be highly effective in inhibition of *S. mutans* growth indicating in minor effect resulted from surface density differences between nano- and micro- tested particles. Experiments to prepare larger microparticles of QA-PEI were failed.

Effect of counter ion: Nitrate, acetate and iodide form QA-PEI nanoparticles demonstrated similar efficiency in bacterial growth inhibition. This behavior can be explained by the fact that antibacterial activity of the QA-PEI particles is depended on the hydrophobic chain and positive charge of the derivative and not on the counter ion. Thus, all tested materials similarly inhibited bacterial growth. Another explanation for this behavior is the fact that counter ion can affect antibacterial properties where it alters the solubility of the biocides; whereas QA-PEI nanoparticles are crosslinked. Thus, counter ions showed minor effect on the antibacterial activity of the QA-PEI nanoparticles. Based on foregoing data, it was decided to focus on the iodide form QA-PEI due to simplicity of the synthesis and further study physical, chemical and biological properties of the restorative composite resins incorporating QA-PEI particles.

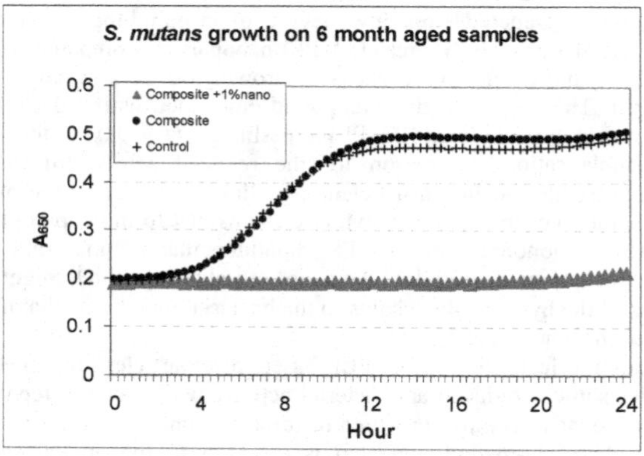


**Figure 3:** antibacterial effect of quaternary ammonium polyethyleneimine (QA-PEI) nanoparticles embedded in restorative composite resin at 1 % w/w aged for 6 months, against *S. mutans*. No bacterial growth was found.

QA-PEI based nanoparticles were synthesized in high reproducibility and showed high chemical stability against environmental changes including exposure to different oxidizers and storage conditions [17].

QA-PEI particles incorporated in restorative composite resin at 1 % w/w demonstrated high antibacterial effect, owing to their hydrophobic nature and cationic density, table 1. Absorption of positively charged polymers onto negatively charged bacterial surface is thought to be responsible for the increase of cell permeability and may disrupt bacterial cell membranes. Furthermore, antibacterial ratio can reach 100 % already at very low dosage (15 mg/l) and a short contact time (4 min) with bacteria cells [18]. Release profile of QA-PEI particles from restorative composite resin was observed using agar diffusion test which demonstrated no antibacterial zone as evidenced by normal bacterial growth indicating probably no or undetectable levels of the antibacterial components leached out from the modified composite resins [19]. Moreover, no residual QA-PEI nanoparticles eluted from the composite materials following polymerization of the resins were revealed by chromatography and spectroscopy studies [19]. These results confirm that composite resins supplemented with QA-PEI are stable and do not excrete sufficient antibacterial substances to the surrounding media to affect bacterial growth. Since no eluted QA-PEI particles were detected from the cured modified composite resins, the inhibition effect on bacterial growth is considered to be due to direct contact of the bacterial cells with the incorporated biocide particles. In addition, no release of the QA-PEI from the composite matrix appears to be an important determinant of the long lasting antibacterial activity. Consequently, bacterial growth is kept inhibited and thus could be long-standing and effective once the restoration is placed. According to experimental results, restorative composite resins embedded with 1 % w/w QA-PEI inhibited totally *S. mutans* growth in commercial composite resins used in the daily dental practice for at least 6 months.

Table 1: Bacterial growth rate (in percentage) of three Gram-positive bacteria (*S.aureus, E.faecalis, S.epidermidis*) and two Gram-negative bacteria (*E.coli, P.aeruginosa*) following direct contact with resin composite with 0, 1, and 2% w/w incorporated PEI nanoparticles. Growth rate determined in four separate experiments in which samples were aged for 1 h, 48 h, 1 week and 4 weeks. The results are expressed as the mean ± SD.

| % Growth Rate | S.Aureus | S.Epidermidis | P.Aeruginosa | E. Faecalis | E.Coli |
|---|---|---|---|---|---|
| Control 1 hour | 100.0 ± 0.1% | 100.0 ± 0.0% | 100.0 ± 10.0% | 100.0 ± 0.0% | 100.0 ± 10.0% |
| Control 48 hours | 100.0 ± 0.0% | 100.0 ± 5.0% | 100.0 ± 5.0% | 100.0 ± 0.1% | 100.0 ± 10.1% |
| Control 1 week | 100.0 ± 10.0% | 100.0 ± 0.2% | 100.0 ± 2.2% | 100.0 ± 2.0% | 100.0 ± 1.8% |
| Control 4 weeks | 100.0 ± 0.3% | 100.0 ± 3.0% | 100.0 ± 7.1% | 100.0 ± 2.8% | 100.0 ± 3.0% |
| | | | | | |
| Composite 1 hour | 92.2 ± 2.0% | 88.6 ± 4.3% | 91.7 ± 1.3% | 72.6 ± 0.0% | 98.9 ± 1.3% |
| Composite 48 hours | 107.5 ± 5.0% | 128.3 ± 10.1% | 100.3 ± 0.7% | 79.3 ± 0.1% | 68.7 ± 0.1% |
| Composite 1 week | 85.1 ± 8.3% | 100.7 ± 2.3% | 90.8 ± 3.3% | 128.7 ± 7.3% | 130.3 ± 8.6% |
| Composite 4 weeks | 120.2 ± 7.0% | 100.2 ± 4.1% | 150.2 ± 11.3% | 131.6 ± 6.1% | 100.0 ± 4.3.1% |
| | | | | | |
| Composite + 1% QA-PEI 1 hour | 3.5 ± 8.0% | 10.2 ± 3.5% | 39.2 ± 9.5% | 2.1 ± 6.0% | 8.2 ± 2.7% |
| Composite + 1% QA-PEI 48 hours | 1.2 ± 4.1% | 6.3 ± 7.2% | 41.2 ± 11.2% | 6.3 ± 1.1% | 17.5 ± 0.2% |
| Composite + 1% QA-PEI 1 week | 8.0 ± 0.1% | 19.3 ± 8.3% | 92.3 ± 4.6% | 1.1 ± 5.7% | 14.3 ± 4.3% |
| Composite + 1% QA-PEI 4 weeks | 5.0 ± 8.0% | 21.3 ± 6.7% | 107.6 ± 12.4% | 0.9 ± 7.1% | 36.8 ± 3.7% |
| | | | | | |
| Composite + 2% QA-PEI 1 hour | 1.2 ± 1.9% | 4.3 ± 2.0% | 9.8 ± 2.5% | 5.2 ± 7.1% | 4.3 ± 4.1% |
| Composite + 2% QA-PEI 48 hours | 0.2 ± 2.1% | 0.0 ± 4.7% | 11.4 ± 4.9% | 4.6 ± 5.1% | 0.0 ± 0.1% |
| Composite + 2% QA-PEI 1 week | 6.0 ± 0.1% | 3.8 ± 3.9% | 8.7 ± 1.6% | 0.0 ± 3.1% | 0.2 ± 0.1% |
| Composite + 2% QA-PEI 4 weeks | 2.0 ± 6.0% | 5.2 ± 1.0% | 12.1 ± 14.3% | 0.0 ± 2.0% | 0.0 ± 0.1% |

Surface chemical analysis of the restorative composite resin containing QA-PEI depicted surface modification of higher hydrophobicity and presence of quaternary amino groups on the surface of the modified composite resins compared to the corresponding commercial material although only one percentage of the particles was added. In particular, the water contact angles were increased following the addition of the QA-PEI nanoparticles, raising the hydrophobicity of the material surface [20]. The presence of active antibacterial components on the surface of the restorative composite materials may also offer an additional explanation for the long lasting antibacterial properties of the materials following incorporation of QA-PEI. These findings add another aspect to the belief that the effective antibacterial outcome of these components is through lethal direct contact with bacteria.

Incorporation of the QA-PEI at low concentrations in restorative composite resin did not change its biocompatibility when compared to the commercial composite resin *in vitro*. No distinct difference in the cell viability of the restorative composite and composites supplemented with 1 % w/w and 2 % w/w of QA-PEI was detected using cytotoxicity studies of the eluted supernatants. Local toxicity studies, presented by implantation of modified dental composite materials, indicated a normal immune response compared to the unmodified restorative composite resin. No adverse reaction upon application of the tested implants embedded with up to 2 % w/w of QA-PEI. In addition, no evidence of any active inflammatory reaction or necrosis was noted. Following *in vitro* cytotoxicity tests and histopathological evaluation, addition of a small percent (up to 2 % w/w) of QA-PEI nanoparticles was biocompatible.

Since dental biofilm has been defined as a complex microbial community directly associated with both caries and periodontal diseases, QA-PEI incorporation is a possible approach to achieve antibacterial dental composites which do not release any antibacterial components, enabling long-term effectiveness. Preliminary clinical studies approved by Helsinki committee of the hospital, were proceeded with control and test disks containing 1 % w/w of QA-PEI, which were held for 4 hours in the volunteers' mouth (10 volunteers). Clinical studies demonstrated a change in the biofilm thickness as well as significantly more dead bacterial stain in the composite resins supplemented with 1 % w/w QA-PEI.

## CONCLUSIONS

Quaternary ammonium polyethyleneimine nanoparticles synthesized by crosslinking polyethyleneimine and subsequent alkylation of the amino groups with octyl iodide and methyl iodide, demonstrated high antibacterial effect *in vitro*. These quaternary ammonium polyethyleneimine particles embedded in restorative composite resin at 1 % w/w resulted in complete inhibition of *Streptococcus mutans* growth as well as other common bacteria and the effect was lasted for weeks.

## REFERENCES


1. E. R. Kenawy, *J.Appl. Polym. Sci.*, **82(6)**, p. 1364-1374, (2001).
2. E. R. Kenawy, S. D. Worley and R. Broughton, *Biomacromolecules* **8(5)**, p. 1359-1384, (2007).
3. E. R. Kenawy, F. I. Abdel-Hay, A. El-Raheem, R. El-Shanshoury and M. H. El-Newehy, *J. Controlled Release*, **50(1-3)**, p. 145-152, (1998).
4. T. Tashiro, *Macromol. Mater. Eng.*, **286(2)**, p. 63-87, (2001).

5. I. L. Shih, M. H. Shen and Y.T. Van, *Bioresource Technol.*, **97(9)**, p. 1148-1159, (2006).
6. N. Beyth, I. Yudovin-Farber, M. Perez-Davidi, A. J. Domb and E. Weiss, *PNAS*, **107(51)**, p. 22038-43285, (2010).
7. J. Lin, S. Qiu, K. Lewis and A. M. Klibanov, *Biotechnol. Bioeng.,* **83(2)**, p. 168-172, (2003).
8. J. Hiraki, *Fine Chem.*, **29**, p. 25-28, (2000).
9. J. Hiraki, *J. Antibact. Antifungal Agents*, **23**, p. 349-354, (1995).
10. G. Li and J. Shen, *Abstr. Papers Am. Chem. Soc.*, **216**, p. U97-U97, (1998).
11. E. S. Park, H. S. Kim, M. N. Kim and J. S. Yoon, *Eur. Polym. J.*, **40(12)**, p. 2819-2822, (2004).
12. S. Imazato, N. Ebi, Y. Takahashi, T. Kaneko, S. Ebisu and R. Russell, *Biomaterials*, **24(20)**, p. 3605-3609, (2003).
13. S. Imazato, Y. Kinomoto, H. Tarumi, S. Ebisu, F. R. Tay, *Dent. Mater. J.*, **19(4)**, p. 313-319, (2003).
14. S. Imazato, *Dent. Mater. J.*, **28(1)**, p. 11 - 19, (2009).
15. H. Sashiwa and S.I. Aiba, *Prog. Polym. Sci.*, **29**, p. 887–908 (2004).
16. Z. L. Shi, K. G. Neoh, E. T. Kang and W. Wang, *Biomaterials*, **27(11)**, p. 2440-2449, (2006).
17. S. Farah, W. Khan, I. Farber, D. Kesler-Shverob, N. Beyth, E. Weiss and A. J. Domb *Polym. Adv. Technol.*, (2013).
18. B. Gao, *J. Biomater. Sci. Polymer Ed.*, **18(5)**, p. 531-544, (2007).
19. N. Beyth, I. Yudovin-Farber, R. Bahir, A. J. Domb, E. Weiss, *Biomaterials*, **27(21)**, p. 3995-4002, (2006).
20. N. Beyth , I. Yudovin-Farber, A. J. Domb , and E. I. Weiss, *Dent. Mater. J.*, (2008).

**Mater. Res. Soc. Symp. Proc. Vol. 1569 © 2013 Materials Research Society**
**DOI: 10.1557/opl.2013.800**

# DNA Transfection Efficiency of Antimicrobial Peptide as Revealed by Molecular Dynamics Simulation

Namsrai Javkhlantugs[1, 2], Janlav Munkhtsetseg[3], Chimed Ganzorig[1], and Kazuyoshi Ueda[2]

[1]Center for Nanoscience and Nanotechnology & Department of Chemical Technology, School of Chemistry and Chemical Engineering, National University of Mongolia, Main building, University street 1, Ulaanbaatar 14201, Mongolia
[2]Department of Advanced Materials Science, Graduate School of Engineering, Yokohama National University, 79-5 Tokiwadai, Hodogaya ku, Yokohama 240-8501, Japan
[3]Department of Biochemistry and Laboratory, School of Biomedicine, Health Sciences University of Mongolia, Zorigiin gudamj, Ulaanbaatar 14210, Mongolia

## ABSTRACT

The peptide-DNA complex was investigated by using molecular dynamics simulation to analyze the transfection efficiency of cationic amphipathic peptide. Previously, the cationic peptide, LFampinB, with positively charged amino acid residues of Lysines was used to investigate the orientation and interaction energies for entering the cell though disruption of the endosomal membrane. The same interactions were obtained for N-terminus of the LFampinB peptide with membrane and with plasmid DNA. The N-terminus of LFampinB can bind at minor groove of DNA to make complexation of the peptide with DNA.

## INTRODUCTION

The investigations of the efficiency of DNA-transfection to eukaryotic cells are important as a basic research in potential applications of gene delivery. Cationic peptides with positively charged amino acid residues can bind to polyanionic DNA to form complexes with a net positive charge which can enter the cell through disruption of the endosomal membrane. One of such peptide is bovine lactoferrampin (LFampinB) which has strong antimicrobial activity and kills a wide variety of *Candida albicans* yeast and a number of bacteria such as *Escherichia coli*. This peptide was found in bovine lactoferrin and corresponding to 268-284 amino acid residues in the protein [1]. Solid-state nuclear magnetic resonance (ssNMR) method with $^{13}C$ chemical shift oscillation analysis and molecular dynamics (MD) simulation were performed to reveal the local structure and orientation of LFampinB in the molecular level with mimetic bacterial membrane [2]. The interactions and the dynamic structure of LFampinB in mimetic bacterial membranes were demonstrated by ssNMR and MD simulation. LFampinB specifically interacted with acidic phospholipids in bacterial membranes. LFampinB may kill bacteria by causing leakage of potassium ions from the cell membranes and breaking off the concentration gradient across the cell membrane as revealed by quartz crystal microbalance and ion selective electrode measurements [2]. In this work, we investigated the interactions of LFampinB with DNA to reveal the DNA transfection efficiency.

## MATERIALS AND METHODS

### Construction of peptide and DNA

LFampinB is a cationic amphipathic peptide which consists of 17 amino acid residues with a sequence of Trp-Lys-Leu-Leu-Ser-Lys-Ala-Gln-Glu-Lys-Phe-Gly-Lys-Asn-Lys-Ser-Arg [1]. Trp1 at the N-terminus was modeled with protonated form of $-NH_3^+$ and Arg17 at the C-terminus was formed with the structure of $-COO^-$ which are the dominant states at neutral pH. As a total, the peptide gives a net charge of +5. To build the initial conformation of LFampinB, the main chain of LFampinB assumed to have right handed α-helical structure [3, 4]. As for the side chain conformation, we searched the lowest energy conformation for considering all possible orientations of the side chain by rotating every torsion angles with a 120° interval using the iteration procedure [5]. Energy minimization was performed using CHARMM [6] with all-atom force field [7]. Figure 1(a) shows the lowest energy conformation of LFampinB after taking the minimization procedure. The obtained LFampinB conformation was used as the initial structure of the following calculations.

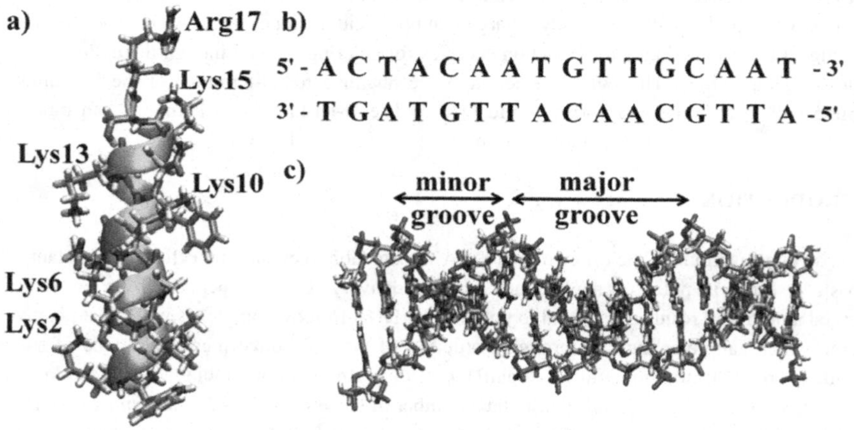


**Figure 1.** Snapshots of lowest energy conformation of LFampinB (a) and DNA (c). The carbon, nitrogen, oxygen, phosphorus and hydrogen atoms, and α-helical conformation of the peptide backbone are shown by light blue, blue, red, orange, and gray licorice, and green cartoon representations, respectively. DNA sequence is shown schematically (b). See online version for color.

As a model of the B-DNA polymer, 32 base pair double-stranded B-DNA, which was derived from the *Drosophila* fat body enhancer [8] and characterized at 1.6 Å atomic resolution crystallographic analysis (PDB entry 3BSE) [9] (Figure 1(b) and (c)), was used in this study. As far as we know, this is the largest double helical structure of B-DNA in PDB database.

**Simulation procedure**

The procedure of docking simulation was the same to our previous work [10]. After docking simulation, the orientations with the highest interaction energy between peptide and DNA were selected to perform the molecular dynamics (MD) simulation. All the initial conformations were placed in water boxes (6 nm × 6 nm × 6 nm) to calculate the MD simulation. Three-dimensional periodic boundary condition was assigned to the box. MD simulations were performed on those boxes for 10 ns without any constraints using CHARMM34 [6] software with its all atom force field [7]. Isobaric-isothermal (NPT) ensemble was carried out with 1 fs time step. The details of the simulation procedure were the same as our previous simulation [5, 11]. That is, non-bonded interactions were calculated using a group-based cutoff with a switching function [6] and were updated every 5 time steps. The switching function was turned on at 1.2 nm and turned off at 1.35 nm. All the bonds containing hydrogen were constrained using SHAKE BONH algorithm [12]. Electrostatic interaction was calculated using Particle Mesh Ewald (PME) summation method [13, 14] with a real space cutoff 1.35 nm, $k = 0.04\ nm^{-1}$. The dielectric constant was set at 1.0. The temperature was set at 298 K which was controlled by a Nose-Hoover thermostat [15, 16]. All data visualizations were done using VMD1.8.7 [17].

## RESULTS AND DISCUSSION

**Docking simulation of peptide-DNA complex**

Docking simulation was performed to obtain the orientation of the strong interaction between LFampinB and DNA by investigating various kinds of relative orientations according to the procedure of docking simulation. The results were sorted according to the interaction energies between the LFampinB and DNA surface. From this procedure, two orientation patterns were found as the conformation with high interaction energies of LFampinB, one is at the position of DNA major and the other is at the minor grooves (Figure 2). In both cases, it is found that the negative charges of the LFampinB interact with positive charges of the phosphate groups of DNA which exist at the bank of the grooves. To investigate the stability of these orientations, MD simulations were performed using these two structures of LFampinB-DNA complex as the initial conformations.

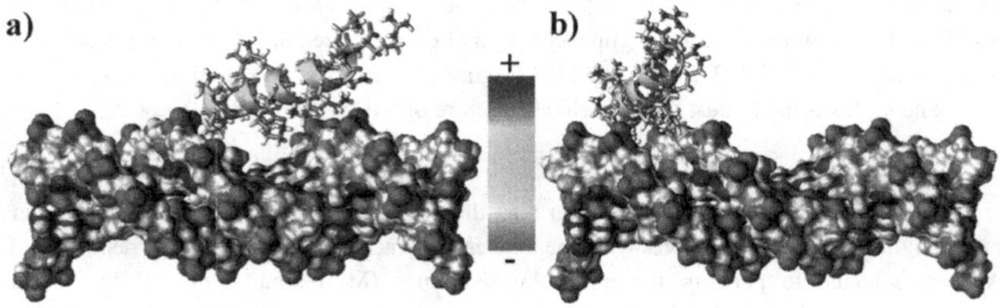


**Figure 2.** Snapshots of the structure LFampinB-DNA complex with high interaction energies using docking simulation. The positions of peptide are at major (a) and minor grooves (b) of DNA, respectively. The atoms of DNA are shown by charge density with color. See online version for color.

**Interactions between peptide and DNA**

Time courses of the interaction energies between peptide and DNA molecule at two different groove positions were analyzed and shown in Figure 3(a). It is found that the interaction at minor groove position of DNA was higher than that of major groove positions in all time regions investigated. It also found that the peptide gradually left from the interaction site at the major groove at around 6 ns and never come back within the simulation time of this work.

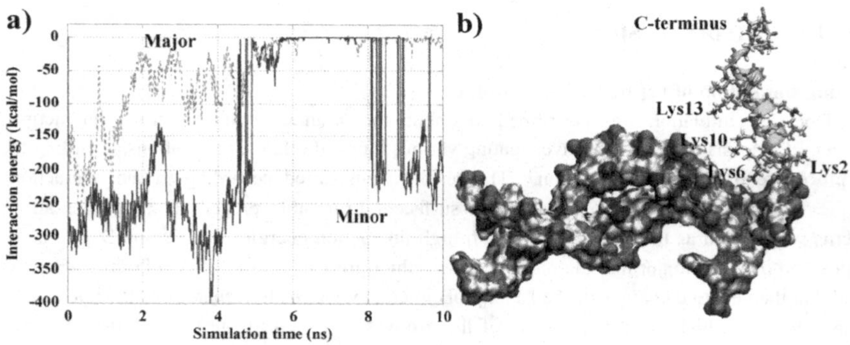


**Figure 3**. Interaction energy changes of peptide (a) with DNA molecule during the entire simulation time. Snapshot of peptide and DNA after simulation at peptide position of minor groove of DNA (b).

Instead, the peptide initially existed at the minor groove site was once left from the interaction site at around 5 ns, but returned to the interaction site from 8 ns. These observations indicate that the peptide would interact with DNA more strongly at the minor groove than the major groove. A snapshot of the interaction behavior with minor groove at the last trajectory of simulation was show in Figure 3(b). It was shown that the N-terminus of peptide interacts with DNA molecule. As the surface of the DNA is surrounded by the negative charges of the phosphate groups, the possible interacting residues of the peptide with positive charges at the N-terminal side were selected and the time trajectories of the interaction energies between these peptides and the DNA were investigated. The results were shown in Figure 4. It can be seen that the residues of Lys2, Lys6 and protonated form of Trp1 (N-ter) strongly interact with DNA compared with other Lys residues which locate far from the N-terminal side. It is noteworthy that these interaction sites are similar to those found between peptide and membrane in our previous work [2]. Although the more detail discussions and the investigation will be needed, the binding behavior of the LFampinB and DNA found in this study may be a basic information on the consideration of the transfection efficiency.

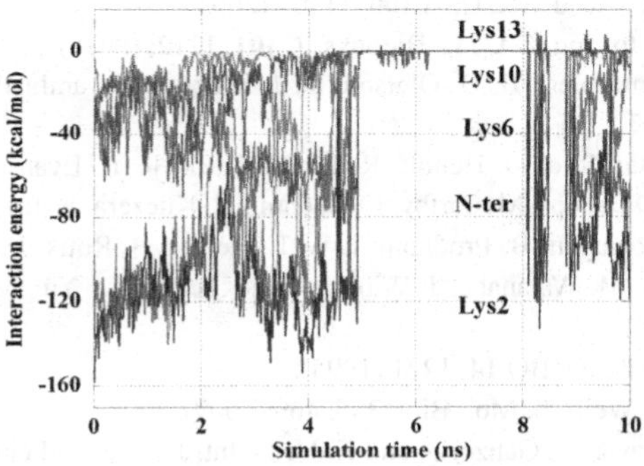


**Figure 4**. Interaction energy changes of the positively charged residues of peptide with DNA molecule during the entire simulation time.

## CONCLUSION

To elucidate the role of the LFampinB peptide on gene delivery, the interaction of peptide with DNA need to be understood in detail. In this work, we performed the docking simulation to investigate the interaction between LFampinB and DNA. The results showed that the N-terminus of LFampinB can bind at minor groove of DNA to make complexation of the peptide with DNA. In future work, the interaction of multiple peptides with DNA needs to be investigated using theoretical and experimental methods.

## ACKNOWLEDGEMENT

This research was supported by The Asia Research Center in Mongolia and the Korean Foundation for Advanced Studies within the framework of the Project #1 (2013-2014). The authors thank the Research Center for Computational Science, Okazaki, Japan for the use of a computer to perform part of the calculation.

## REFERENCES

1. M. I. A. van der Kraan, J. Groeink, K. Nazmi, E. C. I. Veerman, J. G. M. Bolscher, A. V. Nieuw Amerongen, Peptides **25**, 177 (2004).
2. A. Tsutsumi, N. Javkhlantugs, A. Kira, M. Umeyama, I. Kawamura, K. Nishimura, K. Ueda, and A. Naito, Biophys. J. **103**, 1735, (2012).
3. B. L. Pauling, R. B. Corey, H. R. Branson, Proc. Natl. Acad. Sci. USA **37**, 205 (1951).
4. D. Eisenberg, Proc. Natl. Acad. Sci. USA **100**, 11207 (2003)
5. N. Javkhlantugs, A. Naito, and K. Ueda, Biophys. J. **101**, 1212, (2010).
6. B. R. Brooks, R. E. Bruccoleri, B. D. Olafson, D. J. States, S. Swaminathan, M. Karplus, J. Comp. Chem. **4**, 187 (1983).
7. A. D. Mackerell, D. Bashford, D. Bellott, R. L. Dunbrack, R. L. Evanseck, M. J. Field, S. Fischer, J. Gao, H. Guo, D. Joseph-McCarthy, L. Kuchnir, K. Kuczera, F. T. K. Lau, C. Mattos, S. Michnick, T. Ngo, D. T. Nguyen, B. Prodhom, I. W. E. Reiher, B. Roux, M. Schlenkrich, J. C. Smith, R. Stote, J. Straub, M. Watanabe, J. Wiorkiewicz-Kuczera, D. Yin, M. Karplus, J. Phys. Chem. B. **102,** 3586 (1998).
8. W. An and P. C. Wensink, EMBO **14**, 1221 (1995).
9. N. Narayana and M. A. Weiss, J. Mol. Biol. **385**, 469 (2009).
10. N. Javkhlantugs, H. Bayar, C. Ganzorig, and K. Ueda, Int. J. Nanomedicine, in press (2013).
11. H. I. Watanabe, M. Kamihira, N. Javkhlantugs, R. Inoue, Y. Itoh, H. Endo, S. Tuzi, H. Saito, K. Ueda, and A. Naito, Phys. Chem. Chem. Phys. in press (2013)
12. J. P. Ryckaert, G. Cicotti, H. J. C. Berendsen, J. Comput. Phys. **23**, 327 (1977).
13. T. Darden, D. York and L. Pedersen, *J. Chem. Phys.,* 1993, **98**, 10089-10092.
14. U. Essmann, L. Perera, M. L. Berkowitz, T. Darden, H. Lee and L. G. Pedersen, *J. Chem. Phys.*, 1995, **103**, 8577-8593.
15. S. Nose, J. Chem. Phys. **81,** 511 (1984).
16. W. G. Hoover, Phys. Rev. A. **31**, 1695 (1985).
17. W. A. Humphrey, A. Dalke, K. Schulten, J. Mol. Graphics **14**, 33 (1996).

Mater. Res. Soc. Symp. Proc. Vol. 1569 © 2013 Materials Research Society
DOI: 10.1557/opl.2013.802

# Superhydrophobicity-Enabled Interfacial Microfluidics on Textile

Siyuan Xing, Jia Jiang and Tingrui Pan
Micro-nano Innovation (MiNI) Laboratory, University of California, Davis, CA 95616, U.S.A.

## ABSTRACT

Capillary-driven microfluidics, utilizes the capillary force generated by fibrous hydrophilic materials (e.g., paper and cotton) to drive biological reagents, has been extended to various biological and chemical analyses recently. However, the restricted capillary-driving mechanism persists to be a major challenge for continuous and facilitated biofluidic transport. In this abstract, we have first introduced a new interfacial microfluidic transport principle to automatically and continuously drive three-dimensional liquid flows on a micropatterned superhydrophobic textile (MST). Specifically, the MST platform utilizes the surface tension-induced Laplace pressure to facilitate the liquid motion along the fibers, in addition to the capillary force existing in the fibrous structure. The surface tension-induced pressure can be highly controllable by the sizes of the stitching patterns of hydrophilic yarns and the confined liquid volume. Moreover, the fluidic resistances of various configurations of connecting fibers are quantitatively investigated. Furthermore, a demonstration of the liquid collection ability of MST has been demonstrated on an artificial skin model. The MST can be potentially applied to large volume and continuous biofluidic collection and removal.

## INTRODUCTION

Interfacial microfluidics, where one or more gas-liquid interfaces exists as a boundary condition, are emerging directions in microfluidics from which several new and flexible operations have been established, including self-propelled motion, three-dimensional connectivity, open sample accessibility, direct reactivity and readability [1, 2]. Recently, the interfacial microfluidic concept has been extended to textile-based structures and surfaces (e.g., yarns and fabrics) [3]. The textile-based microfluidics utilizes the wicking force produced by hydrophilic yarns (e.g., cotton yarns) to direct biological reagents along the fiber which provides a low-cost and scalable solution based on well-established textile manufacturing techniques. Although the current development of capillarity-enabled interfacial microfluidics holds great promises to biofluidic manipulation, its intrinsic driven mechanism persists to be a major challenge for continuous and facilitated biofluidic transport. For instance, an external fluidic driver (e.g., capillary or syringe pumps) still remains necessary to provide continuous flow on the fabric network [3]. Alternatively, the latest efforts of implementing surface tension-driven flow by harvesting the Laplace pressure gradients on the interfacial microfluidics enable a new approach to automated microflow operations [5]. Importantly, the surface tension-driven micropump can be achieved simply by lithographically defining hydrophilic flow paths on a superhydrophobic substrate to provide extensive pumping capacity, flexible pumping rates, as well as bidirectional control [6].

In this report, we propose a novel interfacial microfluidic transport principle on micropatterned superhydrophobic textiles (MST) in an autonomous, controllable and continuous manner. The interfacial microfluidic transport utilizes the surface tension-induced pressure gradient along the flow path defined by extreme wetting contrast to facilitate the liquid motion in the MST network in addition to the capillarity presented in the fibrous structure. The transport duration on MST can be highly-controlled by the dimensions of the hydrophilic patterns (inlet and outlet). The gravitational influence on the transport duration has also been investigated quantitatively. Finally, a radical pattern has been demonstrated to automatically collect droplets from the surface of an artificial skin model without any external pumping source.

**EXPERIMENT**

Commercially available mercerized cotton yarns with a nominal diameter of 500 μm were purchased online (100% Mercerized ELS Cotton, Star®). However, these cotton yarns came with a layer of wax on the surface and cannot be easily primed with aqueous solutions. Therefore, a hydrophilic pre-treatment (degumming) process was applied to remove the outer wax shielding. Detailed procedure can be found elsewhere [3].

Superhydrophobic textile substrates were prepared from off-the-shelf cotton fabrics. The woven cotton fabrics (Fabric Flair®) composed of two layers of mesh with 31 threads per inch were purchased from a local tailor store. Commercially available superhydrophobic coating (Fluoropel™ M1604V, Belsville) was obtained from Cytonix company and spray-coated onto the cotton fabrics of a 3 $cm^2$ surface area by an airbrush at the pressure of 50 kPa. The coating thickness can be controlled by the volume of the treatment solution per unit surface area. To achieve the complete coverage with optimal superhydrophobic performance, 2 mL of the coating solution was applied and coated onto the substrate followed by a thermal drying process.

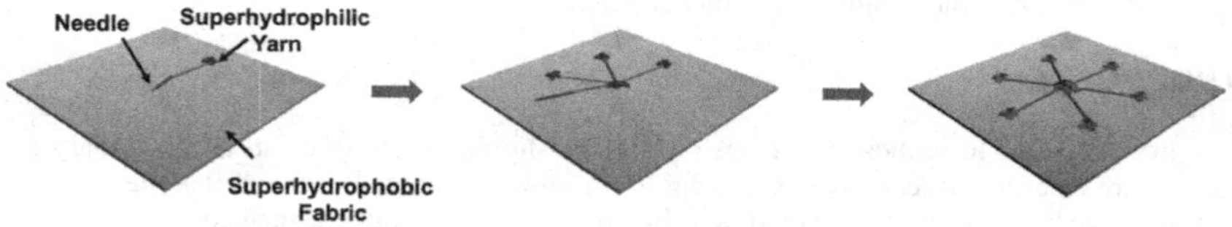


**Figure 1.** The fabrication process of MST.

Micropatterned fluidic networks were formed by stitching treated hydrophilic yarn on the superhydrophobic textile substrates (Figure 1). The cotton yarns used was commercially available mercerized cotton yarns purchased online with an average diameter of 500 μm. A 28 gauge stitching needle (of 0.362 mm in diameter) was used to sew cotton yarn to form the fluidic networks. In order to achieve the 3D flow profile, two-stage stereo-stitching technique, involving both penetrating and parallel stitching, was utilized to build the 3D hydrophilic structures in the inlet and the outlet. Specifically, an inlet was fabricated by closely stitching the thread to penetrate through the fabric layers in a defined area until the substrate was fully covered by the yarn. Alternatively, an outlet was fabricated by stitching the yarn in parallel through the meshes only presented in the top layer.

The artificial skin model was devised on a 3-layer microfluidic structure, of which the bottom two pieces were cut with microchannels and the top layer was perforated with pores. The diameter and density of the pores were 60 μm and 114 pores/$cm^2$, respectively, similar to the opening and density of human sweat glands on human backs (160±30 glands/cm$^2$). The device was fabricated from PDMS using a simple laser micromachining PDMS approach. A syringe pump was connected through tubes and micro droplets were generated on the surface to mimic the perspiration process at different flow rates.

## DISCUSSION

### Transport analysis

The MST utilizes hydrophilic yarns micro-stitched on the superhydrophobic textile which possesses a very high CA (typically greater than 140°) and inherently low hysteresis. Once hydrophilic micropatterns are formed on the superhydrophobic textile, the extreme wetting contrast enables immobilization of the triple line of gas/liquid/solid phases along the pattern boundaries. By connecting two circular patterns (reservoir) through a hydrophilic pathway, e.g., a cotton yarn or glass fiber, a simple surface-tension driven microfluidic system can be devised, as shown in Figure 2a. Once the surface tension-induced pressure gradient is established along the outflow path, the liquid at the inlet (of the diameter $D_i$) would be automatically directed towards the outlet (of the diameter $D_o$) via the hydrophilic connection (i.e., the yarn) (Figure 2b). As its laminar nature, the flow rate is approximately proportional to the Laplace pressure difference between the inlet and outlet, but inversely proportional to flow resistance of the yarn.

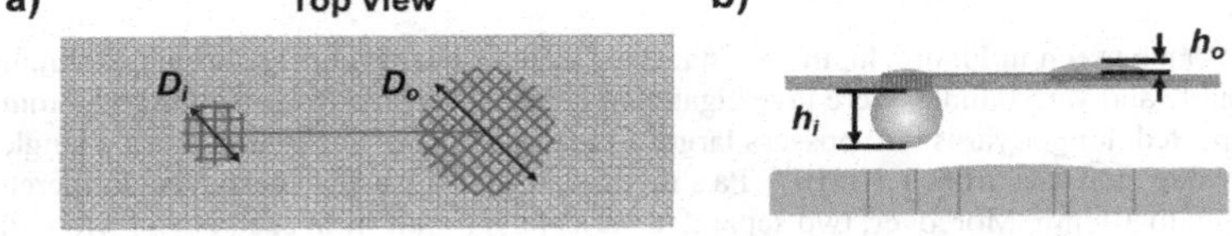


**Figure 2**. Illustrations of a) the design of MST and b) the transport process on MST.

### Micropatterned superhydrophobic material

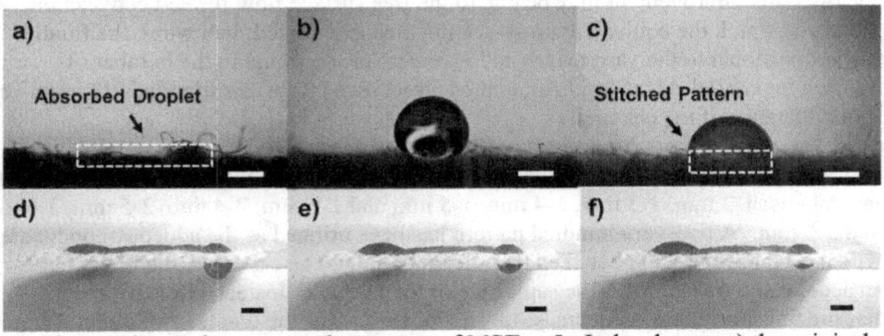


**Figure 3**. The wetting and transport phenomena of MST: a 5 μL droplet on a) the original hydrophilic substrate, b) superhydrophobic treated substrate and c) a 2 mm hydrophilic micropattern (scale bars: 1 mm); d-f) 3D Laplace pressure-driven transport on MST (scale bars: 2 mm).

Figure 3a shows a 5 μL water droplet that has been rapidly absorbed into the original hydrophilic fabric. To alter wettability on the substrate, we utilized a superhydrophobic surface coating made of fluoro-microparticles to modify the chemical inertness and microscopic topology. After the spray-coating, the microparticles are physically absorbed onto the textile surface and render the surface to be superhydrophobic with a water CA of 140±3°. As can be seen in Figure 3b, a water droplet of 5 μL maintains nearly a spherical shape on the superhydrophobic substrate. As described previously, hydrophilic yarns with a CA close to 0° have been stitched on the superhydrophobic-modified fabric and formed highly wettable micropatterns. As shown in Figure 3c, a 5 μL droplet is deposited onto a 2 mm circular hydrophilic pattern and forms a semi-spherical shape as its boundary is pinned along the wetting-contrast pattern. Furthermore, the 3D bi-droplet surface-tension driven transport has been demonstrated on MST as illustrated in Figure 3d-f.

**Control of transport duration**

According to our theoretical analysis, the pumping rate of MST is determined by the Laplace pressure difference as well as the channel resistance between the inlet and outlet. In general, various parameters including the yarn diameter, porosity, length, bundle arrangement could all result in the difference of the fluidic resistance. Since we would like to confine our material to be off-the-shelf (the diameter and porosity is fixed), we only varied the yarn length and bundle arrangement (single or multiple threads) to investigate their influence on the flow resistance.

As can be seen in Figure 4a, the resistance of different yarns configurations, including single, double and yarn bundle, were investigated at different lengths from 10 mm to 30 mm. As can be expected, longer yarns will possess larger fluidic resistance. For example, for a single yarn, the value increases from $1.10\times10^{13}$ Pa s m$^{-3}$ to $2.86\times10^{13}$ Pa s m$^{-3}$ as the length increases from 10 mm to 30 mm. Moreover, two separated yarns (e.g. 5 mm from each other) will reduce the resistance to half of the single yarn's value at the same length. Furthermore, a yarn buddle with two yarns in a compact form significantly increase the fluidic conductance compared with the two separated yarn. For instance, at the same yarn length, the flow resistance of the bundled configuration is measured at $1.04\times10^{12}$ Pa s m$^{-3}$, which is only one fifth of that of the two separated yarns. This improvement may be due to the free surface flow formed between the two adjacent fibres, of which the equivalent cross-section area is enlarged. In a word, the fluidic resistance is proportional to the yarn length and inversely proportional to the number of connecting fibre in parallel. Meanwhile, a bundled structure of fibre can significantly improve the fluidic conductance of the channel.

Moreover, the transport durations have been measured for various inlet-outlet configurations, given a fixed droplet volume (of 4 μL). Figure 4b shows the selected inlet-outlet dimensions in pairs: 1-2 mm, 1-3 mm, 1-4 mm, 1-5 mm and 1-6 mm; 2-4 mm, 2-5 mm, 2-6 mm; 3-5 mm and 3-6 mm. A two-yarn bundled pattern has been utilized as the aqueous-conducting channel with a 10 mm separation between the inlet and outlet. As expected, the larger outlet leads to an accelerated transport for the same size of the inlet. For instance, for the inlet of 1 mm in diameter, the transport duration decreases from 53 s to 11 s as the diameter of outlet extends from 2 mm to 6 mm. This is attributed by the fact that the same liquid droplet possess a lower profile on a larger size of reservoir, and thereby, a smaller internal pressure. In contrast, a larger reservoir at inlet will induce a slower transport given the same outlet dimension. As can be seen, the pump duration increases from 15 s to 75 s when the inlet reservoir size rises from 1 mm to 3 mm. These results have clearly demonstrated the potential of the MST in manipulating flow rates by the micro-stitched geometries.

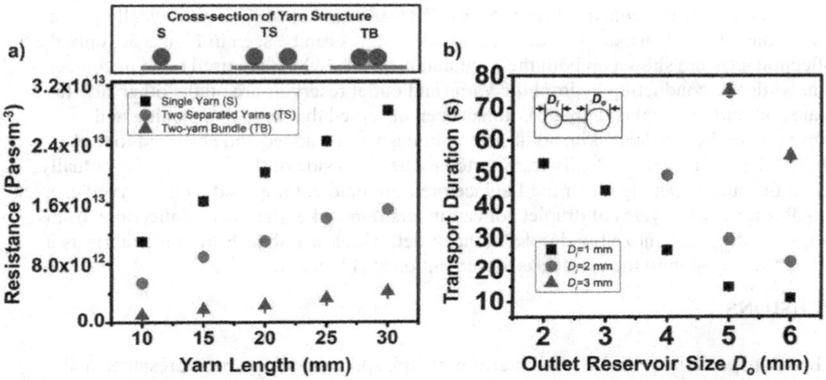


**Figure 4.** a) The fluidic resistance of different yarn configurations (the inset shows the cross-section of yarn structures); b) different inlet-outlet dimensions

__Liquid collection demonstration__

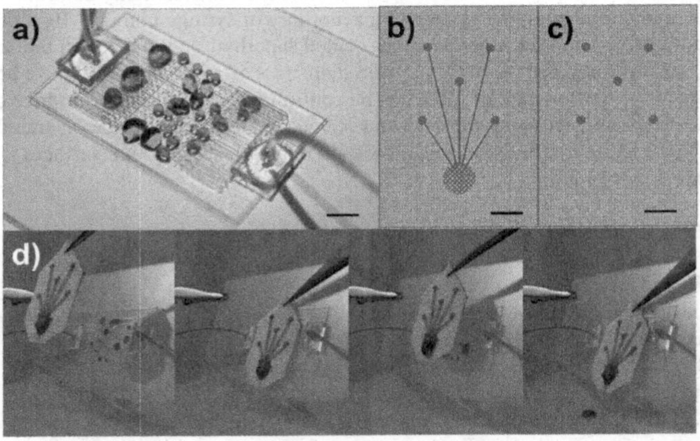


**Figure 5.** a) The artificial skin model; illustration of b) the front side and c) the backside of the MST for sweat collection and removal (scale bar 5 mm); d) the liquid collection of MST on the surface of the skin model.

An artificial skin model has been fabricated for demonstration of the MST by the method aforementioned for manageable simulation of perspiration on skin (Figure 5a). As the human body keeps up-right positions throughout most of the day, it is reasonable to conduct our sweating collection from the MST surface by placing the artificial skin vertically (of 75°). A micro-stitched pattern with 1 mm inlet and 6 mm outlet in a compact 5-inlet-1-outlet has been utilized for efficient sweat removal (Figure 5b-c). The substrate of MST is periodically put in contact with the artificial sweating surface during the test. As can be seen in Figure 5c, only the 5 inlet collection sites are shown on both the front and backside, which are used to be in contact with skin, while the conducting hydrophilic yarns and outlet reservoir are on the other side for liquid transport and removal (Figure 5b). It has been observed that in each attaching and detaching cycle to the artificial skin, additional volumes will be added and absorbed to the hydrophilic inlet reservoirs, gradually transported to the other side of the fabric, and eventually, collected at the outlet opening under the Laplace pressure gradient imposed by the curved droplet surface. After repetitive cycles of droplet collection and removal at the inlets, collections from the multiple inlets merge into a big droplet at the outlet, which will detach from the fabric as it overweighs the adhesion to the surface as aforementioned (Figure 5d).

## CONCLUSIONS

In this paper, a novel interfacial microfluidic transport model has been presented and implemented on micropatterned superhydrophobic textiles (MST). The MST platform, utilizing the extreme wetting contrast between very hydrophilic yarns and a superhydrophobic textile substrate, confines the aqueous flow pathway and enables autonomous surface tension-driven microflow on the textile surface. As compared to the previously reported textile-based microfluidic systems, it is capable of achieving a continuous flow in a highly controllable manner without an external pumping system (e.g., capillary or syringe pumps). By a unique two-stage stereo-stitching approach on the spray-coated superhydrophobic cotton fabric (CA=140±3°), various micropattern designs (from diameters of 1 mm to 6 mm) have been fabricated and tested. The flow rate of MST can be controlled by the fluidic resistance of the channel and the dimensions between the inlet and outlet. Finally, we have demonstrated its applicability toward facilitated and controlled biofluid removal, such as skin surfaces experiencing heavy perspiration.

## REFERENCES


1. Glenn, M.W. and J.B. David, *Lab Chip*, vol. **2**, pp. 131-134 (2002).
2. Hong, L. and T. Pan, *Lab Chip*, vol. **10**, pp. 3271-3276 (2010).
3. Li, X., J. Tian, and W. Shen, *ACS Appl. Mater. Interfaces*, vol. **2**, pp. 1-6 (2010).
4. Safavieh, R., G.Z. Zhou, and D. Juncker, *Lab Chip*, vol. **11**, pp. 2618-2624 (2011).
5. Hong, L. and T. Pan, *Microfluidics Nanofluidics*, vol. **10**, pp. 991-997 (2011).
6. Xing, S., R. Harake, and T. Pan, *Lab Chip*, vol. **11**, pp. 3642-3648 (2011).

**Responsive Multifunctional Biomaterials**

Mater. Res. Soc. Symp. Proc. Vol. 1569 © 2013 Materials Research Society
DOI: 10.1557/opl.2013.799

# Comparison of memory effects in multiblock copolymers and covalently crosslinked multiphase polymer networks composed of the same types of oligoester segments and urethane linker

Li Wang[1, 2], Ulrich Nöchel[1], Marc Behl[1], Karl Kratz[1], Andreas Lendlein[1, 2]
[1] Institute of Biomaterial Science and Berlin-Brandenburg Centre for Regenerative Therapies (BCRT), Helmholtz-Zentrum Geesthacht, 14513 Teltow, Germany
[2] Institute of Chemistry, University Potsdam, 14476 Potsdam, Germany

* Corresponding Author: andreas.lendlein@hzg.de

## ABSTRACT

Phase-segregated multiblock copolymers (MBC) as well as covalently crosslinked multiphase polymer networks, which are composed of crystallizable oligo(ε-caprolactone) (OCL) and oligo(ω-pentadecalactone) (OPDL) segments have been recently introduced as degradable polymer systems exhibiting various memory effects. Both types of copolyesterurethane networks can be synthesized via co-condensation of the respective hydroxytelechelic oligomers and 2,2(4),4-trimethyl-hexamethylene diisocyanate (TMDI) as aliphatic linker.

In this work the dual-shape properties as well as the temperature-memory capability of thermoplastics and covalently crosslinked copolyesterurethanes containing OCL and OPDL domains are explored. Both copolyesterurethane networks exhibited excellent dual-shape properties with high shape fixity ratios $R_f \geq 93\%$ and shape recovery ratios in the range of 92% to 100% determined in the 2nd and 3rd test cycle, whereby the dual-shape performance was substantially improved when covalent crosslinks are present in the copolymer.

A pronounced temperature-memory effect was achieved for thermoplastic as well as crosslinked copolyesterurethanes. Hereby, the switching temperature $T_{sw}$ could be adjusted via the variation of the applied deformation temperature $T_{deform}$ in the range from 32 °C to 53 °C for MBC and in the range from 29 °C to 78 °C for multiphase polymer networks.

## INTRODUCTION

In recent years different memory effects such as dual-, triple-, or multi-shape effect or temperature-memory effect (TME), where a material can memorize the temperature at which it was deformed before, have been reported for various thermo-sensitive polymers [1, 2, 4, 16]. These memory effects are characterized by a controlled shape change in a predefined way, when the response temperature of a material is exceeded, and therefore these materials are of great technological interest for the realization of actively moving intelligent devices [3, 4]. Polymers, which are capable to memorize defined shapes or temperatures, have two structural elements in common: netpoints, which stabilize the original, permanent shape and reversible crosslinks related to a thermal transition (e.g. a glass or melting transition) acting as switching domains and in this way enabling on the one hand side the fixation of temporary shapes as well as the entropy driven recovery of the original shape. The netpoints in thermo-sensitive polymers can be either of chemical (covalently crosslinked polymer networks) or physical (thermoplastics) nature. It could be shown that the dual-shape capability of covalently crosslinked polymer networks is superior to that obtained for thermoplastics, where creeping/relaxation effects have to be considered, which result in reduced shape-memory performance [5]. For quantification of the dual-shape properties of polymers cyclic, thermomechanical tensile tests are typically applied, which allow the determination of characteristic values such as the shape fixity ratio ($R_f$) and the shape recovery ratio ($R_r$) as well as the response temperatures (e.g. the switching temperature $T_{sw}$ representing the temperature of maximum recovery rate $v_{rec}$) from the obtained stress-temperature-strain data [6].

In thermoplastic multiblock copolymers composed of two crystallizable segments, the crystalline domains associated to the highest melting temperature $T_{m,perm}$ determine the permanent shape of the material and therefore solely the second set of crystallisable domains can be utilized as switching domains for achieving a dual-shape capability [7-10]. In contrast in multiphase polymer networks the covalently crosslinks determining the permanent shape and both crystallisable domains can act as switching domains resulting in a triple-shape capability [11-15].

In this study we explored the dual-shape properties as well as the temperature-memory performance of thermoplastic copolyesterurethane networks (named PDLCL) and covalently crosslinked copolyesterurethanes networks (named $^3$PDL-$^4$PCL), via specific cyclic, thermomechanical tensile test protocols. Both copolymers PDLCL as well as $^3$PDL-$^4$PCL were synthesized from hydroxytelechelic oligo($\varepsilon$-caprolactone) (OCL) and oligo($\omega$-pentadecalactone) (OPDL) segments by co-condensation with 2,2(4),4-trimethyl-hexamethylene diisocyanate (TMDI) as linker (see Experimental Details). All investigated PDLCL and $^3$PDL-$^4$PCL copolyesterurethanes exhibited two well separated melting transitions in differential scanning calorimetry (DSC) experiments, with a $T_{m,PCL}$ around 36 to 44 °C and a $T_{m,PDL}$ in the range of 64 to 86 °C (see Experimental Details).

## EXPERIMENTAL DETAILS

### Materials

For synthesis of PDLCL multiclock copolymer equal weight amounts of oligo($\varepsilon$-caprolactone)diol ($M_n$ = 3000 g⋅mol$^{-1}$, $T_m$ = 43 °C) and oligo($\omega$ –pentadecalactone)diol $M_n$ = 5600 g⋅mol$^{-1}$, $T_m$ = 91 °C) were reacted with a mixture of the two isomers 1,6-diisocyanato-2,2,4-trimethylhexane and 1,6-diisocyanato-2,4,4-trimethylhexane (TMDI, > 99%, Fluka, Taufkirchen, Germany) in 1,2-dichlorethane at 80 °C and afterwards precipitated in methanol at -10 °C as described previously in [8]. PDLCL050 with $M_n$ = 59000 g⋅mol$^{-1}$ and a polydispersity of PD = 2.0 (as determined by gel permeation chromatography) was achieved with a yield of 92 %. DSC showed two melting transitions at $T_{m,PCL}$ = 44 °C and $T_{m,PDL}$ = 86 °C [8].

The covalently crosslinked multiphase polymer networks were prepared by co-condensation of star-shaped poly($\omega$-pentadecalactone)triol ($^3$PDL; $M_n$ = 4500 g⋅mol$^{-1}$) and poly($\varepsilon$-caprolactone)tetrol ($^4$PCL, $M_n$ = 9600 g⋅mol$^{-1}$) with TMDI according to the method reported in [10]. Copolymer networks with two different compositions 40 wt% and 75 wt% $^4$PCL in the starting reaction mixture of the hydroxytelechelic oligomers have been synthesized named $^3$PDL-$^4$PCL040 and $^3$PDL-$^4$PCL075, respectively. In DSC analysis two distinct melting transitions were observed for both copolymer networks at $T_{m,PCL}$ = 42 °C and $T_{m,PDL}$ = 78 °C for $^3$PDL-$^4$PCL040) and $T_{m,PCL}$ = 36 °C and $T_{m,PDL}$ = 64 °C ($^3$PDL-$^4$PCL075).

### Dual-shape and temperature-memory experiments

Cyclic, thermomechanical tensile tests for quantification of the dual-shape and temperature-memory properties were performed on tensile tester Zwick Z2.5 (Zwick-Roell, Ulm, Germany) equipped with a thermo chamber and a temperature controller (Eurotherm e2408, Limburg, Germany).

Dual-shape tests consisted of three repetitive cycles with identical test parameters composed of a programming module, where the temporary shape was created followed by a recovery module under stress-free conditions. In particular the test specimen was deformed at $T_{deform}$ = 55 °C to $\varepsilon_m$ = 150%, afterwards cooled to $T_{low}$ = 0 °C either under strain ($^3$PDL-$^4$PCL075) or stress control (PDLCL050). Finally the stress or strain was removed and temporary shape $\varepsilon_u$ was obtained. For recovery of the original shape the sample was reheated to $T_{high}$ = 50 °C ($^3$PDL-$^4$PCL075) or 65 °C (PDLCL050). As characteristic values for quantification of the dual-shape effect, the shape fixity ratio $R_f$, which describes how well the applied deformation $\varepsilon_m$ can be fixed, and the shape recovery ratio $R_r$, which is a measure for the recovery of the original shape are calculated according the following equations (1) to (4). Where $\varepsilon_u(N)$ is the resulting strain after completion of the programming module in the N$^{th}$ test cycle and $\varepsilon_m$ as well as $\varepsilon_l$ are the maximum strains achieved either during strain controlled and stress controlled cooling to $T_{low}$. Here, $\varepsilon_p$ characterizes the recovered shape.

$$\text{Strain controlled: } R_f = \frac{\varepsilon_u(N)}{\varepsilon_m} \tag{1}$$

$$\text{Stress controlled: } R_f = \frac{\varepsilon_u(N)}{\varepsilon_l} \tag{2}$$

$$\text{Strain controlled: } R_r = \frac{\varepsilon_m - \varepsilon_p(N)}{\varepsilon_m - \varepsilon_p(N-1)} \tag{3}$$

$$\text{Stress controlled: } R_r = \frac{\varepsilon_l - \varepsilon_p(N)}{\varepsilon_l - \varepsilon_p(N-1)} \tag{4}$$

The temperature-memory capability of both polymers was explored in cyclic, thermomechanical experiments by variation of $T_{deform}$ in each subsequent cycle. For PDLCL050 $T_{deform}$ = 25 °C, 35 °C, 45 °C and 55 °C were chosen whereby $\varepsilon_m$ = 150%, $T_{low}$ = 0 °C and $T_{high}$ = 70 °C were kept constant in each cycle [4]. The TME of $^3$PDL-$^4$PCL040 was investigated at $T_{deform}$ = 30 °C, 60 °C and 90 °C with $\varepsilon_m$ = 100%, $T_{low}$ = 0 °C and $T_{high}$ = 115 °C.

## RESULTS AND DISCUSSION

The dual-shape capability and the temperature-memory properties (TME) of two types of phase segregated copolyesterurethane networks comprising either physical (PDLCL) or covalent netpoints ($^3$PDL-$^4$PCL) were investigated by different cyclic, thermomechanical tensile tests protocols.

### Dual-shape properties of copolyesterurethane networks

For evaluation of the dual-shape effect three subsequent test cycles with identical test parameters were applied to PDLCL050 and $^3$PDL-$^4$PCL075 and analysed. According to the $T_{m,PCL}$ obtained in DSC experiments a $T_{deform} > T_{m,PCL}$ of 50 °C or 55 °C was selected. A stress-controlled programming module was utilized for investigation of PDLCL050, while $^3$PDL-$^4$PCL075 was explored via a strain-controlled programming process (see Experimental Details and Figure 1). The resulting dual-shape characteristics are summarized in Table 1. An almost complete fixation of the applied deformation with $R_f \geq 92\%$ was achieved for both thermoplastic as well as covalently crosslinked copolymers, whereby slightly higher $R_f$-values were found for $^3$PDL-$^4$PCL075 compared to PDLCL050 (Table 1). From Figure 1B it becomes obvious that for the covalently crosslinked $^3$PDL-$^4$PCL075 almost identical stress-temperature-strain curves were obtained in the subsequent cycles, which demonstrates the excellent dual-shape properties with high shape-recovery ratios of $R_f \geq 95\%$, a $T_{sw}$ of 37 °C and a small recovery temperature interval of $\Delta T_{rec} \approx 20$ °C. In contrast, the thermoplastic PDLCL050 exhibited a great difference (fatigue behaviour) in the stress-temperature-strain of the 1st ($R_f \geq 64\%$) and the 2nd test cycle, which can be attributed to the creeping/relaxation effects, while the 2nd and 3rd cycle were almost similar with $R_f \geq 92\%$ (Figure 1A). A $T_{sw}$ of 53 °C was determined for PDLCL050 with a significant broader $\Delta T_{rec} \approx 30$ K compared to $^3$PDL-$^4$PCL075.

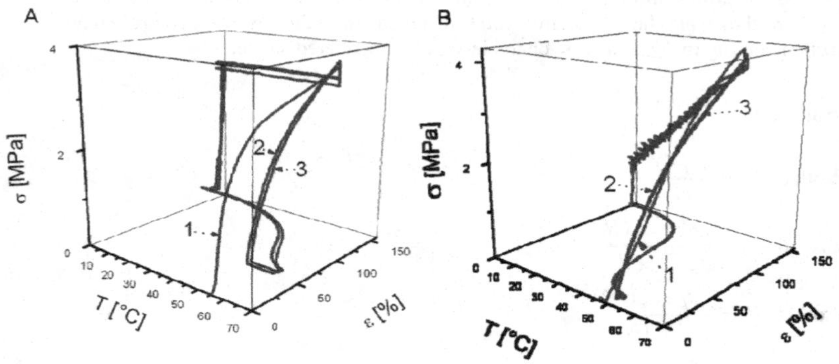


**Figure 1**. Stress-strain-temperature curves obtained in three subsequent dual-shape test cycles for **A)** PDLCL050 (stress controlled programming module) and **B)** $^3$PPD-$^4$PCL075 (strain controlled programming module).

**Table 1**. Dual-shape properties of PDLCL050 and $^3$PPD-$^4$PCL075

| Sample ID | $R_f(1)$ [%] | $R_r(1)$ [%] | $R_f(2)$ [%] | $R_r(2)$ [%] | $R_f(3)$ [%] | $R_r(3)$ [%] |
|---|---|---|---|---|---|---|
| PDLCL050 | 94 [a] | 64 | 93[a] | 92 | 93[a] | 97 |
| $^3$PPD-$^4$PCL075 | 99 [b] | 95 | 99 [b] | 99 | 99 [b] | 100 |

a) Test parameter: $T_{deform}$ = 55 °C, $\varepsilon_m$ = 150%, $T_{low}$ = 0 °C $T_{high}$ = 65 °C; programming at constant stress.
b) Test parameter: $T_{deform}$ = 50 °C, $\varepsilon_m$ = 150%, $T_{low}$ = 0 °C $T_{high}$ = 50 °C; programming at constant strain.

**Temperature-memory properties of copolyesterurethane networks**

The temperature-memory capability of thermoplastic and covalently crosslinked copolyesterurethanes was explored in cyclic, thermomechanical experiments, whereby the deformation temperature $T_{deform}$ was increased in each subsequent cycle (see Experimental Details). An almost linear correlation of the applied $T_{deform}$ = 25 °C or 35 °C or 45 °C and 55 °C and the resulting $T_{sw}$ of 32 °C or 42 °C or 49 °C and 53 °C was observed for PDLCL050 (Figure 2) with similar shape recovery ratios around $R_r$ = 90 to 92% (Table 2). Furthermore, $R_f$ was found to increase from $R_f$ = 70% at 25 °C (representing a cold-drawing situation exhibiting an enhanced relaxation during programming) to $R_f$–values ≥ 90% when

$T_{deform} > T_{m,PCL}$. At $T_{deform}$ = 55 °C also low melting PPDL crystallites can support a temperature-memory effect (TME) of PDLCL as recently reported [4].

In contrast to their thermoplastic counterparts for $^{3}$PDL-$^{4}$PCL040 both thermal transitions $T_{m,PCL}$ and $T_{m,PPDL}$ could be utilized to achieve a temperature-memory capability. Here $T_{sw}$ could be adjusted in the range of 29 °C to 78 °C when $T_{deform}$ was increased from 30 °C to 90 °C. $^{3}$PDL-$^{4}$PCL040 exhibited an excellent shape recovery with $R_r$ 97 to 100% when programmed at 30 °C or 60 °C or 90 °C (Table 2 and Figure 2). The obtained $R_f$ values increased from 82% ($T_{deform}$ = 30 °C) to 99% ($T_{deform}$ = 90 °C). High shape recovery ratios of $R_r \geq 97\%$ were obtained for $^{3}$PDL-$^{4}$PCL040 independent from the applied $T_{deform}$, whereby a decrease in $\Delta T_{rec}$ from 40 °C to 15 °C with increasing $T_{deform}$ was observed.

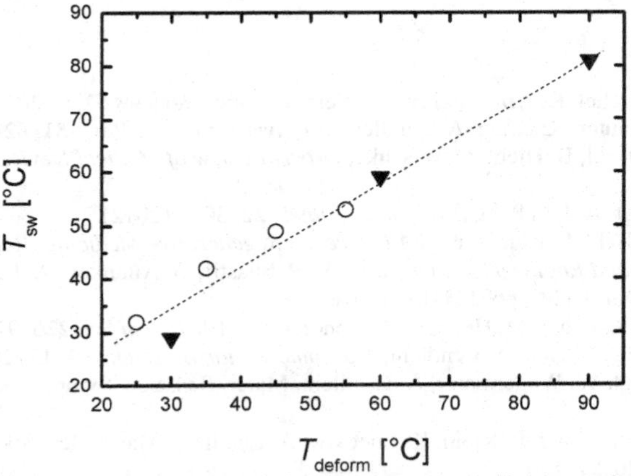


**Figure 2.** Temperature-memory plot (switching temperature $T_{sw}$ versus deformation temperature $T_{deform}$) for multiblock copolymer PDLCL050 (open circles) programmed at $T_{deform}$ = 25 or 35 or 45 or 55 °C and the multiphase copolymer network $^{3}$PPD-$^{4}$PCL040 ($T_{deform}$ = 30 or 60 or 90 °C).

**Table 2.** Temperature-memory properties of PDLCL050 and $^{3}$PPD-$^{4}$PCL040

| **Sample ID** | $R_f$ [%] | $R_r$ [%] | $T_{sw}$ [°C] | $R_f$ [%] | $R_r$ [%] | $T_{sw}$ [°C] | $R_f$ [%] | $R_r$ [%] | $T_{sw}$ [°C] |
|---|---|---|---|---|---|---|---|---|---|
| PDLCL050 | $T_{deform}$ = 25 °C | | | $T_{deform}$ = 45 °C | | | $T_{deform}$ = 55 °C | | |
| | 70 [a] | 91 | 32 | 90 [a] | 96 | 49 | 92 [a] | 97 | 53 |
| $^{3}$PPD-$^{4}$PCL040 | $T_{deform}$ = 30 °C | | | $T_{deform}$ = 60 °C | | | $T_{deform}$ = 90 °C | | |
| | 82 [b] | 98 | 29 | 93 [b] | 97 | 59 | 99 [b] | 100 | 78 |

a) Test parameter: $\varepsilon_m$ = 150%, $T_{low}$ = 0 °C $T_{high}$ = 65 °C; programming at constant stress.
b) Test parameter: $\varepsilon_m$ = 100%, $T_{low}$ = 0 °C $T_{high}$ = 50 °C; programming at constant strain.

## CONCLUSION

Thermoplastic as well as covalently crosslinked copolyesterurethane networks comprising crystallizable oligo($\varepsilon$-caprolactone) and olig($\omega$-pentadecalactone) segments have been investigated regarding their dual-shape and temperature-memory properties. It could be demonstrated that the dual-shape performance was substantially enhanced when covalent crosslinks are present in the copolymer, resulting in higher $R_f$ as well as $R_r$ values and a small recovery temperature interval. In contrast, the thermoplastic copolymers exhibited a fatigue behaviour, which can be attributed to the creeping/relaxation processes. For both copolyesterurethane networks a pronounced temperature-memory effect (TME) could be observed. While for the thermoplastic material the TME is mainly limited to the temperature range of the PCL melting transition, multiphase copolymer networks allow to utilize both melting transitions for the physical adjustment of the switching temperature.

## REFERENCES


1. T. Sauter, M. Heuchel, K. Kratz and A. Lendlein, *Polymer Reviews* **53**, 6 (2013).
2. M. Heuchel, T. Sauter, K. Kratz, A. Lendlein, *J Polym Sci Polym Phys*. **51**, 621(2013).
3. A. Lendlein, M. Behl, B. Hiebl, C. Wischke, *Expert Review of Medical Devices* **7**, 357 (2010).
4. K. Kratz, U. Voigt, A. Lendlein, *Adv. Funct. Mater.* **22**, 3057 (2012).
5. A. Lendlein, M. Behl, S. Kamlage, in *Advances in Regenerative Medicine: Role of Nanotechnology and Engineering Principles,* V. P. Shastri, G. Altankov, A. Lendlein, Eds. (Springer, Dordrecht), pp. 131-156 (2010).
6. W. Wagermaier, K. Kratz, M. Heuchel, A. Lendlein, in *Adv Polym Sci.* **226**, 97 (2010).
7. Y. Feng, M. Behl, S. Kelch, A. Lendlein, *Macromolecular Bioscience* **9**, 45 (2009).
8. K. Kratz, U. Voigt, W. Wagermaier, A. Lendlein, Mater. Res. Soc. Symp. Proc. **1140**, 17 (2008)
9. S. A. Madbouly, K. Kratz, F. Klein, K. Luetzow, A. Lendlein, Mater. Res. Soc. Symp. Proc. **1190**, 99 (2009).
10. H. Matsumoto, T. Ishiguro, Yuichi Konosu, M. Minagawa, A. Tanioka, K. Richau, K. Kratz, A. Lendlein, *European Polymer Journal* **48**, 1866 (2012).
11. I. Bellin, S. Kelch, R. Langer, A. Lendlein, *Proceedings of the National Academy of Sciences of the United States of America* **103**, 18043 (2006).
12. U. N. Kumar, K. Kratz, M. Behl, A. Lendlein, *Express Polymer Letters*, **6**, 26 (2012).
13. J. Zotzmann, M. Behl, D. Hofmann, A. Lendlein, *Adv. Mater.* **22**, 3424 (2010).
14. J. Zotzmann, M. Behl, Y. K. Feng, A. Lendlein, *Adv. Funct. Mater.* **20**, 3583 (2010).
15. J. Zotzmann, M. Behl, A. Lendlein, *Macromol. Symp*. **309/310**, 147 (2011).
16. K. Kratz, S. A. Madbouly, W. Wagermaier, A. Lendlein, *Adv. Mater*. **23**, 4058 (2011).

**Mater. Res. Soc. Symp. Proc. Vol. 1569 © 2013 Materials Research Society**
**DOI: 10.1557/opl.2013.1060**

# Tailoring the recovery force in magnetic shape-memory nanocomposites

M. Behl, M.Y. Razzaq, and A. Lendlein
Institute for Biomaterial Science and Berlin Brandenburg Center for Regenerative Therapies, Helmholtz-Zentrum Geesthacht, Kantstr. 55, 14513 Teltow, Germany

## ABSTRACT

Composites from magnetic nanoparticles in a shape-memory polymer (SMP) matrix allow a remote actuation of the shape-memory effect by exposure to alternating magnetic fields. At the same time the incorporation of MNP may affect the thermal properties and the structural functions of the SMP.

Here, we explored the adjustability of the recovery force as an important structural function in magnetic shape-memory nanocomposites (mSMC) by variation of the programing temperature ($T_{prog}$) and nanoparticle weight content. The nanocomposites were prepared by coextrusion of silica coated magnetite nanoparticles (mNP) with an amorphous polyether urethane (PEU) matrix. In tensile tests in which $T_{prog}$ was varied between 25 and 70 °C and the particle content from 0 to 10 wt% it was found that the Young's moduli ($E$) decreased with temperature and particle content. Cyclic, thermomechanical experiments with a recovery module under strain-control conditions were performed to monitor the effect of mNP and $T_{prog}$ on the recovery force of the composites. During the strain-control recovery the maximum stress ($\sigma_{m,r}$) at a characteristic temperature ($T_{\sigma,max}$) was recorded. By increasing the mNP content from 0 to 10 wt% in composites, $\sigma_{m,r}$ of 1.9 MPa was decreased to 1.25 MPa at a $T_{prog}$ = 25 °C. A similar decrease in $\sigma_{m,r}$ for nanocomposites with different mNP content could be observed when $T_{prog}$ was increased from 25 °C to 70 °C. It can be concluded that the lower the deformation temperature and the particle content the higher is the recovery force.

## INTRODUCTION

The incorporation of magnetic (nano)particles in shape-memory (SMP) polymers as a matrix enabled nanocomposites whose shape-memory effect could be actuated remotely-controlled on demand.[1-5] This principle could be extended to triple-shape nanocomposites.[6] However, the influence of the mNP on the mechanical properties is not well understood. In shape-memory polymers and composites an important parameter in addition to the maximum deformability is the recovery stress as this influences the recovery ratio $R_r$ under constraint conditions such as in an implant situation. The recovery stress can be determined from cyclic, thermomechanical experiments consisting of a deformation followed by recovery module under strain-control conditions. [7] The recovery under constant strain conditions is quantified by a characteristic temperature $T_{\sigma,max}$, at which the maximum recovery stress $\sigma_{m,r}$ occurs.

Recently polymers with a broad transition temperature ($\Delta T_{trans}$) have enabled a temperature-memory effect (TME), which is the capability to remember the temperature $T_{prog}$, where the polymer was deformed recently. [8, 9] We have asked us whether such a matrix with a programmable temperature ($T_{prog}$) at which the deformation is applied can be utilized to study the effect of deformation temperature on the recovery force. We report about the adjustability of the $\sigma_{max}$ in mSMC with amorphous switching domain as a function of mNP content when $T_{prog}$ was applied in the range of $\Delta T_{trans}$. We selected an amorphous PEU as matrix material to which 5 to 10 wt% of silica coated mNP were incorporated to enable mSMCs. PEU has a phase-segregated multiblock structure. It contains $H_{12}MDI$/1,4-BD hard segments, and $H_{12}MDI$/PTMG soft segments. In addition, the polymer exhibits a homogeneously mixed phase with a broad a broad glass transition temperature ($\Delta T_{g,mix}$), which is acting as switching phase. After the thermal and morphological characterization of the composites cyclic, thermomechanical tensile tests, consisting of a programming procedure, where $T_{prog}$ was altered followed by recovery under constant strain conditions, were applied for the evaluation of $\sigma_{m,r}$.

## EXPERIMENTAL DETAILS

### *Material details*

PEU with the trade name of Tecoflex EG60D was purchased from Noveon (Wilmington, MA, USA). Nanoscaled particles AdNano® MagSilica 50 were from Degussa Advanced Nanomaterials (Degussa AG, Hanau, Germany). The particles consist of an iron(III) oxide ($Fe_2O_3$) core in a silica matrix ($SiO_2$). The mean aggregate size (Photon correlation spectroscopy of an aqueous dispersion) is 90 nm, the mean domain size (X-ray diffraction) is 20–26 nm, additional information can be found in reference.[10]

### *Preparation of mSMCs*

T60D and mNP were coextruded with a HAAKE PolyLab system (PTW 16/25; Thermo Electron Corporation) co-rotating twin extruder at 170 °C with a screw speed between 100 and 150 rpm. The films of the pure T60D and the composites were prepared by press molding the compound (200 E, Dr. Collin, Ebersberg, Germany) for 30 min at 150 ° C and pressure of 100 bar.

### *Characterization methods*

Scanning electron microscopy (SEM) was performed on ultrathin sections of 70 nm cut on a Leica FC6 cryo ultra-microtome at 130 °C (Leica Microsystems GmbH Wetzlar, Germany) with a Zeiss Gemini, Supra 40VP (Zeiss, Jena Germany) at an accelerating voltage of 10 kV. DSC experiments were performed with a Netzsch (Selb, Germany) DSC 204, in the temperature range from -100 to 100 °C with a constant heating and cooling rate of 10 $K{\cdot}min^{-1}$. The determination of the dynamic mechanical properties was performed on a DMTA Eplexor 25N (Gabo, Ahlden, Germany) in temperature-sweep mode from -100 to 150 °C at a constant heating rate of 2 $K{\cdot}min^{-1}$ with an oscillation frequency of 10 Hz.

Tensile tests at varied temperatures and cyclic, thermomechanical tensile tests were performed on a Zwick Z1.0 (Zwick, Ulm, Germany) tensile tester equipped with a temperature controlled thermochamber (Eurotherm Regler, Limburg, Germany). The cyclic, thermomechanical tensile consisted of a preconditioning step to remove the thermal history of the samples followed by a programming procedure, where the temporary shape was created, and a recovery module under constant strain conditions. After completion of the recovery module a waiting period of 10 min was applied before the next cycle started.

*Preconditioning step*: The sample was cooled from $T_{high}$ = 80 °C to $T_{prog}$ = 70 °C and deformed to a strain $\varepsilon_m$ =150%. After a waiting period of 10 min, the sample was cooled to $T_{low}$ = 0 °C and equilibrated at $T_{low}$ for 10 min. Finally, the stress was reduced to zero resulting in the temporary

shape $\varepsilon_u$. The recovery of the sample to $\varepsilon_p$ under stress free conditions was carried out by heating from $T_{low}$ to $T_{high}$ with a heating rate of 1 °C·min$^{-1}$.
*Cyclic, thermomechanical experiments*: During the step 1 the specimen was cooled from $T_{high}$ = 80 °C to $T_{prog}$ (25, 40, 55, or 70 °C) and deformed to strain $\varepsilon_m$ =150%, at $T_{prog}$ (step 2). Then the strain was kept constant for 10 min to allow relaxation (step 3). Afterwards, the sample was cooled to the fixation temperature $T_{low}$ = 0 °C and equilibrated for 10 min at $T_{low}$ (step 4). Finally, the stress was reduced to zero stress $\sigma_0$ (step 5), resulting in the fixed strain $\varepsilon_u$ representing the temporary shape. Four subsequent cycles were performed at fixed elongation of $\varepsilon_m$ = 150%, where $T_{prog}$ increased with each cycle from 25 to 70 °C.
*Recovery under strain-control conditions*: The obtained strain level after completion of the programming procedure was kept constant while the sample was heated up to $T_{high}$ = 80 °C with a heating rate of 1 °C·min$^{-1}$ (step 6). The recovery stress built up with raising temperature to $\sigma_{m,r}$ at $T_{\sigma,max}$ until the softening of the material allowed stress relaxation. After reaching $T_{high}$, the stress was removed (step 7), allowing the sample to recover its original shape, before the next cycle started.

## RESULTS AND DISCUSSION

In this study, we investigated magnetic nanocomposites from amorphous PEU containing 5 (PEU5) or 10 wt% (PEU10) of mNP. The distribution of mNP within the mSMCs was investigated with a scanning electron microscope. In Figure 1a the SEM back scattered image for PEU10 with 10 wt% mNP is exemplarily shown and an almost homogeneous distribution of the mNP in the form of small clusters in the sub micrometer length-scale could be observed. The existence of mNP clusters in mSMCs is in good agreement with the previously reported results.[11-13]
DSC thermograms revealed only one glass transition temperature ($T_{g,mix}$) at around 24 °C (with the onset at 0 °C and $\Delta T_{g,mix} \approx 50$ °C), which was attributed to the mixed phase. No significant change in the $T_{g,mix}$ by the addition of the mNP was observed. Two additional transition temperatures attributed to the soft domains ($H_{12}$MDI/PTMEG segment) $T_{g,sd}$ at -42 °C and the softening temperature of the $H_{12}$MDI/1,4-BD hard domains $T_{g,hd}$ at 130 °C were detected in the tan-$\delta$ temperature curves. The broad peak with a maxima at 55 °C reflects the $T_{g,mix}$ and ranges from 20 °C to 80 °C.[7] By the addition of mNP in PEU5 and PEU10, no changes in the thermo-mechanical transitions were observed.

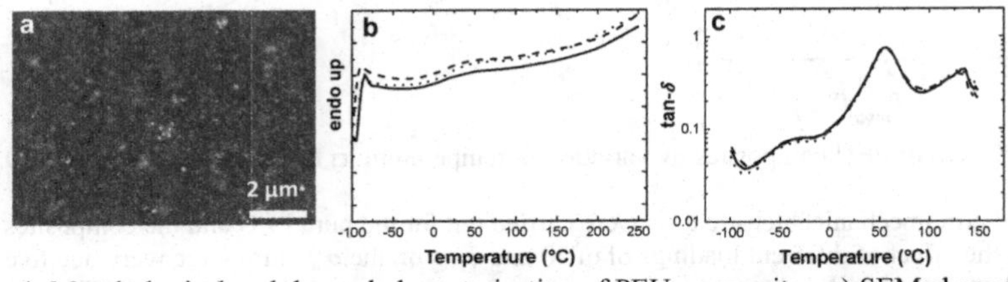


**Figure 1.** Morphological and thermal characterization of PEU composites. a) SEM characterization of a PEU10 composite from the cross-section of the sample; b) DSC curves, second heating run; c) DMTA curves; solid line PEU0, dashed line PEU5, dotted line PEU10.

**Table 1:** Mechanical properties of the PEU and PEU composites determined by tensile tests at varied temperatures

| Sample ID | *T* (°C) | *E* (MPa) | $\sigma_{max}$ (MPa) | $\sigma_b$ (MPa) | $\varepsilon_b$ (%) |
|---|---|---|---|---|---|
| PEU0 | 25 | 101 ± 1 | 75 ± 1 | 18 ± 1 | 470 ± 5 |
| PEU5 | 25 | 95 ± 7 | 70 ± 10 | 63 ± 9 | 390 ± 20 |
| PEU10 | 25 | 110 ± 10 | 69 ± 3 | 67 ± 4 | 390 ± 20 |
| PEU0 | 40 | 50 ± 20 | 24 ± 3 | 23 ± 3 | 430 ± 20 |
| PEU5 | 40 | 68 ± 1 | 46 ± 1 | 45 ± 2 | 510 ± 10 |
| PEU10 | 40 | 64 ± 4 | 50 ± 10 | 50 ± 10 | 438 ± 20 |
| PEU0 | 55 | 24 ± 2 | 16 ± 1 | 14 ± 4 | 620 ± 30 |
| PEU5 | 55 | 26 ± 2 | 15 ± 1 | 15 ± 1 | 550 ± 11 |
| PEU10 | 55 | 23 ± 6 | 11 ± 1 | 21 ± 4 | 580 ± 11 |
| PEU0 | 75 | 1 ± 0.1 | 0.8 ± 0.2 | 0.7 ± 0.1 | 460 ± 60 |
| PEU5 | 75 | 9 ± 1 | 3.7 ± 0.1 | 3.5 ± 0.1 | 900 ± 50 |
| PEU10 | 75 | 10 ± 3 | 4.0 ± 0.6 | 2 ± 1 | 600 ± 160 |

In tensile tests at temperatures between 25 and 70 °C the elastic properties of the PEU and the PEU composite were explored (Table 1). The elongation at break ($\varepsilon_b$) increased with temperature and decreased with particle weight content. In contrast, the Young's moduli (*E*) decreased with increasing temperature, while at lower temperatures it was almost not influenced by the particle content. Only at $T_{prog}$ = 75 °C the nanocompsites provided a Young's modulus, which was significantly higher that the Young's modulus of PEU, which can be attributed to a reinforcing effect of the mNP. $\sigma_{max}$ determined from the tensile test showed a similar tendency (Fig. 2). Slightly higher values were obtained for PEU10, which might be attributed to the higher margin of error.

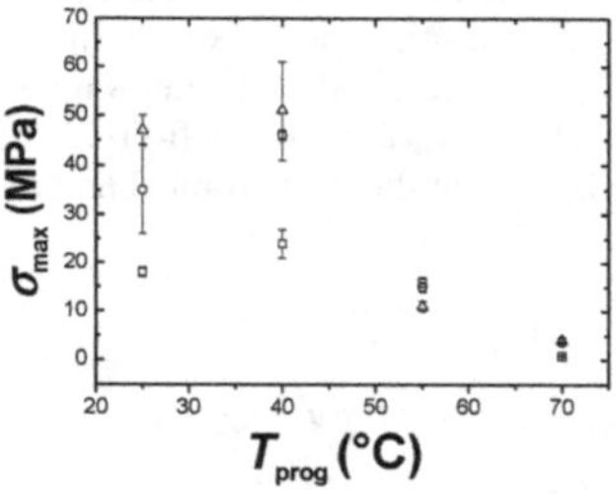


**Figure 2.** $\sigma_{max}$ of PEU composites by variation of temperature: □ PEU0, ○ PEU5, △ PEU10.

Cyclic, thermomechanical tensile tests were carried out for the pure PEU and the composites to monitor the effect of different loadings of mNP and $T_{prog}$ on the $\sigma_{m,r}$ during recovery and to explore whether the results obtained from the uniaxial tensile test could be confirmed. The programming procedure during thermomechanical experiments consisted of five steps (experimental details) and recovery under strain control conditions. Cooling from $T_{high}$, which was above $T_{g,mix}$, to $T_{prog}$ (step 1) led to a partial vitrification of the mixed domains which support the stability of the permanent shape. After deformation to $\varepsilon_m$ at $T_{prog}$ (step 2) and relaxation for 10 min at $T_{prog}$ (step 3), the switching segments, which fix the temporary shape, were oriented. In step 4, the sample was cooled to $T_{low}$, resulting in the partial fixation of the pre-oriented switching segments which were in viscoelastic state at $T_{prog}$. Then, the stress was released to $\sigma_0$ (step 5) and the fixed temporary shape $\varepsilon_u$ was obtained. Finally, the sample was reheated from $T_{low}$ to $T_{high}$ with fixed strain. The stress increased initially with rising temperature and reached a maximum $\sigma_{m,r}$, where the characteristic temperature $T_{\sigma,max}$ was determined, while at higher temperatures the stress decreased dramatically because of the softening of the material.

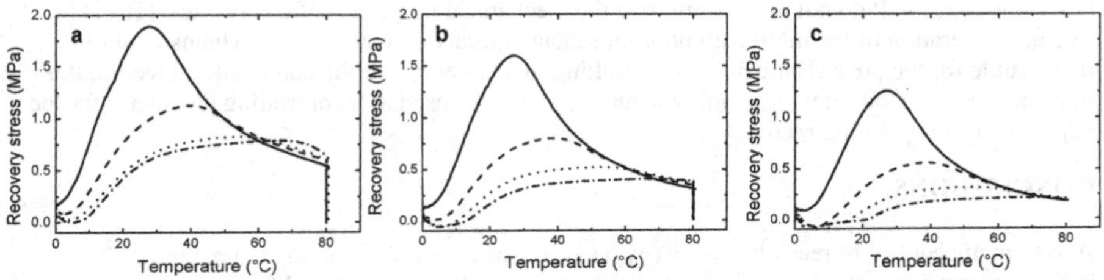


**Figure 3.** Stress generated for (a) PEU0, (b) PEU5, (c) PEU10, programmed at four different $T_{prog}$ (solid line: $T_{prog}$ = 25 °C, dashed line: $T_{prog}$ = 40 °C, dotted line: $T_{prog}$ = 55 °C, dash dot line: $T_{prog}$ = 70 °C).

During programming, the $T_{prog}$ was varied from 25 °C to 70 °C, which is in the range of $\Delta T_{g,mix}$. During the recovery under strain control conditions, a strong correlation between $T_{\sigma,max}$ and $T_{prog}$ was observed for pure PEU and nanocomposites as shown in Figure 3. The values of $T_{\sigma,max}$ were increased linearly by increasing the $T_{prog}$ in all the samples. Such a linear relationship was maintained even after the addition of mNP. However, a significant decrease in the values of $\sigma_{m,r}$ by increasing the $T_{prog}$ was observed. In pure PEU the $\sigma_{max}$ of 1.9 MPa achieved at $T_{prog}$ = 25 °C was decreased to $\sigma_{m,r}$ = 0.8 MPa for $T_{prog}$ = 70 °C. The same decrease in $\sigma_{m,r}$ by increasing $T_{prog}$ from 25 °C to 70 °C was observed for PEU5 and PEU10. This observation can be explained by the fact that for a given $\varepsilon_m$ higher deformation-stress ($\sigma_d$) values were required to achieve at lower $T_{prog}$ because of the increasing stiffness of the materials at lower temperatures. For example for PEU0 to reach $\varepsilon_m$ = 150% required a $\sigma_d$ of 5.4 MPa at $T_{prog}$ = 25 °C, which was decreased to 1.2 MPa at a $T_{prog}$ = 70 °C for the same $\varepsilon_m$. For PEU5 and PEU10, a relatively lower value of $\sigma_d$ at each $T_{prog}$ as compared to PEU0 was required. At low $T_{prog}$ the higher volume fraction of the mixed domain in glassy state results in higher stiffness of the polymer. By increasing the $T_{prog}$, the major fraction of the switching domain is in viscoelastic state, which results a tremendous decrease in stiffness and also the $\sigma_{m,r}$ during strain-control recovery.[14] At the same time the $\sigma_{m,r}/\sigma_d$ ratio increased for each sample when $T_{prog}$ was increased. For PEU10 at $T_{prog}$ of 25 °C $\sigma_{m,r}/\sigma_d$ = 0.30 increased to $\sigma_{m,r}/\sigma_d$ = 0.63 at a $T_{prog}$ of 70 °C. This increase of $\sigma_{m,r}/\sigma_d$ ratio with increasing $T_{prog}$ was attributed to a lower degree of irreversible deformation at lower $T_{prog}$ of the hard domains, which act as netpoints and determine the permanent shape. However, it is particularly important to note that the $\sigma_{m,r}/\sigma_d$ ratio was almost constant at each $T_{prog}$ for PEU0 and the composites.

In addition to $T_{prog}$, also the effect of mNP content on the $\sigma_{m,r}$ during strain control recovery conditions can be discussed. The maximum value of $\sigma_{m,r}$ at each $T_{prog}$ was determined for the pure PEU while for the PEU nanocomposites under the same deformation and recovery conditions lower values of $\sigma_{m,r}$ were obtained. At a $T_{prog}$ = 25 °C, the $\sigma_{m,r}$ = 1.9 MPa observed for PEU0 was decreased to 1.25 MPa for PEU10. A similar decrease in $\sigma_{m,r}$ by increasing the mNP content at higher $T_{prog}$ was observed. However, sharp stress recovery peaks at lower $T_{prog}$ of 25 °C were converted to broad peaks at higher $T_{prog}$ of 70 °C. This decrease in $\sigma_{m,r}$ by increasing the mNP content was attributed to the weak elastic properties of the composites as compared to the pure PEU. In general mNP should have a reinforcing effect resulting in enhanced mechanical properties. However, mNP do not always improve the mechanical properties of composites. [10, 11] The agglomeration of the mNP with poor interfacial interaction with polymer chains might be responsible for the poor elastic behavior resulting in a lower $\sigma_{m,r}$ of the composites. Overall, this indicates that in addition to $T_{prog}$, mNP content can also be used as a controlling parameter for the adjustment of $\sigma_{m,r}$ during recovery.

## CONCLUSIONS

A systematic control of recovery stress ($\sigma_{m,r}$) as a function of programming temperature ($T_{prog}$) and magnetite nanoparticle (mNP) content was successfully demonstrated for magnetic nanocomposites with an amorphous polyetherurethane (PEU) matrix. A significant decrease in $\sigma_{max}$ by increasing the $T_{prog}$ was observed for the pure PEU and the nanocomposites in uniaxial tensile tests. A similar decrease in $\sigma_{m,r}$ by increasing the mNP content from 0 to 10 wt% was also observed for the cyclic, thermomechanical measurements under strain control conditions. For a $T_{prog}$ of 25 °C, the $\sigma_{m,r}$ of 1.9 MPa for PEU0 was decreased to 1.25 MPa for PEU10 with 10 wt% mNP content. It can be concluded that in such amorphous composites the recovery stress can be maximized by increasing the weight fraction of the plastically deformed material while the weight content of mNP should be minimum.

## REFERENCES


1. M.Y. Razzaq, M. Behl, and A. Lendlein, *Nanoscale*, **4**, 6181 (2012).
2. I.A. Rousseau, *Polymer Engineering and Science*, **48**, 2075 (2008).
3. C. Liu, H. Qin, and P.T. Mather, *Journal of Materials Chemistry*, **17**, 1543 (2007).
4. I.S. Gunes and S.C. Jana, *Journal of Nanoscience and Nanotechnology*, **8**, 1616 (2008).
5. T. Sauter, et al., *Polymer Reviews*, **53**, 6 (2013).
6. U.N. Kumar, et al., *Express Polymer Letters*, **6**, 26 (2012).
7. J. Cui, K. Kratz, and A. Lendlein, *Smart Materials & Structures*, **19**, 065019 (2010).
8. T. Xie, K.A. Page, and S.A. Eastman, *Advanced Functional Materials*, **21**, 2057 (2011).
9. K. Kratz, et al., *Advanced Materials*, **23**, 4058 (2011).
10. R. Mohr, et al., *Proceedings of the National Academy of Sciences of the United States of America*, **103**, 3540 (2006).
11. T. Weigel, R. Mohr, and A. Lendlein, *Smart Materials & Structures*, **18**, 025011 (2009).
12. M.Y. Razzaq, M. Behl, and A. Lendlein, *Advanced Functional Materials*, **22**, 184 (2012).
13. U.N. Kumar, et al., *Journal of Materials Chemistry*, **20**, 3404 (2010).
14. X. Chen and T.D. Nguyen, *Mechanics of Materials*, **43**, 127 (2011).

Mater. Res. Soc. Symp. Proc. Vol. 1569 © 2013 Materials Research Society
DOI: 10.1557/opl.2013.840

# ABA triblock copolymer based hydrogels with thermo-sensitivity for biomedical applications

Lucile Tartivel[1,2], Marc Behl[1], Michael Schroeter[1] and Andreas Lendlein[1,2]

[1]Institute of Biomaterial Science and Berlin-Brandenburg Center for Regenerative Therapies (BCRT), Helmholtz-Zentrum Geesthacht, Kantstr. 55, 14513 Teltow, Germany.
[2]Institute of Chemistry, University of Potsdam, 14476 Potsdam, Germany.

## ABSTRACT

Oligo(ethylene glycol)-oligo(propylene glycol)-oligo(ethylene glycol) (OEG-OPG-OEG) triblock copolymers are hydrogel forming and extensively investigated in the field of drug release due to their biocompatibility and thermo-sensitivity. Here the synthesis and characterization of OEG-OPG-OEG based polymer networks from methacrylated oligomers by photo-irradiation are reported. Two precursors were selected to have comparable hydrophilicity (80 wt% OEG content) but different molecular weights of $M_n$ = 8400 g·mol$^{-1}$ and 14600 g·mol$^{-1}$. The precursor solutions were prepared in concentration 10 to 30 wt%. The resulting polymer networks prepared from high $M_n$ precursors exhibited higher swellability at equilibrium (up to 3400%) and mechanical properties in the range of $G' \sim 0.1$ to 1 kPa at 5 °C compared to networks based on low $M_n$ precursors. A more significant thermo-sensitive behavior in terms of swellability, volumetric contraction and mechanical transition, starting at 30 °C could also be observed for the networks based on high $M_n$ precursors, thus promoting future application in the field of drug release.

## INTRODUCTION

Hydrogels are an emergent class of polymer materials especially because their properties such as swellability and mechanical properties, and function such as stimuli-sensitivity can be adjusted to the requirements of specific applications to mimic e.g. the extracellular matrix [1]. Hydrogels based on synthetic compounds such as poly(vinyl alcohol) or poly(ethylene glycol) or natural polymers such as gelatin have been extensively explored [2, 3] due to their biocompatibility. However, other polymers such as poly(*N*-isopropylacrylamide) (PNIPAm) or block copolymers such as ABA triblock copolymer based on oligo(ethylene glycol) segments A and an oligo(propylene glycol) segment B (OEG-OPG-OEG) exhibit thermal transition when in aqueous solution, creating a physical gel at higher temperatures. They have attracted strong interest for pharmaceutical technologies, because of their biocompatibility and thermo-sensitivity [4]. In particular, the micellization of a OEG-OPG-OEG monomer solution upon its critical micellization temperature (cmt) can induce a self-gelation [5]. In this context, a drug can be incorporated in a solution and injected at room temperature and form a physical gel at body temperature. However, such a gel based on physical crosslinks can quickly dissolve when surrounded by body fluids, which could cause a burst release of the active compounds. On the other hand, a covalently crosslinked hydrogel with thermo-sensitivity used as template for drug loading might induce a more gradual release of the drug because its diffusion might be slowed down by the polymer network. In addition, the covalent crosslinking enables the variation of composition parameters and therefore tailorable swellability and elastic properties can be achieved, which might interfere with the micellization process. In this context, we explore the crosslinking efficiency, swellability, and mechanical strength of hydrogels prepared by photocrosslinking of OEG-OPG-OEG functionalized precursors, also in the context of thermo-sensitivity. Two different block copolymer were used, with $M_n = 8400\ \text{g}\cdot\text{mol}^{-1}$ or $M_n = 14600\ \text{g}\cdot\text{mol}^{-1}$, both with comparable hydrophilicity 80 wt% OEG, and aqueous precursors solutions were prepared in concentration 10 to 30 wt%.

## EXPERIMENTAL DETAILS

### Synthesis

The functionalization of OEG-OPG-OEG of $M_n = 8400\ \text{g}\cdot\text{mol}^{-1}$ and $M_n = 14600\ \text{g}\cdot\text{mol}^{-1}$ with 2-isocyanate ethyl methacrylate (IEMA) was performed as described in reference [6]. The hydrogel networks were prepared from OEG-OPG-OEG-IEMA monomer solutions, in concentration 10 wt%, 20 wt% and 30 wt% in water for injection (WFI). A 33.3 $\text{mg}\cdot\text{mL}^{-1}$ photoinitiator solution (Irgacure 2959, BASF, Germany) was prepared and added in concentration 15 wt% to the monomer solution. The mixture was carefully injected with a syringe between two quartz glass plates (100 mm x 100 mm, separated by a 1 mm thick Teflon spacer) whereby the formation of air bubbles was avoided. The mold was placed under an Excimer Laser (Bluelight PS 30P excimer laser, $\lambda = 308$ nm, Heraeus Noblelight, Germany) at a distance of 15 cm on the surface of ice cold water/crushed ice at 5 °C, and irradiated for 5 min on each side. 3 sample pieces were cut from the hydrogel after crosslinking for the determination of gel content ($G$), the remaining polymer network was carefully detached from the quartz glass plates, extracted 24 hours in 25 vol% Ethanol, and then stored in sterile water in a fridge until further characterization.

### Characterization methods:

For the determination of $G$, three sample pieces were extracted in a 25 vol% Ethanol solution for 24h, the mass of the dried samples before extraction ($m_d$) and after extraction ($m_{d,ext}$) were recorded and $G$ was calculated according to equation 1:

$$G = \frac{m_{d,ext}}{m_d} \cdot 100 \quad (\%) \tag{1}$$

The degree of swelling ($Q$) was determined in water at 5 °C and 50 °C, as well as in phosphate buffered saline without $Ca^{2+}$ and $Mg^{2+}$ (PBS, Dulbecco). Swelling kinetics were performed in distilled water at 5 °C. 25 mm diameter chips were punched from the hydrogels at 5 °C, and immersed in a water bath at 5 °C or 50 °C. The surface area was determined from photographs by the use of the program ImageJ. Rheological investigations in time sweep mode were performed on a HAAKE rheometer Mars II (Thermo Scientific), with a plate-plate geometry on punched samples of 2 cm diameter. After verifying that all measurements were performed in the linear viscoelastic region for all samples, a constant deformation of $\gamma = 0.002$ was applied for all frequency- and temperature-dependent measurements. Temperature sweeps were performed between 1 and 60 °C with a frequency $f = 0.15$ Hz and a constant force of $F = 0.10$ N. Other characterization methods were performed as described in reference [6].

## RESULTS AND DISCUSSION

Samples are denoted as xxK(yy), were xxK refers to the $M_n$ of the selected OEG-OPG-OEG block copolymer in kDa, and (yy) to the monomer concentration in wt% of the precursor solutions. The hydrogels appeared as homogeneous, transparent, and air bubble free films. With the exception of the sample of composition 15K(10), crosslinked films samples could be obtained.

### Gel content and swellability

The gel content ($G$) was measured by gravimetric methods to assess the efficiency of crosslinking. Values ranged between 80% and 95% for 8K(yy) and between 74% and 88% for 15K(yy). $G$ was higher for 8K(yy) samples than for 15K(yy) ones and increased with the monomer concentration. In general, $G$ was lower for these networks, synthesized at 5 °C, than those described in our previous work [6], which were synthesized at room temperature. This difference can be attributed to the thermo-sensitive behavior of OEG-OPG-OEG block copolymers solutions which already form micelles at room temperature.

**Table 1:** Composition of the hydrogels and properties

| *Sample Id*[a)] | $G$[b)] | in water | | in PBS | | Contraction |
|---|---|---|---|---|---|---|
| | | $Q$ 5 °C[c)] | $Q$ 50 °C[d)] | $Q$ 5 °C[e)] | $Q$ 50 °C[f)] | 5↔50 °C[g)] |
| | [%] | [%] | [%] | [%] | [%] | [%] |
| **8K(10)** | 80 ± 1 | 1740 ± 30 | 830 ± 50 | 1280 ± 60 | 730 ± 20 | 51 |
| **8K(20)** | 94 ± 1 | 1270 ± 40 | 600 ± 20 | 1230 ± 60 | 540 ± 10 | 44 |
| **8K(30)** | 95 ± 2 | 1000 ± 50 | 530 ± 10 | 1040 ± 60 | 490 ± 10 | 35 |
| **15K(10)** | n.c. | n.m. | n.m. | n.m. | n.m. | n.m. |
| **15K(20)** | 74 ± 2 | 3390 ± 50 | 1150 ± 70 | 3130 ± 60 | 1100 ± 200 | 55 |
| **15K(30)** | 88 ± 1 | 2300 ± 120 | 770 ± 20 | 2200 ± 130 | 760 ± 20 | 51 |

a) Samples are denoted as xxK(yy), were xx refers to the $M_n$ of the selected OEG-OPG-OEG block copolymer, and (yy) to the monomer concentration in wt% of the precursor solutions, b) gel content $G$, n.c. = not crosslinked, c) Degree of swelling $Q$, measured in distilled water at 5 °C, d) $Q$ at 50 °C, e) $Q$ in PBS at 5 °C, f) $Q$ in PBS at 50 °C, g) volumetric shrinking calculated from the change in surface area between 5 °C and 50 °C in water, n.m. = not measurable

The swelling kinetics was calculated by determining $Q$ at 5 °C at different time intervals (see Figure 1). The equilibrium degree of swelling was determined at 5 °C and 50 °C in distilled water as well as in PBS (See Table 1). 8K(yy) networks reached their equilibrium $Q$ within 2 h and 15K(yy) in 3 h, probably due to the greater volume of water absorbed. 15K(yy) samples could swell up to 3400% and to a higher extent than 8K(yy) ones, up to 1740%. The swellability in PBS was slightly lower than in water, but in a comparable range. However, an increasing monomer concentration of the precursor solution directly influenced the swellability, with decreasing values, due to the resulting tightening of the networks. In addition, at 5 °C, the value of $Q$ was for all samples twice as high as the value at 50 °C.

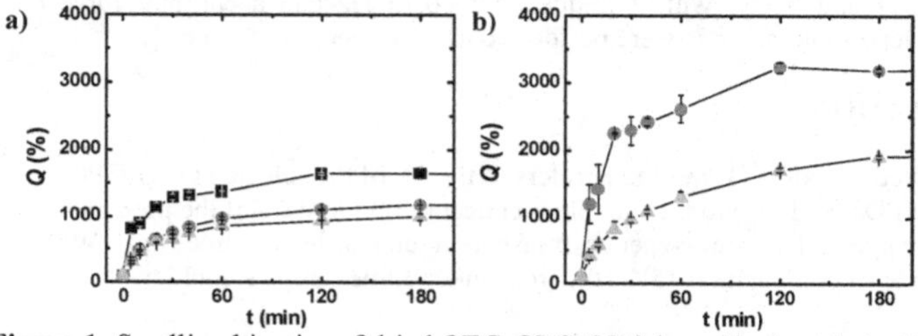


**Figure 1:** Swelling kinetics of dried OEG-OPG-OEG based hydrogels in distilled water at 5 °C. a) 8K(yy), ■ 8K(10), ● 8K(20), ▲ 8K(30), b) 15K(yy), ● 15K(20), ▲ 15K(30).

The difference of swellability of factor 2 measured by $Q$ was also translated in a volumetric contraction (Figure 2). The relative contractions calculated from the surface area are recorded in Table 1. As this phenomenon has been reported for thermo-sensitive hydrogels, e.g based on PNIPAm [7], it is assumed that the volumetric contraction is the result of loss of hydrophilicity of OPG-blocks at increased temperatures, inducing the formation of hydrophobic domains and tightening the network. It could be observed that the volumetric shrinking was higher for polymer networks or hydrogels prepared from lower monomer concentration and for 15K based hydrogels. The hydrogels from precursors with a higher $M_n$ resulted in a loose network with higher flexibility in terms of swellability.

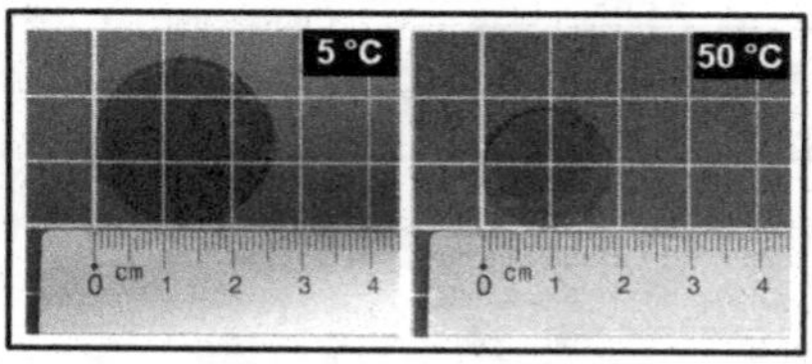


**Figure 2:** Photographs of a hydrogel of composition 8K(10), after immersing in a water bath at 5 °C or 50 °C.

## Mechanical properties

The mechanical strength of the hydrogels was investigated by rheology, from which the storage modulus $G'$ and loss modulus $G''$ were determined as a function of frequency and temperature. Figure 3 shows that all hydrogels exhibited a typical behavior of covalently crosslinked hydrogels, with $G' > G''$ and both moduli showing rather independent values from the frequency. In addition, a systematic correlation between mechanical properties and monomer concentration with increasing moduli with increasing precursor content could be observed, e.g. for 15K(30) $G'$ appeared in the range of 1 kPa while for 15K(20), values were found at ~0.2 kPa. The same tendency was observed for 8K(yy) networks. A systematic comparison between $G'$ and $G''$ of hydrogels prepared from OEG-OPG-OEG block copolymer from different $M_n$ but with similar precursors concentration showed that 15K(yy) samples were of lower mechanical strength than 8K(yy) gels, which is in good correlation with its higher swellability as well as with our previous study [6].

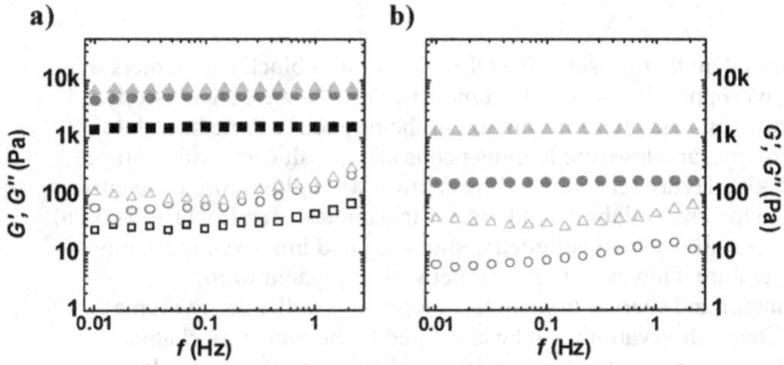


**Figure 3:** Rheological investigation, frequency dependencies of *G'* and *G''* of a) 8K and b) 15K-based hydrogels, measured at 5 °C. Solid symbols: *G'*, open symbols: *G''*, ■ 8K(10), ● 8K(20), ▲ 8K(30), b) 15K(yy), ● 15K(20), ▲ 15K(30).

As the linear oligomers are thermo-sensitive, it was explored whether this thermo-sensitivity could also influence the mechanical properties of the crosslinked hydrogel networks (Figure 4). All samples exhibited an increase of *G''* with augmenting temperature. On the other hand, the transition of *G'* with temperature could not be so clearly observed. This transition corresponds to the formation of hydrophobic domains when the temperature is increasing, thus inducing a deswelling and a reduction of flexibility of the networks, as described previously [8]. Such behavior has also been observed in PNIPAm-based hydrogels [9]. All samples displayed a transition of *G'* with augmenting temperature of about one order of magnitude, e.g. for 8K-based hydrogels, *G''* went from a 10-100 Pa range to approximately 1 kPa. In general, compared to 8K(yy) hydrogels, a sharper transition at higher temperatures could be observed for 15K(yy) gels, which can be correlated to a higher degree of flexibility in the material, also noticed from the more significant volumetric contractions of those networks.

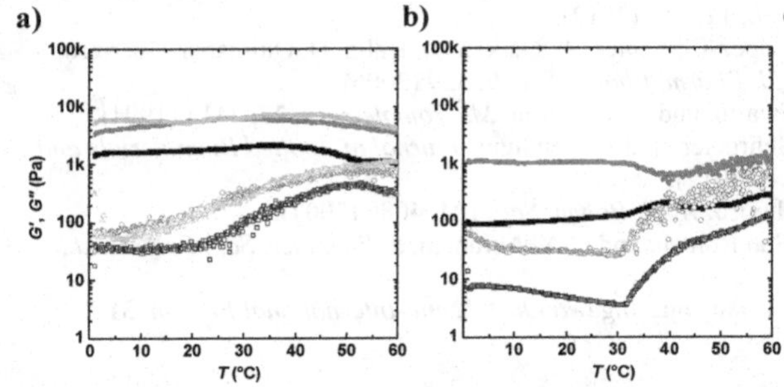


c)

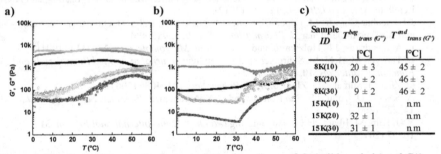

| Sample ID | $T^{beg}_{trans\ (G'')}$ [°C] | $T^{end}_{trans\ (G')}$ [°C] |
|---|---|---|
| **8K(10)** | 20 ± 3 | 45 ± 2 |
| **8K(20)** | 10 ± 2 | 46 ± 3 |
| **8K(30)** | 9 ± 2 | 46 ± 2 |
| **15K(10)** | n.m | n.m |
| **15K(20)** | 32 ± 1 | n.m |
| **15K(30)** | 31 ± 1 | n.m |

**Figure 4:** Rheological investigation as temperature dependency of *G'* (solid symbols) and *G''* (open symbols). a) 8K(yy) networks, ■ 8K(10), ● 8K(20), ▲ 8K(30), b) 15K(yy), ● 15K(20), ▲ 15K(30). c) Table of the temperatures corresponding to the beginning and the end of the transition of *G''*, n.m. = not measurable.

## CONCLUSIONS

Hydrogels networks based on thermo-sensitive OEG-OPG-OEG block copolymers of two different $M_n$ and similar hydrophilicity by a OEG content of 80 wt% were successfully synthesized in several concentrations via photo-irradiation of the respective dimethacrylate telechelic oligomers. The resulting samples were homogeneous and translucent, with storage moduli in the range of $G' \sim 1$ kPa. Hydrogels prepared from lower $M_n$ and/or more concentrated precursor solutions displayed a lower swellability and greater mechanical stability (1 to 10 kPa). All samples exhibited thermo-sensitivity with volumetric shrinking and improved mechanical stability with increasing temperature. However, hydrogel networks prepared from $M_n = 14600$ g·mol$^{-1}$ showed larger and sharper transitions in terms of swellability, volume shrinking and loss modulus. These observations can be attributed to the higher mechanical flexibility of those hydrogels, and therefore to a higher degree of freedom of the crosslinked monomers to form hydrophobic domains when the temperature is increased, which complements the results of our previous study [6]. Such hydrogels fabricated from biocompatible monomers and with demonstrated tunable properties and thermo-sensitivity at physiological temperatures can be considered as relevant candidates for thermo-sensitive drug delivery system.

## ACKNOWLEDGMENTS

LT acknowledges Berlin-Brandenburg School for Regenerative Therapies (BSRT – DFG-GSC 203) for financial support.

## REFERENCES

1. R. Langer and J.P. Vacanti, *Science*, **260**, 920 (1993).
2. S. Piluso, B. Hiebl, S.N. Gorb, A. Kovalev, A. Lendlein, and A.T. Neffe, *Int. J. Artif. Organs*, **34**, 192 (2011).
3. B.F. Pierce, E. Pittermann, N. Ma, T. Gebauer, A.T. Neffe, M. Hoelscher, F. Jung, and A. Lendlein, *Macromol. Biosci.*, **12**, 312 (2012).
4. J.J. Escobar-Chavez, M. Lopez-Cervantes, A. Naik, Y.N. Kalia, D. Quintanar-Guerrero, and A. Ganem-Quintanar, *J. Pharm. Pharm. Sci.*, **9**, 339 (2006).
5. P. Alexandridis, J.F. Holzwarth, and T.A. Hatton, *Macromolecules*, **27**, 2414 (1994).
6. L. Tartivel, M. Behl, M. Schroeter, and A. Lendlein, *Journal of Applied Biomaterials and Functional Materials*, **10**, 6 (2012).
7. Q. Zhao, J. Sun, and Q. Zhou, *J. Appl. Polym. Sci.*, **104**, 4080 (2007).
8. A. Sosnik, D. Cohn, J.S. San Roman, and G.A. Abraham, *J. Biomater. Sci., Polym. Ed.*, **14**, 227 (2003).
9. Y.-J. Kim, M. Ebara, and T. Aoyagi, *Angewandte Chemie-International Edition*, **51**, 10537 (2012).

Mater. Res. Soc. Symp. Proc. Vol. 1569 © 2013 Materials Research Society
DOI: 10.1557/opl.2013.838

# Hysteresis modeling of porous SMA for drug delivery system designed and fabricated by the laser-assisted sintering

Igor V. Shishkovsky
P.N. Lebedev Physics Institute of Russian Academy of Sciences, Samara branch,
Novo-Sadovaja st. 221, Samara 443011, Russia. shiv@fian.smr.ru

## ABSTRACT

In this report we develop a complete mathematical model for a porous scaffold from nitinol (NiTi – intermetallic phase) with a shape memory effect (SME), fabricated layerwise via the selective laser sintering (SLS) process. The operation of the SME bio-fluidic MEMS involves such physical process as a heat transfer, a phase transformation with a temperature hysteresis, stress-strain and electrical resistance variations accompanied the phase transformation. The simulations were conducted for the electro- and a thermo- mechanical hysteresis phenomenon, during the SME in the porous nitinol structures of the cylinder shape, which allow to formulate a recommendations for SLS. Previously done the temperature evolution of electrical resistivity was compared with our present calculations as a function of the laser-processing parameters for three dimensional nitinol samples. This model can be used for an estimation of a drug delivery system route during a porous phase volume changing.

## INTRODUCTION

The intermetallic NiTi (named as *nitinol*) has a great potential as in implantlogy as during a fabrication of the active bio fluidic drug delivery systems (DDS), lab - on a chip devices [1,2] due to intrinsical properties even in a porous state a biocompatibility, a corrosion resistance, damping characteristics and a unique shape memory effect (SME) [3].

In general, the fluid propulsion can be effectuated mechanically, electrically, or thermally [1, 4, 5]. In this study the thermal methods of propulsion will be discussed. Under the reversible austenite –martensite transformations in the NiTi, the driving pressure arises from the volume change due to the crystal lattice reconfiguration from one state to another under the temperature variation. An essential component in the micro-fluidic system is a control ability to stop and start the fluid flow. So, in the field of the controlled drug delivery, self- expanding (occurred in our case owing to the SME) in vivo devices that use diagnostic measurements to control the drug release and that can be initiated at the temperature of a living organism will be widely developed. The laser additive approaches and the Selective Laser Sintering method in particularly have a great potential for engineering the implants, scaffolds, DDS with a prespecified and reproducible external and internal surface morphology that could be used in the bio engineering applications [3, 6-7]. In the papers [6-7] it were proposed the SLS of a biopolymer implant with a bioactive drug. After implantation in a living tissue a drug delivery process took place and the implant was resorbed during time.

In our earlier studies it was shown that the NiTi scaffolds fabricated by the SLS process typically had high values of a corrosion resistance and porosity, a significant water absorption ability and a well-developed surface structure [8]. A controllable internal structure of the interconnected porous channels allows providing of a framework for the bone that ensures its growing into the matrix of the material, thus increasing the interfacial area between the implant and the tissue and thereby reducing the shift of the implant in the tissue. Also, some suitable channels could be available for the drug release in the blood, aimed to reach the connective ingrown tissue thereby preventing necrosis of the penetrated cells. Early the general modeling scheme of nitinol DDS was developed in our report [2]. The present report is dedicated to theoretical modeling of porous DDS made of nitinol by the method of laser-assisted-manufacturing. Our modeling scheme is used on Dutta *et al.* and Ikuta *et al.* approaches [9, 10]. Some numerical estimation about the thermo mechanical and electro-thermal hysteresis phenomenon will be made.

## THEORETICAL MODEL

The phase transformation in nitinol is characterized by the transformation start and finish temperatures: $A_s$ is the austenite start temperature, $A_f$ is the austenite finish temperature, $M_s$ is the martensite start temperature, and $M_f$ is the martensite finish temperature. The amount of *A - M* transformation is characterized by the martensite fraction $V_M$. Martensite fraction is defined as the volume fraction of martensite present in the shape memory alloy (SMA) at any instant, therefore $0 \leq V_M \leq 1$. The austenite fraction can also estimate by the rule $V_A = 1 - V_M$ (where $V_M, V_A \in [0, 1]$) as the volume fraction of austenite present in the SMA. In general, at any given temperature, the phases *A* and *M* coexist. Obviously, the thermal, mechanical and electrical properties of SMAs can be predicted if the $V_M$ and the stress history of the SMA are known.

Differential Scanning Calorimetry (DSC) is a well-known approach for analyzing phase transformations, and a temperature hysteresis evaluation during the *A - M* phase transformation [9, 11]. But Antonucci *et al.* brought out clearly in the [11] that an electrical resistivity (ER) measurement has proved to be a good probe for the identification of various phases in SMA, resulting more sensitive than the conventional DSC technique in the phases evaluation and of its start and finish transition temperatures. As in our study [3], Antonucci *et al.* showed by the ER method the presence of the intermediate *R-* phase on the cooling stage during *A-M* transformation. Therefore the summation of fractions becomes always 1 written in formula $V_A + V_M + V_R = 1$.

Ikuta *et al.* [10] proposed a variable sublayer model to derive expressions for both strain and electrical resistance of the SMAs. The variable sublayer model hypothesizes that the SMA consists of the parallel connected sublayers of the different phases with the different mechanical characteristics. Although the main three phases (*A - R - M*) are distributed randomly in the bulk metal from the microscopic point of view, it can be considered that their sublayers corresponding to an each phase are connected parallel. Yet one interesting approach was done in the Likhachev paper [12], where it was proposed a general differential model for representing the SMA hysteresis minor loops, which is probably the first differential model of SMA hysteresis in the literature.

Let us take the porous cylindrical sample of nitinol synthesized by SLS method. The $L_0$ is the length and $d_0$ is the diameter of the undeformed DDS cooled to the 100% martensite state and at a zero pretension. If the $\varepsilon_0$ is the strain caused by an initial deformation, while the DDS is still cooled to the 100% martensite state, so the $\varepsilon_r$ is the recoverable strain caused by *M-A* phase transformation. The total strain will be $\varepsilon = \varepsilon_0 + \varepsilon_r$.

The balance of the heat energy governs the temperature of the SMA DDS. This must be essentially a transient heat transfer problem. We will consider that the internal resistance of the DDS to heat conduction is negligible compared to the convective heat transfer with the environment. We also assume that only a natural convection occurs. Then, the temperature *T* of the porous SMA DDS is governed by the following convective heat transfer equation:

$$\delta * c * \frac{\pi d_0^2 L_0}{4} * \frac{dT}{dt} = \pi * d_0 * [d_0 / 2 + L_0 * (1 + \varepsilon / 2)] * h(T - T_{amb}) \pm \Delta q * \eta_V, \qquad (1)$$

where $\delta$ - is a mass density of the porous SMA DDS, *c* – is a specific heat, *h* – a convection heat transfer coefficient, $T_{amb}$ - a temperature at the normal "healthy" state (~ 36.6 $^0$C), $\Delta q$ – a latent heat of the A $\leftrightarrow$ M transformations, and $\eta_V$ – is a specific molar volume of a new phase. We will consider that the above equation assuming that both *h* and *c* are constants. However in general, its temperature-dependent parameters and in future it could be use a polynomial interpolation for the temperature-dependent thermo-physical properties of the porous SMA DDS also. It is known that the $M - A$ transformation under a heating is endothermic, while the $A-M$ transformation under a cooling is exothermic. This observation is also taking into account of the heat transfer model (1).

It is known that the blood flow has the Newtonian nature into the large arteries and rheological type of flow into the narrow capillaries [13]. The rheological Darcy equation allows to calculate a permeability factor – $K$ during a drug delivery process. With regard to a contribution of a deformation a liquid flow (blood flow) can be estimated by the next equation [14]:

$$Q = \frac{K * \delta_L * g * \Delta H * S * P}{\zeta_L * L * (1 \pm \varepsilon)^2}, \qquad (2)$$

where $Q$ – is a liquid flow, $\Delta H$ – a liquid level, $g$ – a gravity acceleration, $S$ – a sectional area of a filter element (the DDS in our case), $\delta_L$ – is a liquid density, $\zeta_L$ – a liquid dynamic viscosity and $P$ – is a sample porosity. Under an alternating deformation of the DDS the liquid pulsation due to a pressure difference in its permeable structure will proceed. From the equation (2) it is visible that the pulsation rate will depend not only the pressure difference but on the size of the pores and the liquid viscosity also.

Fig. 1 shows numerical estimations the temperature changing according to the equation (1) and the flow rate of the blood plasma under the strain changing (the equation (2) and blood parameters from [14]). As is evident, a diameter change of the porous matrix section accordance with the capillary diameters (the Fig. 1a) affects to the nature of the temperature relaxation under its increasing up to 40 $^0$C. From the other side, a change in the particle sizes (pores) into the porous DDS (the Fig. 1b) significantly reduces the flow speed and can lead down the thrombosis.

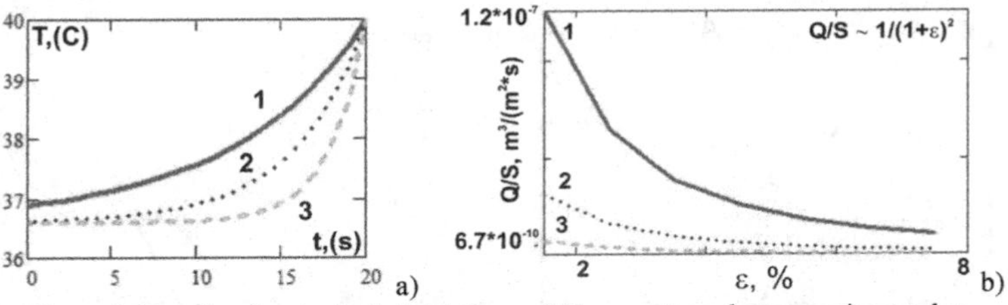


**Figure 1.** Results of numerical simulations: a) Temperature change vs time under different capillary sizes $d_0$ = 1.0, 0.5, 0.25 mm; b) - Blood plasma flow rate vs strain in porous NiTi with different pore sizes: 1.- 500; 2. 250; 3. – 100 μm.

The crystalline phase transformation between the martensite and austenite, and hence the relationship between the martensite fraction and temperature has a hysteretic character. The typical $V_M$ - $T$ hysteretic relationship for the complete $A \leftrightarrow M$ transformation was shown schematically in the Fig. 1. Such hysteresis loop corresponding to the complete $A \leftrightarrow M$ transformation is called a major hysteresis loop. The area enclosed by the major hysteresis loop is called the hysteresis region. An incomplete $A$ - $M$ transformation yields minor hysteresis loops within the major hysteresis loop. The underlying assumption is that the shape and transformation temperatures of the minor loops are the same as those of the major loop [9, 10]. The shape of the minor loops is pertinent to hysteresis modelling of the SMA DDS because forward and reverse phase transformations can occur at any temperature within the hysteresis region while the DDS is an operation. We will use the special functions for the general differential hysteresis model based on the Likhachev approach [12]. We follow a phenomenological approach to the hysteresis modeling also, whish was proposed by Dutta *et al* [9]. However in a real situation the transformation temperatures $M_f$, $M_s$, $A_s$, and $A_f$ are not constant throughout and must be change with a degree of porosity, presence of alloying elements in the nitinol [3].

Dutta *et al.* [9] proposed to use the Gaussian probability distribution functions (PDFs) $g(T)$ and $f(T)$ as the slope functions of the major hysteresis loop. The Gaussian PDF is characterized by the mean $\mu$ and the variance $\sigma^2$:

$$g_{+/-}(T) = \frac{1}{\sigma_{+/-}\sqrt{2\pi}} * exp(-\frac{(T-\mu_{+/-})^2}{2\sigma^2{}_{+/-}}),\quad f_{+/-}(T) = \frac{1}{2} * \left[1 + erf(\frac{T-\mu_{+/-}}{\sigma_{+/-}\sqrt{2}})\right],$$

$$\frac{dV_M}{dT} = \begin{bmatrix} \dfrac{f_-(T)+V_M-1}{f_+(T)-f_-(T)} * g_+(T), T \geq 0, \\ \dfrac{f_+(T)+V_M-1}{f_-(T)-f_+(T)} * g_-(T), T < 0, \end{bmatrix}, \qquad (3)$$

where $V_M(0) = 1$ and subscripts + and − denote increasing and decreasing curves, respectively. Scaling constants for the minor hysteresis loops given in (3) were originally proposed by Likhachev [12]. Equation (3) is justly for $V_R$ instead of $V_M$. Fig. 2 shows the hysteresis loops obtained by the temperature profile (1). Under construction, we have assumed the martensite fraction at time $t = 0$ to be $1$. This is a reasonable assumption, because the initial temperature of the DDS is "normal state" temperature, which means that the SMA DDS comprises 100% martensite at time $t = 0$. Hence, the solution of the last equation from the (3) gives the complete martensite fraction - temperature differential hysteresis model.

Hence, the differential equation solution in the system (3) allows to completely describe temperature- phase hysteresis under the DDS work in our porous NiTi matrix.

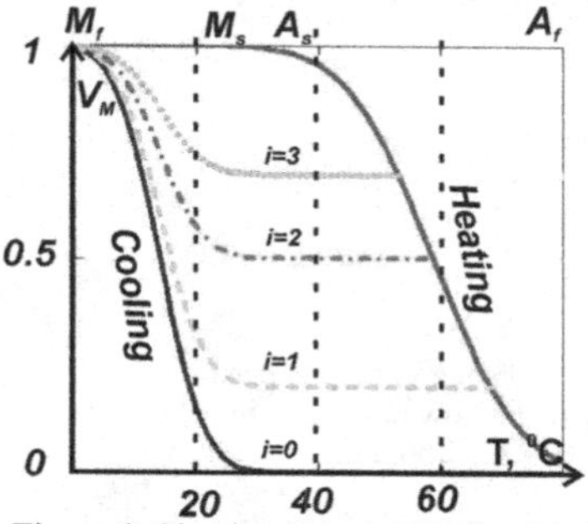


**Figure 2.** Simulated martensite fraction vs. temperature hysteresis from eq. 3 with minor loops ($i = 1..3$). Vertical dot lines show minor hysteresis loop start-finish points (compare with fig. 4).

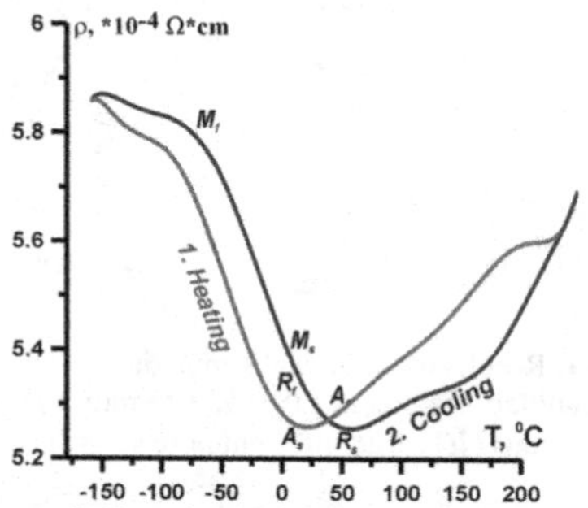


**Figure 3.** Experimental dependence of electrical resistivity - ρ vs. temperature in porous nitinol [3]: 1) curve – is heating; 2) curve – is cooling stages. Laser regime of the SLS: $P = 21.5$ W; $v = 2.5$ cm/s; a beam diameter and a hatch distance were 50 μm.

Above mentioned equations (3) have a phenomenological basis [9, 10]. However, we can propose the own method for the phase fraction estimation by the experimental data from [3]. Converted fractions of the each phase have been evaluated also using resistivity results (Fig. 3), where a resistivity is a function of the temperature. In order to get an expression of converted phase as a function of the $\rho(T)$, the mixing rule has been used, where the total resistivity has been considered as the sum of the resistivity of the each phase. In order to find an expression of the converted percentage for the each phase, at first the experimental resistivity data have been linearly interpolated in the temperature ranges where a single phase exists. Then, on the basis of the mixing rule for the resistivity, the following relationships for the phase fractions have been considered [11]:

$$V_i(T) = \frac{\rho(T) - \rho_j(T)}{\rho_i(T) - \rho_j(T)}, \qquad (4)$$

where $\rho(T)$ - is the experimental ER, $i$ = ($M$ or $R$ phases), $j$ = ($A$ or $R$ phases), respectively. The equation (4) will express the martensite fraction decreasing (if $i = M, j = A$) during a heating passing from *1* at a temperature lower than the $A_s$ to *0* at a temperature higher than the $A_f$. This equation will express respectively the martensite fraction increasing from 0 to 1 (if $i = M, j = R$), during a cooling from a $R$- phase and the $R$- phase fraction increasing from 0 to 1 (if $i = R, j = A$), during the cooling from the austenite phase.

The martensite and austenite have very different stress-strain characteristics [10]. Therefore, the SMA DDS strain $\varepsilon$ is a function of both the stress $\sigma$ and the martensite fraction $V_M$. However, the stress is caused by the DDS tension and can be expressed as a function of the strain. We eliminate the $\sigma$ from the $\varepsilon$ - $\sigma$ - $R_m$ relationship and derive $\varepsilon$ (note that $\varepsilon = \varepsilon_0 + \varepsilon_r$.) as an explicit function of the $R_m$.

Mechanical properties of the austenite phase are modelled by a simple complete elastic body. Because the practical SMA should be used under a yield stress (i.e. an elastic limit) not to introduce a fatigue and will described the Hooke's law. The $M$- phase has an elastic area under a low elastic limit and a plastic area caused by a "twin deformation". It should be noted that the mechanism of the twin deformation completely differs from that of a regular plasticity based on the dislocation. Though dislocation is a kind of a slip mechanism and non-reversible, the twin deformation in the SMA is reversible one. Based on above reason, the mechanics of the $M$-phase can be described by the serial connection of the elastic part and plastic one with the limit strain of the twin deformation. Mechanical function is written in following equations:

$$\sigma_i(\varepsilon, \varepsilon_i^p) = E_i\varepsilon : \ |\varepsilon| < |\varepsilon_i^p + \varepsilon_i^y| \ (elastic\ area), \qquad (5a)$$
$$= E_i\varepsilon_i^y : |\varepsilon| > |\varepsilon_i^p + \varepsilon_i^y| \ (plastic\ area)$$

where $E_i$ – a modulus of elasticity an according phase ($i$=$M$, $R$), $\varepsilon_i^y$ is a elastic limit strain (an yield strain) of the $i$- phase. The similar approach could be use for the mechanical property of the $R$- phase also. Since the each sublayer is connected parallel, the total stress is divided by three phases as equation (5):

$$\sigma = V_A * \sigma_A(\varepsilon) + V_M * \sigma_M(\varepsilon, \varepsilon_M) + V_R * \sigma_R(\varepsilon, \varepsilon_R) \qquad (5)$$

where $\sigma_A$, $\sigma_M$ and $\sigma_R$ mean the stress functions of the $A$-, $M$- and $R$ phases.

There is known existence the "Clausius - Clapeyron like relationship" between the stress and the transformation temperature, which expressed a next form [10]:

$$\frac{d\sigma}{dT} = \frac{\rho\Delta H}{T_0 * \Delta\varepsilon}, \qquad (6)$$

where $\rho$ – is a density, $\Delta H$ – is a latent heat, $T_0$ – a transformation temperature under no stress, $\Delta\varepsilon$- a strain due to the phase transformation. It should be noted that a common stress $\sigma^*$ which can contribute shift of the transformation temperatures does not depend on the direction of the stress macroscopically [10]. Hence the $\sigma^*$ can be expressed by a following equation:

$$\sigma^* = V_A * |\sigma_A(\varepsilon)| + V_M * |\sigma_M(\varepsilon, \varepsilon_M^P)| + V_R * |\sigma_R(\varepsilon, \varepsilon_R^P)| \qquad (7)$$

The ER of the austenite and martensite are different [3, 9-11]. Thus the electrical resistance of the SMA DDS in any given state depends on the martensite fraction $V_M$ in that state. At the same time, the resistivity of the each phase is also the function of the temperature (Fig. 3) [3, 9]. We use the variable sublayer model [10] to model the electrical resistance of SMA DDS:

$$R^A(T)=\frac{4L_0(1+2\varepsilon)}{\pi d_0^2}*\rho_A(T), R^M(T)=\frac{4L_0(1+2\varepsilon)}{\pi d_0^2}*\rho_M(T), \quad (8)$$

where $R^A$ – the electrical resistance of the 100% austenite, $R^M$ – the electrical resistance of the 100% martensite, $\rho_A(T)$ – the ER of the austenite, and $\rho_M(T)$ – the ER of the martensite. Using the variable sublayer model [10], the two phases can be considered to be in parallel. Then the common electrical resistance $R$ of the DDS is given by

$$\frac{1}{R}(T)=\frac{(1-V_M)}{R^A}+\frac{V_M}{R^M} \quad (9)$$

Substituting for the $R^A$ and $R^M$ from (8), we obtain the following expression for the $R$:

$$\frac{1}{R}(T)=\frac{\pi d_0^2}{2L_0(1+2\varepsilon)}*\left[\frac{1-V_M}{\rho_A(T)}+\frac{V_M}{\rho_M(T)}\right] \quad (10)$$

Total resistivity of the SMA can be estimated as a linear summation of resistivities of the each phase as following equation:

$$\rho = V_A*\rho_A + V_M*\rho_M + V_R*\rho_R \quad (11)$$

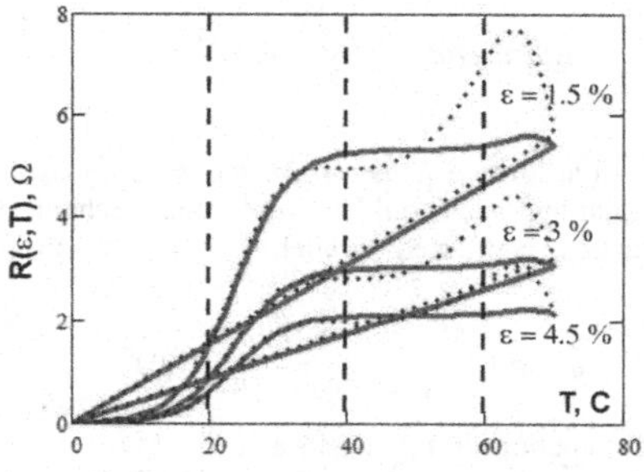


**Figure 4.** Simulated electrical resistance - R vs. temperature – T for different strain – ε (ε = 1...8 %). Model simulations: broken curve - are the cooling; solid curves – are the heating stages. Vertical dot lines show minor hysteresis loop start-finish points (compare with fig. 2).

where $\rho_A$, $\rho_M$ and $\rho_R$ mean the ER of *A*- phase, *M*- phase and *R*- phase respectively. Hence, the austenite and martensite resistivities could be obtained from equations (8-10). Consider the Fig. 2, which shows the $V_M$ as the function of $T$. Suppose the $T$ increases from the $M_f$ to the $A_s$. We know that the $V_M = 1$ in this region. Hence, the electrical resistance $R$ would only depend on the $\rho_M$, and $\rho_M$ should take a form which would yield the shape of the increasing curve in the Fig. 3. Similarly, suppose the $T$ decreases from the $A_f$ to the $M_s$. Then the $V_M = 0$, and the $\rho_A$ should take the form which would the yield the shape of the decreasing curve in the Fig. 3. The above reasoning explains qualitatively why the temperature dependence of the resistivities of the each phase is as shown in the experimental dependence (Fig. 3). Having obtained the $R$ as the function of the $V_M$ and the $T$, we will use the electro-thermal hysteresis model developed us to model the SMA porous cylinder resistance vs the temperature relationship.

Fig. 4 shows the simulation of the common resistance for the NiTi porous cylinder by equations (8-10). It is seen that with the strain increasing the resistance falls, and a hysteresis loop squares are decreased. Developed us the hysteresis phenomena model allows to make one additional important assertion. In order to reach the required pore dimension, i.e., to govern the drug delivery, it is possible to apply an electrical potential to the porous NiTi sample from outside and to change its electrical resistance into the calculated range of the parameters. In other words we obtained an effective way of influencing upon the DDS efficiency, which makes the device completely controllable.

## CONCLUSIONS

In the present work on the based of the previously developed common design, the theoretical model for the combine description of the thermo-mechanical and electro-thermal phenomena in the bio fluidic DDS was proposed. The deformation and stress produced in the NiTi must create a useful displacement and occurrence of the force necessary for the drug release from the pores. The model is capable of simulating the temperature, martensite fraction, stress, strain and electrical resistance of the similar lab-on-chip MEMS.

The basic stages of the DDS functioning are connected with the complete or incomplete martensite transformation at the body temperature level. The SME temperature range shifting and a temperature increasing-decreasing during a disease evolution gives the possibility to govern and regulate the distance between the structural components (pore sizes) and to use the drug dosed supply in the blood. The effective way of measurement under DDS is shown. In our opinion, the micro devices of this type can be used repeatedly.

## ACKNOWLEDGMENTS

The research was supported by the grants of the RAS Presidium under the Program "Fundamental Sciences for Medicine – 2010-2011" and RFBR, 13-08-00208-a Povoljie project.

## REFERENCES

1. J.Z. Hilt, N.A. Peppas. *International Journal of Pharmaceutics*, **306**, 15, (2005).
2. I.V. Shishkovsky. *MRS Proceedings*, **1415,** mrsf11-1415-ii03-10, (2012).
3. I. Shishkovsky, V. Sherbakoff et al. *Proc. Instn. Mech. Engrs., Part C,* **226** (12)**,** 2882, (2012).
4. J.L. Moore, A. McCuiston et al. *Microfluidics and Nanofluidics,* **10** (4), 877, (2010).
5. E. Gultepe, D. Nagesha, S. Sridhar et al. *Advanced Drug Delivery Reviews,* **62**, 305, (2010).
6. C.M. Cheah, K.F. Leong, C.K. Chua et al. *Proc. Instn. Mech. Engrs. Part H,* **216**, 369, (2002).
7. C.L. Liew, K.F. Leong, C.K. Chua, Z. Du. *Int. J. Adv. Manuf. Technol.,* **18**, 717, (2001).
8. Y.I. Arutyunov, L.V. Zhuravel et al. *Physics of metals and metallography,* **93** (2), 185, (2002).
9. S. Dutta, F. Ghorbel. *IEEE/ASME Transactions on Mechatronics,* **10**/2, 1, (2005).
10. K. Ikuta, M. Tsukamoto, S. Hirose. *IEEE Micro Electro Mechanical Systems,* 103, (1991).
11. V. Antonucci, G. Faiella, M. Giordano et al. *Thermochimica Acta,* **462,** 64, (2007).
12. A. Likhachev. *Scripta Metallurgica et Materialia,* **32**/4, 633, (1995).
13. A.E. Medvedev. *Mathematical Biology and Bioinformatics,* **6** (2),228, (2011).
14. V.E. Gunter. *Delay law a new class of materials and implants in medicine.* Northampton, MA: STT Publishing, 2000, 432 p.

# Particulate Biomaterials

Mater. Res. Soc. Symp. Proc. Vol. 1569 © 2013 Materials Research Society
DOI: 10.1557/opl.2013.766

# In situ Preparation of Gold Nanoparticles in Alginate Gel Matrix by Solution Plasma Sputtering Process

Anyarat Watthanaphanit [1,2,*], Gasidit Panomsuwan [3], and Nagahiro Saito [2,3]
[1] EcoTopia Science Institute, Nagoya University, Nagoya 464-8603, Japan
[2] Green Mobility Collaborative Research Center, Nagoya University, Nagoya 464-8603, Japan
[3] Department of Materials, Physics and Energy Engineering, Graduate School of Engineering, Nagoya University, Nagoya 464-8603, Japan

## ABSTRACT

The rapid synthesis of gold nanoparticles (AuNPs) was done by solution plasma sputtering (SPS) process in the presence of a biopolymer, sodium alginate. We utilized the alginate polymer in order to meet three important requirements: (1) to promote the generation of plasma in liquid environment, (2) to provide colloidal stability, and (3) to render biocompatibility to the AuNPs. The effect of sodium alginate concentration (varied as 0.2, 0.5, and 0.9 %w/v) and plasma sputtering time on the particle size and the physical absorption property of the AuNPs were studied. The results indicated that preparation of AuNPs in alginate gel matrix was successful by the SPS process in one step without any reducing agents. This technique has high potential as a novel strategy to produce AuNPs suspended in alginate aqueous solution which is suitable for biomedical application.

## INTRODUCTION

Recent advances in nanotechnology enable AuNPs to be used in biomedical applications [1]. Examples of the applications are DNA sensors [2], drug delivery [3], and treatment of cancer [4]. From a biological viewpoint, the nanoparticles should be passivated with a layer acting as a biocompatible interface before administration into living organisms, so as to confirm the safety of the nanoparticles to interact with target cells. Some work reported the synthesis of AuNPs by chemical method before dispersing them in a natural biopolymer, such as sodium alginate, and then *in vivo* administration [5,6]. Not only for safety issues, this biopolymer matrix functions to stabilize the AuNPs or prevent the agglomeration and settlement of the particles. Although methods for the preparation of AuNPs using chemicals are being widely applied, they may associate with biological toxicity.

Here we present the preparation of AuNPs in alginate gel matrix in one step by using solution plasma sputtering process [7]. We select alginate polymer (Figure 1a) as a matrix since it is one of the most commonly used material for the encapsulation of biologicals [8,9]. The alginate polymer can both provide colloidal stability to the AuNPs and promote the generation of plasma in the liquid environment [10]. Furthermore, no other chemical is needed and the process can be done at room temperature and atmospheric pressure. Effect of alginate concentration on the size of the AuNPs is studied.

## EXPERIMENTAL DETAILS

### Materials and sample preparation

Sodium alginate was purchased from Kanto Chemical Co., Inc. (Tokyo, Japan). Ultra-pure water from Aquarius water distillation apparatus, RFD250NB, Advantec (Tokyo, Japan) was used for preparation of all alginate aqueous solutions. Sodium alginate aqueous solutions were prepared at concentrations of 0.2, 0.5, and 0.9 %w/v.

### Experimental setup

The reactor for SPS process was made from a 100-mL glass beaker having an inner diameter of 43 mm and a height of 75 mm. Two opposing sides of the beaker consist of holes with conical silicone plugs used for the insertion of gold wire electrodes having a diameter of 1 mm (Nilaco, 99.9%). Experimental setup is shown in Figure 1b. The distance gap between two electrodes is 0.3 mm. The plasma was generated in the reactor containing 90 mL of alginate aqueous solution, using a bipolar-DC pulsed power supply (SPIK 2000A/KT-IDP-1010S). The pulse width and pulse frequency employed during the generation of plasma for all solution concentrations were 2 μs and 15 kHz, respectively. The SPS was operated at atmospheric pressure and room temperature, and solutions were stirred continuously during the operation. Effect of the plasma sputtering time was investigated by varying the time as 1, 2, 5, and 10 min.

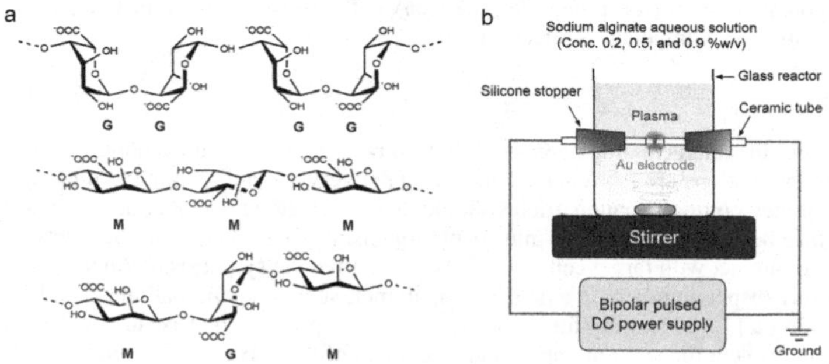


**Figure 1.** (a) Chemical structure of alginate, a linear block co-polymer composing of $\beta$-D-mannuronate (M) and $\alpha$-L-guluronate (G). (b) Experimental setup for the SPS process.

### Characterizations of the AuNPs in alginate aqueous solution

The optical property of AuNPs in alginate aqueous solutions was detected with a UV-vis spectrophotometer UV-2450 (Shimadzu, Japan) in the range of 200–900 nm. Quartz cells (10 mm × 10 mm × 45 mm) were used and pure water was detected as a reference. The shape and microstructure of the AuNPs were observed with a transmission electron microscope (TEM) JEM-2500SE (JEOL, Japan) operated at 200 kV. Samples were prepared by adding ethanol into the AuNPs in alginate aqueous solution. The solution was then dropped onto a copper grid before let it dry by exposure to the air. SemAfore 4.0 program was used to obtain the size and size distribution of the AuNPs.

## RESULTS AND DISCUSSION

### Shape, size, and size distribution of AuNPs

The shape, size, and size distribution of nanoparticles are important factors in some medical applications especially if penetration through a pore structure of a cell membrane is required. Figure 2 shows bright-field (BF) TEM micrographs of AuNPs in alginate aqueous solution with different alginate concentrations, along with their size distributions (Data were collected at a sputtering time of 10 min.). The particles were isotropic in shape, i.e. they seemed to be spherical. Their size and size distribution were dependent on alginate concentration. When the concentration of alginate solution was increased, the size of AuNPs decreased. The average particle diameters of AuNPs were found to be 4.31±0.85, 3.54±1.04, and 2.98±0.82 nm for 0.2, 0.5, and 0.9 %w/v alginate solutions, respectively. These sizes were smaller than those prepared by other methods reported in previous studies [5,11,12–14]. The obtained smaller size of AuNPs is an advantage of the SPS method, since the distribution of AuNPs in tissues/organ is size dependent. Sonavane et al. [5] reported that only AuNPs with the size smaller than 50 nm can pass blood–brain barrier, as evidenced from Au concentration in the brain of mice after 24 h of administration. Thus, the smaller sizes of the AuNPs reported here can indicate their ability to pass through several target cells. Furthermore, the AuNPs suspended in all alginate concentrations exhibited good stability for several months. The electrostatic repulsion forces between $-COO^-$ groups of alginate chains could be an effective driving force of this stability.

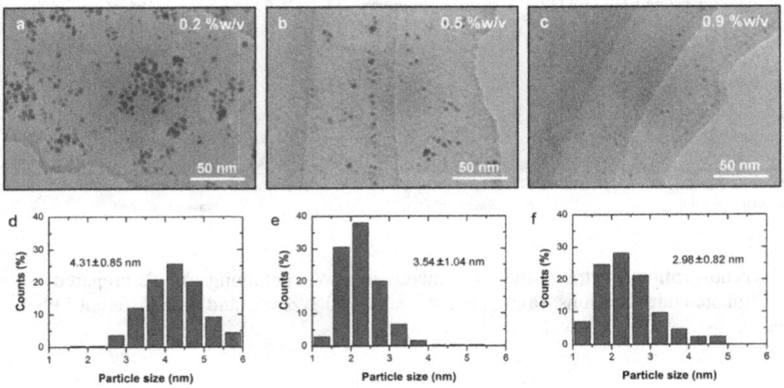


**Figure 2.** BF-TEM micrographs of AuNPs in alginate aqueous solution with different alginate concentrations: (a) 0.2, (b) 0.5, and (c) 0.9 %w/v, along with their size distribution histograms.

## Optical absorption properties of AuNPs

It is well known that AuNPs exhibit unique optical properties. These properties are due to the interaction of light with electrons on the AuNP surface that causes a phenomenon called surface plasmon resonance (SPR) at a specific wavelength of light. The particular wavelength of light where the SPR occurs is strongly dependent on the AuNP size and shape [12]. Figure 3 shows UV-vis absorption spectra of alginate aqueous solution containing AuNPs prepared with different alginate concentrations and SPS discharge times. Spectra of the AuNP suspended in 0.2% and 0.5% alginate aqueous solutions exhibited an absorption band with a maximum absorption wavelength ($\lambda_{max}$) at ~510 nm. Generally, the $\lambda_{max}$ shifts to a shorter wavelength (blue shift) when the size of the nanoparticles decreases. Pal et al. [13] reported that the $\lambda_{max}$ of the AuNPs were found at 527, 530, and 534 nm when the average diameter of the AuNPs were ~8.6, ~12.4, and ~14.4 nm, respectively. In agreement with their reported, the $\lambda_{max}$ observed in our study was at a lower wavelength, since the average sizes of the obtained particles were smaller. On the other hand, the $\lambda_{max}$ was not clearly observed for the AuNPs prepared in 0.9% alginate solution. This result could confirm the small size of the obtained AuNPs since the AuNPs of sizes below 2 nm do not exhibit SPR [12]. Moreover, intensity of the absorption bands remarkably decreased when the alginate concentrations were increased. Since the alginate concentration increment leads to the increase in the solution viscosity, we hypothesized that the viscosity increment may retard the sputtering process of the Au wire. Besides, the more closely alginate chains in alginate aqueous solution, at the higher concentration, could form smaller cavities that may function as a template for the production of AuNPs during sputtering. This is the reason why the size of AuNPs decreased when the alginate concentration was increased.

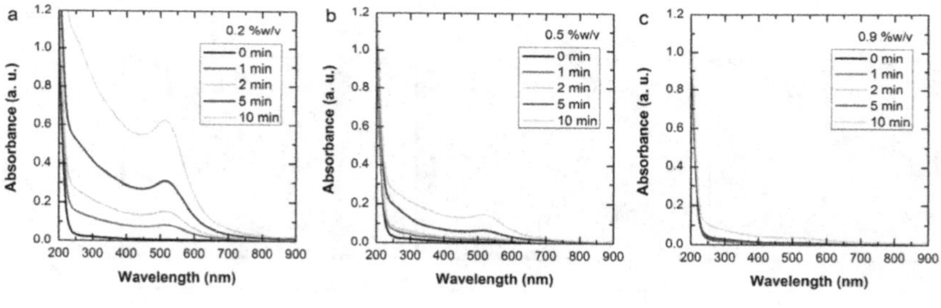


**Figure 3.** UV-vis absorption spectra of alginate aqueous solution containing AuNPs prepared with different alginate concentrations: (a) 0.2, (b) 0.5, and (c) 0.9 %w/v, and with different SPS discharge times.

## CONCLUSIONS

We have demonstrated a one-step method for preparing AuNPs stabilized in alginate gel matrix by the SPS process. The AuNPs produced were spherical and suspended stably in the alginate aqueous solution. The size of the particles depended on the alginate concentration. A gradual size reduction of the AuNPs was observed when the alginate concentration was increased. The average particle diameters of AuNPs produced by this method were smaller than those prepared by other methods reported in previous studies. The method is simple, fast, and chemical free. Therefore, the alginate stabilized AuNPs produced by the SPS process may be useful for biological applications owing to the mild process conditions and the biocompatibility of alginate surrounded the particles.

## ACKNOWLEDGMENTS

We greatly appreciate the financial support from Tokai Region Knowledge Cluster Initiative of Japan Ministry of Education, Cluster, Sports, Science, and Technology (MEXT).

## REFERENCES

1. V. Salata, *J. Nanobiotechnol.* **2**, 1–6 (2004).
2. Y. Cheng, T. Stakenborg, P. V. Dorpe, L. Lagae, M. Wang, H. Chen, and G. Borghs, *Anal. Chem.* **83**, 1307–1314 (2011).
3. S. D. Brown, P. Nativo, J. A. Smith, D. Stirling, P. R. Edwards, B. Venugopal, D. J. Flint, J. A. Plumb, D. Graham, and N. J. Wheate, *J. Am. Chem. Soc.* **132**, 4678–4684 (2010).
4. L. C. Kennedy, L. R. Bickford, N. A. Lewinski, A. J. Coughlin, Y. Hu, E. S. Day, J. L. West, and R. A. Drezek, *Small* **7**, 169–183 (2011).
5. G. Sonavane, K. Tomoda, and K. Makino, *Colloid Surface B* **66**, 274–280 (2008).
6. L. Dykmana and N. Khlebtsov, *Chem. Soc. Rev.* **41**, 2256–2282 (2012).
7. X. Hu, S. P. Cho, O. Takai, and N. Saito, *Cryst. Growth Des.* **12**, 119–123 (2012).
8. S. Chen, Y. Wua, F. Mi, Y. Lin, L. Yu, and H. Sung, *J. Control Release* **96**, 285–300 (2004).
9. T. G. Fernandes, S. Kwon, M. Lee, D. S. Clark, J. M. S. Cabral, and J. S. Dordick, *Anal. Chem.* **80**, 6633–6639 (2008).
10. A. Watthanaphanit and N. Saito, *Polymer Degrad. Stabil.* **98**, 1072–1080 (2013).
11. H. Huang and X. Yang, *Carbohydr. Res.* **339**, 2627–2631 (2004).
12. T. A. El-Brolossy, T. Abdallah, M. B. Mohamed, S. Abdallah, K. Easawi, S. Negm, and H. Talaat, *Eur. Phys. J. Special Topics* **153**, 361–364 (2008).
13. A. Pal, K. Esumi, and T. Pal, *J. Colloid Interface Sci.* **288**, 396–401 (2005).
14. A. T. Anh, D. V. Phu, N. N. Duy, B. D. Du, and N. Q. Hien, *Radiat. Phys. Chem.* **79**, 405–408 (2010).

**Mater. Res. Soc. Symp. Proc. Vol. 1569 © 2013 Materials Research Society**
**DOI: 10.1557/opl.2013.613**

# Advanced Characterization Techniques for Nanoparticles for Cancer Research: Applications of SEM and NanoSIMS for Locating Au Nanoparticles in Cells

Paul J Kempen[1], Chuck Hitzman[2], Laura S Sasportas[3], Sanjiv S Gambhir[3] and Robert Sinclair[1]
[1]Department of Material Science and Engineering, Stanford University, 496 Lomita Mall, Stanford, CA 94305-4034 U.S.A.
[2]Stanford Nanocharacterization Laboratory, Stanford University, 476 Lomita Mall, Stanford, CA 94305-4035 U.S.A.
[3]Molecular Imaging Program at Stanford, Department of Radiology, Stanford University, 318 Campus Drive, Stanford, CA 94305-5427 U.S.A.

## ABSTRACT

The ability of nano secondary ion mass spectrometry (NanoSIMS) to locate and analyze Raman active gold core nanoparticles (R-AuNPs) in a biological system is compared with the standard analysis using the scanning electron microscope (SEM). The same cell with R-AuNPs on and inside the macrophage was analyzed with both techniques to directly compare them. SEM analysis showed a large number of nanoparticles within the cell. Subsequent NanoSIMS analysis showed fewer R-AuNPs with lower spatial resolution. SEM was determined to be superior to NanoSIMS for the analysis of inorganic nanoparticles in complex biological systems.

## INTRODUCTION

With the continuing and growing use of nanoparticles for biological applications [1-3] it is becoming increasingly important to be able to accurately locate and characterize them on and within a cell. Owing to their small size it is necessary to utilize advanced characterization techniques to study them. The scanning electron microscope (SEM), capable of sub-nanometer spatial resolution [4] and good depth of field, is commonly utilized for this purpose [5]. Moreover, it is possible to distinguish materials by atomic number through the use of backscattered electron (BSE) imaging [6]. This atomic number sensitivity creates contrast in the image where higher atomic number inorganic nanoparticles appear bright against a dark low atomic number, organic matrix background. This enables the ready identification of inorganic nanoparticles both on the surface and in the cytoplasm of cell samples [7].

Recently there have been advances in secondary ion mass spectrometry (SIMS), combining SIMS with a focused scanning ion beam, allowing for the creation of high spatial resolution, *ca* 50 nm, compositional maps [8]. The NanoSIMS combines reasonably good spatial resolution with high atomic sensitivity, down to the ppm regime [8] and is powerful because it is capable of distinguishing between isotopes of the same material while analyzing up to seven distinct elements or species at the same time [9]. This technique has been increasingly utilized to study and analyze complex biological systems including the use of isotope labeling [10, 11].

Gold-core silica-shell nanoparticles (R-AuNPs) are promising as a multimodality surface enhanced Raman spectroscopy (SERS) and photoacoustic nanoparticle for diagnostic purposes [12-14]. R-AuNPs, produced by Oxonica Materials Inc (now owned by Cabot Corporation), consist of a 60 nm diameter gold core surrounded by a 30 nm thick silica shell. A monolayer of Raman active dye is adsorbed onto the surface of the gold core providing a surface enhancement effect that can improve the Raman signal intensity by many orders of magnitude [15]. SEM is an established technique for analyzing nanoparticles and has previously been utilized to locate and analyze R-AuNPs in a number of biological systems including T-cells [16] and brain tumors [12]. In this work SEM analysis of RAW264.7 human macrophages incubated with R-AuNPs was compared with NanoSIMS analysis of the same area to determine the viability of NanoSIMS as a technique to locate and characterize inorganic nanoparticles in biological systems.

## EXPERIMENT

R-AuNPs were incubated with RAW264.7 human macrophages grown on glass cover slips to induce cellular uptake of the nanoparticles. The cells were then fixed in a solution of 2.5% glutaraldehyde 2% paraformaldehyde in 0.1 M sodium cacodylate buffer pH 7.4 for 1 hour at room temperature. The samples were then dehydrated in increasing concentrations of ethanol; 50%, 70%, 95% and 100% twice, at 20 minute intervals. The cells were then prepared for SEM and NanoSIMS analysis through critical point drying (CPD), creating a vacuum compatible sample while preserving the structure of the cells [17]. The glass cover slips were then placed on aluminum SEM stubs and coated with a thin layer of Au-Pd to improve conduction. The samples were then cross hatched with a diamond scribe to allow for co-localization between the SEM and the NanoSIMS.

The samples were imaged in the SEM using a FEI Magellan 400 XHR SEM operated at 15 kV with a probe current of 50 pA. After imaging in the SEM, the location was recorded by acquiring a low magnification montage image of the cover slip using the automated software in the Magellan and the sample was placed in the Cameca NanoSIMS 50L. Cesium$^+$ primary ion beam bombardment and negative secondary ion detection were used to optimize the sensitivity for Au. The primary beam current at the sample was on the order of 1pA with an impact energy of 16 keV normal to the surface. Using the secondary electron signal generated by the scanning cesium, $Cs^+$, ion beam the exact location was located. Composition maps for $C_2^-$, $CN^-$, $O^-$, $Si^-$, $P^-$ and $Au^-$ were then generated to show the location of the R-AuNPs within the cell.

## DISCUSSION

### Scanning Electron Microscopy

Scanning electron microscopy is a powerful technique for analyzing R-AuNPs on and inside cells. Secondary electron (SE) imaging provides excellent surface topography information, as shown in figure 1(a-b), while BSE imaging can provide elemental information about the sample as shown in figure 1(c-d). The surface structure is more identifiable in the SE images while the R-AuNPs are can clearly be identified on the surface of the macrophage, appearing as bright areas in the BSE image due to the high atomic number gold core. Figure 1 is representative of the types of images that can be obtained from the SEM. SEM analysis indicated that the R-AuNPs were located across the surface of the macrophages, making this a useful system for comparison with NanoSIMS. A BSE image of the cell that was analyzed using the NanoSIMS is shown in figure 2, with a higher magnification image of the area analyzed for R-AuNPs. This image shows a large number of nanoparticles in the area of interest on or in the cell. BSE imaging can image R-AuNPs that are not at the surface of the sample because the high energy electron beam can penetrate into the sample creating signal several hundred nanometers below the surface of the cell. Some of the R-AuNPs shown in figure 2(B) are located within the cell and not on the surface. This cell was chosen because of the large extension in the upper part region of the cell allowing for easier identification in the NanoSIMS.

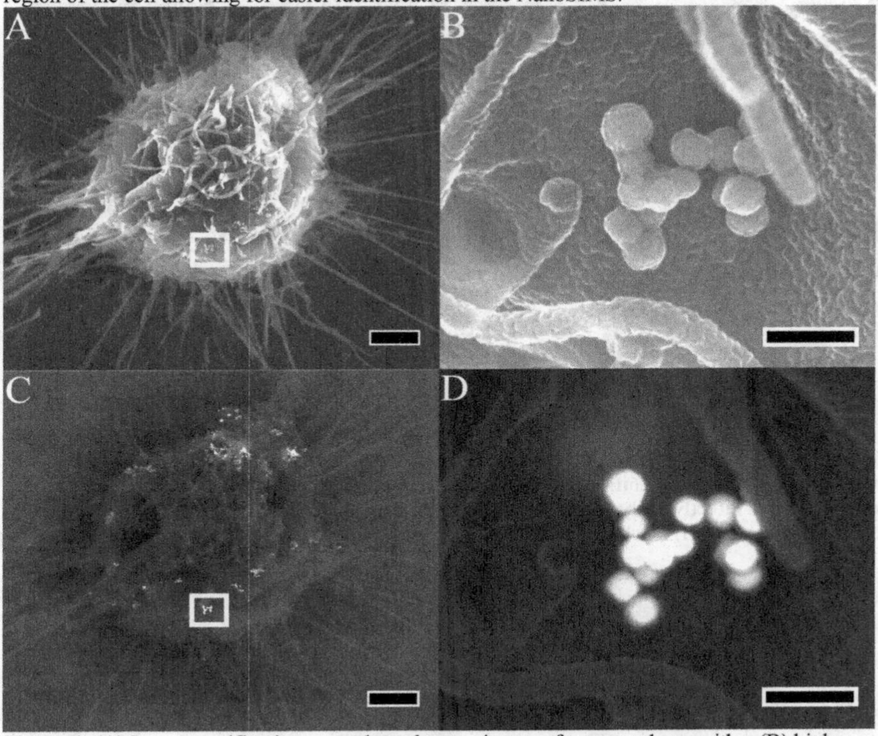


**Figure 1.** (A) Low magnification secondary electron image of a macrophage with a (B) higher magnification image shown R-AuNPs on the surface of the macrophage. (C) Low magnification backscattered electron image of a macrophage showing a number of bright R-AuNPs all over the cell surface with a (D) higher magnification image showing the bright gold core of the R-AuNPs on the surface of a macrophage. Scale (A,C) = 2 μm (B,D) = 200 nm.

**Nano Secondary Ion Mass Spectrometry**

Locating the same cell in the NanoSIMS is a non-trivial endeavor due to differences in image formation and quality. There is an image inversion in the NanoSIMS that must be accounted for when navigating and locating a cell. Once this is taken into account, there are still large differences in image quality due to the ion beam. This beam has a much larger probe diameter, around 50 nm, than the electron beam in the SEM, approximately 1 nm, resulting in decreased surface resolution, shown in figure 3. Additionally, the ion beam quickly sputters the

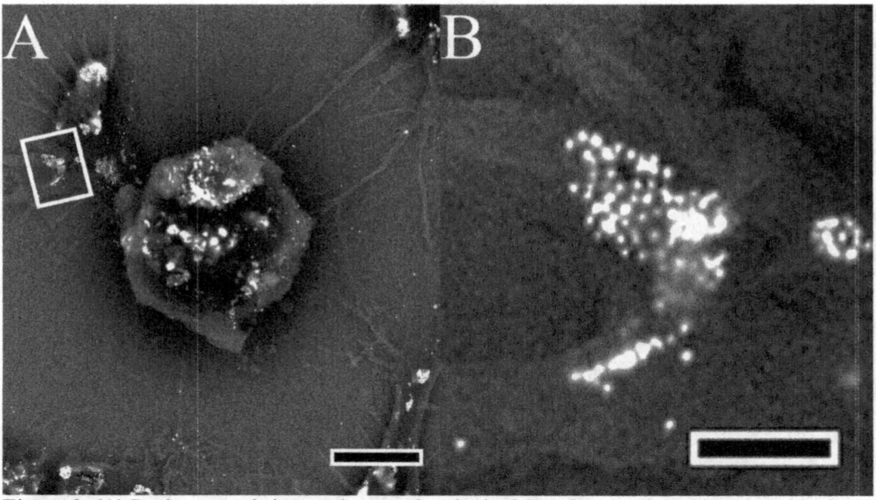


**Figure 2.** (A) Backscattered electron image taken in the SEM showing the cell analyzed in the NanoSIMS. (B) Higher magnification BSE image from the SEM of the boxed area in (A) showing the area directly analyzed for R-AuNPs in the NanoSIMS. R-AuNPs appear as bright areas in both (A) and (B). Scale (A) = 5 μm (B) = 1 μm.

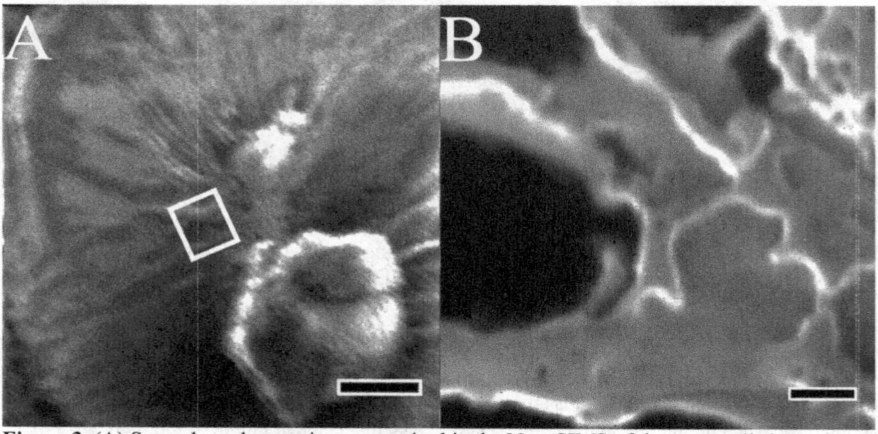


**Figure 3.** (A) Secondary electron image acquired in the NanoSIMS of the same cell shown in figure 2(A). The decreased resolution is due to the use of the ion beam to generate the secondary electrons and the increased charging caused by the glass substrate. (B) Higher magnification secondary electron image of the area shown in figure 2(B) with no R-AuNPs visible. Scale (A) = 5 μm (B) = 500 nm.

thin layer of Au-Pd on the surface of the sample, exposing the underlying cell and glass substrate leading to localized charging. The cell in figure 3(a), the same cell shown in figure 2, appears more blurred as a result of the decreased resolution. There are also bright regions in the image that do not correspond to R-AuNPs but are rather due to charging of the sample. This makes identification of the same region between the two instruments difficult. The unusual shape of this cell was the single identifiable feature between the two techniques. It took about four hours to locate this specific cell in the NanoSIMS.

Once the cell was identified at low magnification, the region of the cell indicated in the white box in figure 3 was analyzed for the presence of $C_2^-$, $CN^-$, $O^-$, $Si^-$, $P^-$ and $Au^-$. Initially no gold was identified in this region. This happened because the R-AuNPs are coated with silica preventing the sputtering of the underlying gold and the nanoparticles were located within the cell in this region requiring the sputtering of the surface of the cell before the R-AuNPs would become visible. NanoSIMS is essentially a surface sensitive technique because ion generation and sputtering can only occur within the first 1-2 nm of the surface. After slowly sputtering the surface away for approximately 30 minutes to expose the R-AuNPs, compositional maps for $C_2^-$, $CN^-$, $O^-$, $Si^-$, $P^-$ and $Au^-$ were created, figure 4.

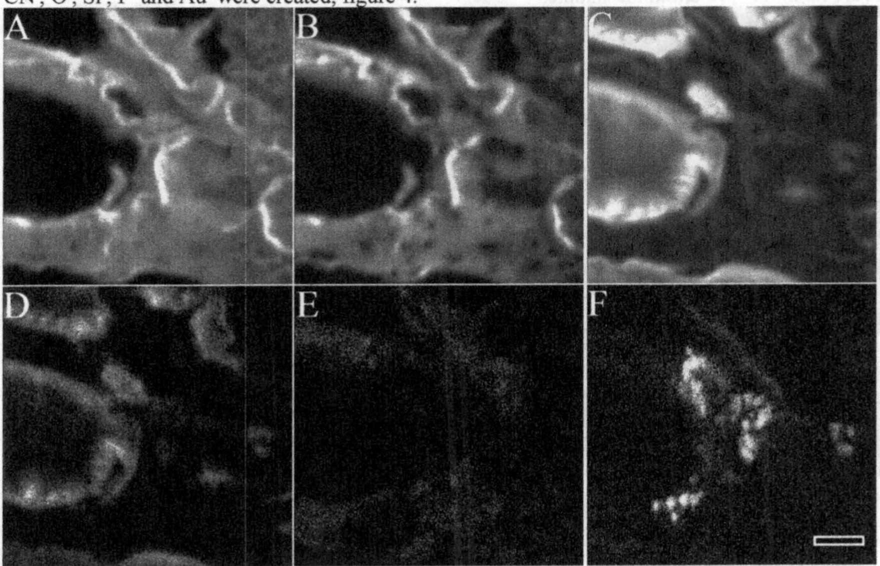


**Figure 4**. Compositional maps for (A) $C_2^-$ and (B) $CN^-$ showing the structure of the cell in this region, (C) $O^-$ and (D) $Si^-$ showing the glass regions exposed by the increased sputtering, (E) $P^-$ showing a weak signal where the cell is located and (F) $Au^-$ showing the presences of the R-AuNPs that have been exposed to the surface. Scale = 500 nm.

The composition maps show the location of the cell in the analyzed region as well as some of the R-AuNPs located in that region. The $C_2^-$, $CN^-$ and $P^-$ maps show the structure of the cell. Both the $C_2^-$ and $CN^-$ maps closely match the secondary electron map shown in figure 3(B) while the $P^-$ signal was too weak to obtain any useful information from it. The $O^-$ and $Si^-$ maps show the presence of the glass substrate in the regions where the ion sputtering has removed the Au-Pd layer exposing the silica underneath. The most intense signals appear closest to the cell possibly due to freshly exposed glass surfaces where the cell previously had covered. These maps were also expected to show the location of the R-AuNPs due to the silica shell but this did not occur.

The $Au^-$ map shows the locations of a number of R-AuNPs in the sample. The nanoparticles do not appear as crisp as in the SEM image, due to decreased resolution and not all of them are visible as shown in figure 5. Comparing the SEM image with the NanoSIMS image, it is possible to locate and identify individual R-AuNPs in the NanoSIMS image but when multiple R-AuNPs are located together their signals appear to blur together making quantification difficult, if not impossible. Additionally, it is clear that not all of the R-AuNPs are visible in the NanoSIMS image, figure 5(A). This could likely occur because the R-AuNPs visible in the SEM image, figure 5(B), could be deeper in the tissue and have not yet been exposed at the surface. Furthermore, the some of the R-AuNPs could have been sputtered away and are no longer present in the sample.

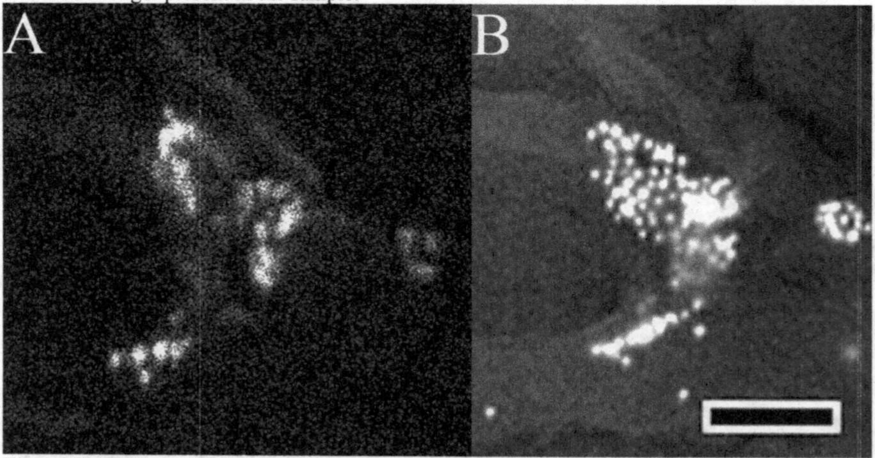


**Figure 5.** (A) NanoSIMS $Au^-$ map showing the location of the R-AuNPs indicating that not all the R-AuNPs located in (B), the SEM image, are visible in the NanoSIMS image. Scale = 1 μm.

## CONCLUSIONS

NanoSIMS is a very useful technique for biological applications. However for locating, imaging and identifying inorganic nanoparticles in complex systems SEM is a superior technique. NanoSIMS is limited by a lower spatial resolution of about 50 nm whereas SEM can achieve resolutions down to 1 nm. The ability to differentiate between isotopes is a unique capability of the NanoSIMS but with inorganic nanoparticles like the R-AuNPs this is not an issue. The ability to identify the R-AuNPs using backscattered electron detection in the SEM allows for higher spatial resolution with increased depth penetration than NanoSIMS.

## ACKNOWLEDGMENTS

Part of this research was conducted using the Cameca NanoSIMS 50L supported by the NSF under award #0922684. The work was funded by NCI Center for Cancer Nanotechnology Excellence Grant CCNE U54 U54CA151459 (S.S.G.).

## REFERENCES

1. C. L. Zavaleta, K. B. Hartman, Z. Miao, M. L. James, P. Kempen, A. S. Thakor, C. H. Nielsen, R. Sinclair, Z. Cheng and S. S. Gambhir, Small **7** (15), 2232-2240 (2011).
2. N. K. Devaraj, E. J. Keliher, G. M. Thurber, M. Nahrendorf and R. Weissleder, Bioconjugate chemistry **20** (2), 397-401 (2009).
3. Z. Liu, S. Tabakman, S. Sherlock, X. Li, Z. Chen, K. Jiang, S. Fan and H. Dai, Nano research **3** (3), 222-233 (2010).
4. L. Y. Roussel, D. J. Stokes, I. Gestmann, M. Darus and R. J. Young, presented at the Proc. SPIE, 2009 (unpublished).
5. A. L. Koh, C. M. Shachaf, S. Elchuri, G. P. Nolan and R. Sinclair, Ultramicroscopy **109** (1), 111-121 (2008).
6. L. Reimer, Measurement Science and Technology **11** (12), 1826 (2000).
7. W. Zhou and Z. L. Wang, *Scanning microscopy for nanotechnology: techniques and applications*. (Springer, 2006), pp. 1-40.
8. B. Wu and J. S. Becker, International Journal of Mass Spectrometry **307** (1–3), 112-122 (2011).
9. J. Pett-Ridge and P. Weber, in *Microbial Systems Biology*, edited by A. Navid (Humana Press, 2012), Vol. 881, pp. 375-408.
10. J. Wells, M. R. Kilburn, J. A. Shaw, C. A. Bartlett, A. R. Harvey, S. A. Dunlop and M. Fitzgerald, Journal of Neuroscience Research **90** (3), 606-618 (2012).
11. M. L. Steinhauser, A. P. Bailey, S. E. Senyo, C. Guillermier, T. S. Perlstein, A. P. Gould, R. T. Lee and C. P. Lechene, Nature (2012).
12. M. F. Kircher, A. de la Zerda, J. V. Jokerst, C. L. Zavaleta, P. J. Kempen, E. Mittra, K. Pitter, R. Huang, C. Campos, F. Habte, R. Sinclair, C.W. Brennan, I.K. Mellinghoff, E.C. Holland and S.S. Gambhir, Nature Medicine **18** (5), 829-834 (2012).
13. J. V. Jokerst, M. Thangaraj, P. J. Kempen, R. Sinclair and S. S. Gambhir, ACS Nano **6** (7), 5920-5930 (2012).
14. S. Keren, C. Zavaleta, Z. Cheng, A. De La Zerda, O. Gheysens and S. Gambhir, Proceedings of the National Academy of Sciences **105** (15), 5844-5849 (2008).
15. S. P. Mulvaney, M. D. Musick, C. D. Keating and M. J. Natan, Langmuir **19** (11), 4784-4790 (2003).
16. P. J. Kempen, Stanford University, 2012.
17. M. J. Dykstra and L. E. Reuss, *Biological electron microscopy: theory, techniques, and troubleshooting*. (Springer, 2003), pp. 1-71.

Mater. Res. Soc. Symp. Proc. Vol. 1569 © 2013 Materials Research Society
DOI: 10.1557/opl.2013.906

# Generation and Surface Functionalization of Electro Photographic Toner Particles for Biomaterial Applications

Christian Speyerer[1], Kirsten Borchers[2], Joachim Storsberg[3], Günter E. M. Tovar[1,2], Thomas Hirth[1,2], Achim Weber[1,2]

[1]Institute of Interfacial Process Engineering and Plasma Technology IGVP, Nobelstr. 12, 70569 Stuttgart, Germany.
[2]Fraunhofer Institute for Interfacial Engineering and Biotechnology IGB, Nobelstr. 12, 70569 Stuttgart, Germany.
[3]Fraunhofer Institute for Applied Polymer Research IAP, Geiselbergstr. 69, 14476 Potsdam, Germany.

## ABSTRACT

In this paper, we provide the investigation about the controlled surface functionalization of acrylic toner particles for electro photography ("laser printing") with sodium hydroxide and the subsequent carbodiimide-mediated coupling of numerous functional amines onto the generated carboxylic group. Various chemically valuable functionalities, comprising of thiol, alkyne and azide, were bound onto the particles' surface and allow for further versatile modifications via huisgen cycloaddition as well as thiol-ene reaction. The functionalization of the acrylic toner surface with alkyne, azide and carboxylic groups increased the cell viability up to 178 % ± 22 % and might offer an interesting path for new applications using common laser printing techniques.

## INTRODUCTION

Electro photography ("laser printing") has emerged to one of the leading print technologies during the last decades. The process enables the simultaneous alignment of multiple toner materials with high spatial resolution (600 dpi, resolution < 100 µm). A new cytocompatible class of toner particles has been developed for the patterning of substrates by common laser printing techniques (Figure 1).

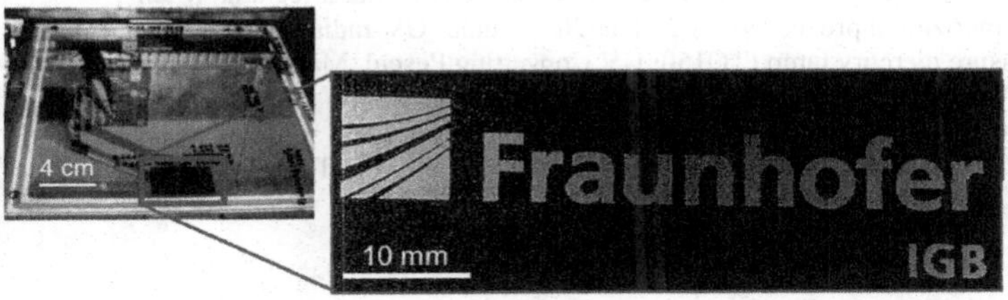


**Figure 1.** Printed institute logo. The surface functionalized particles were printed with an electro photographic printer, which was developed at the Fraunhofer Institute for Manufacturing Engineering and Automation IPA. The image shows the great potential to pattern substrates with the cytocompatible toner particles.

The surface functionalization of the developed microparticles gives rise to further interesting applications [1]. Basic hydrolysis of the poly(methacrylic) surface provides a simple route to highly surface functionalized microparticles via post-synthesis modification without altering their bulk properties. In this article, we provide the investigation about the surface functionalization of acrylic microparticles and the subsequent carbodiimide-mediated coupling of numerous functional amines onto the generated carboxylic groups. Various chemically valuable functionalities, comprising of thiol, alkyne and azide, were bound onto the particles' surface and allow for further versatile modifications via huisgen cycloaddition as well as thiol-ene reaction [2].

**EXPERIMENT**

Propargylamine (98 %), cysteamine (95 %), 11-azido-3,6,9-trioxaundecan-1-amine (90 %), N-(3-dimethylaminopropyl)-N′-ethylcarbodiimide hydrochloride (EDC, 98 %), 2-(N-morpholino)ethanesulfonic acid (MES, 98 %), 2,2′-azobis(2-methylpropionitrile) (AIBN, 98 %), dimethyl sulfoxide (99.9 %), ), propidium iodide (94%) and fluorescein diacetate were purchased from Sigma-Aldrich GmbH (Steinheim, Germany) and used without further purification. Chloroform (99.8 %, Avantor Performance Materials, Griesheim, Germany), acetone (99.3 %, Avantor Performance Materials, Griesheim, Germany), methanol (99.9 %, Avantor Performance Materials, Griesheim, Germany), resazurin sodium salt (AlamarBlue®, Invitrogen Life Technologies GmbH, Darmstadt, Germany), endothelial cell growth medium (Invitrogen Life Technologies GmbH, Darmstadt, Germany) and poly(diallyldimethylammonium chloride) (molecular weight 1,07 x $10^5$ g $mol^{-1}$, BTG Instruments GmbH, Herrsching, Germany) were used as received. Pyrogenic silica (HDK N20 Pharma®) was provided by Wacker Chemie AG (München, Germany). Aqueous sodium hydroxide solutions were created from solid sodium hydroxide pellets (> 99 %, AppliChem GmbH, Darmstadt, Germany) in water (Milli-Q®, Millipore GmbH, Schwalbach, Germany). Methyl methacrylate (Sigma-Aldrich GmbH, Steinheim, Germany) was dried over calcium sulfate for 16 h and distilled under reduced pressure prior to use.

**Polymer synthesis**

The polymeric microparticles were synthesized by an UV-initiated suspension polymerization of methyl methacrylate. The monomer (107 mL, 1000 equivalents) and the initiator 2,2′-azobis(2-methylpropionitrile) (AIBN, 2,45 g, 15 equivalents) were dispersed at 800 rounds per minute in an aqueous silica suspension (0.4 wt.-% $SiO_x$, 900 mL) with a disc turbine stirrer (six blades, outer diameter = 6 cm) in a 1 L-glass reaction vessel (H.W.S. Labortechnik, Mainz, Germany). The polymerization proceeded for 24 h at 20 °C under UV-radiation by a water-cooled medium-pressure mercury lamp (TQ150, UV Consulting Peschl, Mainz, Germany). The synthesized P(MMA) particles were purified by filtration and redispersion in hydrochloric acid (0.1 M) in a methanol/$H_2O$-solution (volume ratio 1:1) for three times and subsequent washing in phosphate buffer (100 mM, pH 7.0) as well as pure water for three times. The particles were dried under reduced pressure ($p < 10$ mbar) for 48 h.

**Surface hydrolysis**

Hydrolysis of the acrylic particles (5.0 g) was performed in an Oak Ridge poly(propylene) copolymer (PPCO)-centrifuge tube by the addition of a10 M sodium hydroxide solution (25 mL). The particles were fully wetted for 20 s using a vortex mixer (Uzusio VTX-3000L, LMS Co., Ltd., Tokyo, Japan) and dispersed afterwards in a heated laboratory shaker (Inkubator 1000, Heidolph Instruments GmbH & Co. KG, Schwabach, Germany) at 37 °C for 72 h. Hydrolysis was stopped by filtration and redispersion in phosphate buffer (c = 100 mM, pH 7.0) for three times, followed by washing the particles in $H_2O$ for three more times. The particles were dried under reduced pressure (p < 10 mbar) for 48 h. The coated soda-lime dishes (ref. to Analysis of the cytocompatibility) were hydrolyzed analogously by the addition of sodium hydroxide solution as well as washing solutions (3 mL each).

**Carbodiimide-mediated coupling**

The carbodiimide-mediated reaction of functional amines was performed at the surface of particles, which had been hydrolyzed with a sodium hydroxide concentration $c_{NaOH}$ of 10 M at 37 °C for 72 h. The hydrolyzed particles (3.0 g) were placed into an Oak Ridge PPCO-centrifuge tube and dispersed in 2-(N-morpholino)ethanesulfonic acid (MES) buffer (15 mL, 100 mM, pH 5.0). In relation to the titrated carboxylic groups at the particles' surface (3.83 µmol), a tenfold molar excess of the functionalizing reagents (propargylamine, cysteamine, 11-azido-3,6,9-trioxaundecan-1-amine) dissolved in dimethyl sulfoxide (DMSO) was added. The coupling was started by the addition of a hundredfold excess of EDC. The particles were dispersed for 48 h at room temperature and subsequently cleaned by filtration and redispersion in phosphate buffer (c = 100 mM, pH 7.0) for three times followed by washing in water for three more times. The particles were dried under reduced pressure (p < 10 mbar) for 48 h. The hydrolyzed coating of the soda-lime dishes (ref. to Surface hydrolysis) was functionalized analogously by the addition of the functionalizing reagents and subsequent washing.

**Analysis of the cytocompatibility**

Primary dermal fibroblasts were used to evaluate the cytocompatibility of the polymeric surfaces. The soda-lime glass dishes (Ø = 3 cm) were coated with a polymeric layer (about 225 µm thick) by the addition of 4 mL of a poly(methyl methacrylate)-solution in chloroform (4 Vol.-% polymer). The solvent was evaporated at room temperature for 5 d and additionally for 48 h at 35 °C under reduced pressure ($5 \times 10^{-3}$ mbar). The glass dishes were γ-sterilized with a radiation dose of $2,5 \times 10^4$ Gy (BBF Sterilisationsservice GmbH, Kernen-Rommelshausen, Germany) prior to the analysis.

Fluorescence staining of cells

Human dermal fibroblasts ($5 \times 10^4$) in endothelial cell growth medium (ECGMmv) were seeded into the coated soda-lime dishes and cultivated for 96 h at 37 °C at a $CO_2$-level of 5 Vol. % and an air humidity of 95 %. Fluorescein diacetate-solution (5 µL, c = 5 µg $mL^{-1}$ in acetone) and propidium iodide-solution (45 µL, c = 0.5 µg $mL^{-1}$ in phosphate buffer) were added. The cells were incubated for 15 min at 37 °C and washed twice with phosphate buffer (pH 7.4). The fluorescence was subsequently analyzed with a fluorescence reader (Synergy 2, Biotek, Bad Friedrichshall, Germany). Fluorescein was excited at the absorption wavelength of 496 nm and detected at the emission wavelength of 530 nm. Propidium iodide was excited at 535 nm and detected at 615 nm.

Cell-viability assay
Human dermal fibroblasts (5 x $10^4$) in endothelial cell growth medium (ECGMmv) were seeded into the coated soda-lime dishes and cultivated for 96 h at 37 °C at a $CO_2$-level of 5 Vol. % and an air humidity of 95 %. The adherent cells were incubated with a resazurin-solution (1 mL, 10 Vol.-% AlamarBlue®-solution in ECGMmv) for 3 h at ambient temperature. The fluorescence was subsequently analyzed with a fluorescence reader (Synergy 2, Biotek, Bad Friedrichshall, Germany). The fluorescent dye was excited at the absorption wavelength of 540 nm and detected at the emission wavelength of 600 nm. The cell viability on the polymeric materials was determined in comparison to a reference value of plasma treated tissue culture dishes.

**Particle charge detector (PCD)**

Specific surface charge was determined potentiometrically with the particle charge detector Muetek PCD 03pH (BTG Instruments GmbH, Herrsching, Germany) connected to a Titrino 702SM (Metrohm AG, Herisau, Switzerland) at room temperature. Dispersion of 400 mg particles in 15 mL phosphate buffer (c = 10 mM, pH 7.0), were titrated with an aqueous poly(diallyldimethylammonium chloride) solution (PDADMAC, c = 1.60 mM).

## DISCUSSION

Basic hydrolysis of P(MMA) microparticles at elevated temperatures of 37 °C for 72 h with a sodium hydroxide concentration $c_{NaOH}$ of 10 M generated carboxylic groups at the poly(methacrylic) surface as well as methanol (Figure 2). It allowed for the further modification of the particles' surface via carbodiimide-mediated coupling with three different functional amines (11-azido-3,6,9-trioxaundecan-1-amine, cysteamine, propargylamine).

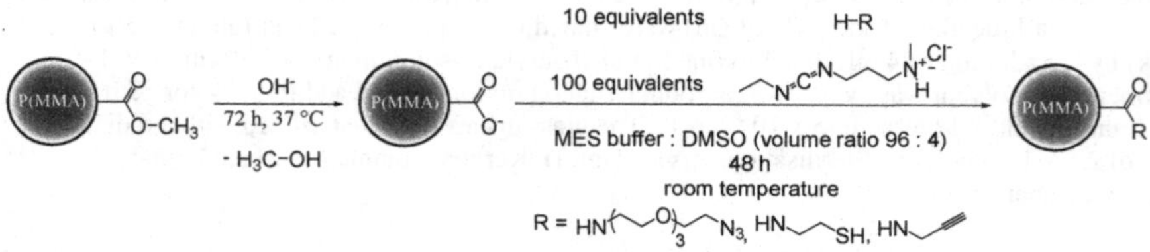


**Figure 2.** Proposed chemical reaction at the microparticles' surface. Basic hydrolysis of the ester groups generates carboxylic groups, which can be subsequently modified with functional amines (11-azido-3,6,9-trioxaundecan-1-amine, cysteamine, propargylamine) via carbodiimide-mediated coupling with N-(3-dimethylaminopropyl)-N'-ethylcarbodiimide hydrochloride (EDC).

Quantitative analysis of the chemical reactions was performed by titration of the carboxylic surface groups. The specific surface charge increased to 1.29 μeq $g^{-1}$ (which is equivalent to

1.28 μmol m$^{-2}$) after the hydrolysis. A high conversion (>95 %) of the generated carboxylic groups was achieved during the coupling of functional amines in the presence of N-(3-dimethyl-aminopropyl)-N'-ethylcarbodiimide hydrochloride (EDC), while the reaction did not yield a large amount of amide bonds without the carbodiimide moiety (Figure 3).

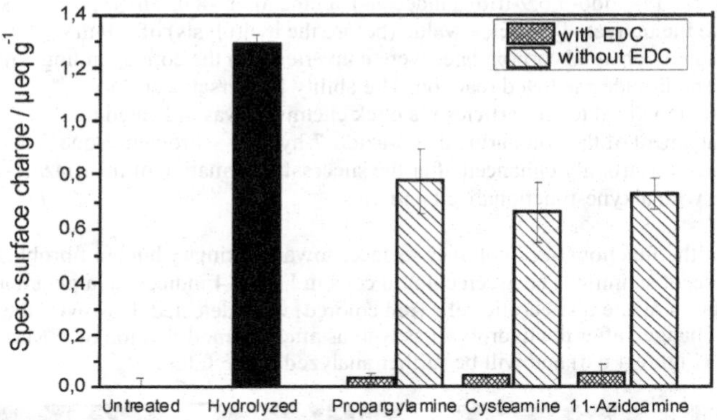


**Figure 3.** Specific surface charge of the polymeric particles by titration of the carboxylic surface groups. The specific surface charge increased after the hydrolysis due to the generation of carboxylic groups. A high conversion (>95 %) was achieved in the presence of N-(3-dimethylaminopropyl)-N'-ethylcarbodiimide hydrochloride (EDC), while the reaction did not yield a large amount of amide bonds without the coupling reagent.

More details to the characteristics of the particles are shown in [3], which we here shortly summarize. In order to determine the surface modifications, electrophoretic light scattering was performed for the determination of the particles' zeta potential in phosphate buffer at pH 7.0. A very quick increase of the particles' negative zeta potential was observed during the first 72 h under basic conditions at 65 °C in a sodium hydroxide / potassium chloride-buffer (3 M, pH = 13.3), while the hydrolytic reaction slowed down afterwards. This qualitative result was confirmed by quantitative titration of the carboxylic groups with a poly(diallyldimethyl-ammonium chloride) solution (PDADMAC solution) in phosphate buffer (pH = 7.0).
We also investigated the pH dependence of the carboxlates at the particles' surface by electrophoretic light scattering in different buffer solutions with the same ionic strength (I = 10 mM). As expected, all carboxylic groups of particles, which had been hydrolyzed in 10 M sodium hydroxide solution for 72 h, were protonated below a pH value of 2 and resulted in a small zeta potential of -10 mV. In contrast, we found a large zeta potential of -81 mV above the pH value of 7, which indicated a fully deprotonated particles' surface under basic conditions.
The reactivity of the generated carboxylic groups at the particles' surface was verified by fluorescent staining. Fluoresceinamine was covalently bound to the carboxylic groups via carbodiimide-mediated coupling in 2-(N-morpholino)ethanesulfonic acid buffer and DMSO (pH = 5.0). The fluorescence intensity of this reagent was enhanced after the covalent bond had been formed due to the reduced electronic donation of the amine group and therefore exhibited a strong response at the particles' surface. Figures are shown in [3].

The kinetics of the carbodiimide-mediated attachment of other functional amines onto the particle's surface was investigated by electrophoretic light scattering in phosphate buffer (pH = 7.0) at 25 °C. We observed that the coupling reaction proceeded quickly during the first 8 h and slowed down afterward for all functional amines. The zeta potential of the functionalized particles asymptotically converged to similar values of -26 mV for propargyl amine, -27 mV for cysteamine and -25 mV for 11-azido-3,6,9-trioxaundecan-1-amine after 48 h. These potentials were only slightly above the untreated particles' value (before the hydrolysis) of -20 mV, because almost all negatively charged carboxylates were converted into the corresponding amide bonds during the the carbodiimide-mediated reaction. The ability for versatile surface modifications of the functionalized toner particles via click chemistry was indicated by copper(I)-catalyzed attachment of the coumarine dye 3-azido-7-hydroxy-chromen-2-one. The fluorescence of this molecule strongly enhanced after the successful formation of the triazole-ring during the reaction with alkyne-functionalized particles.

The cytocompatiblity of the functionalized polymer surfaces towards primary human fibroblasts was analyzed by fluorescent staining. The green colored cells in Figure 4 indicate a large amount of vital cells on all surfaces, while no necrotic cells (red colored) were detected. However, the fibroblasts' phenotype changed after the hydrolysis as well as after the modification of P(MMA) with functional amines (Figure 4 c/d) and will be further analyzed in the future.

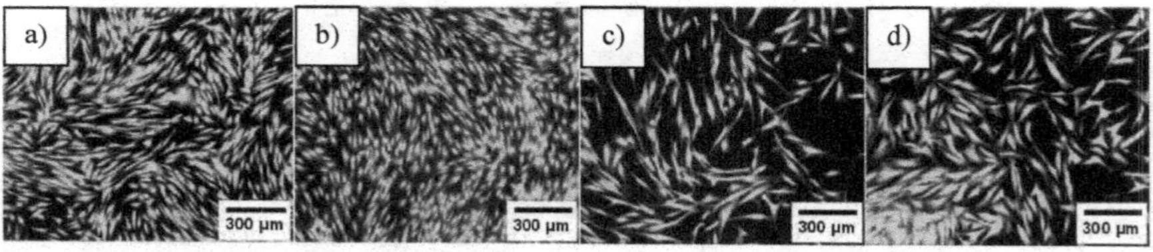


**Figure 4.** Fluorescent live/dead staining of fibroblasts. The green color indicates a large amount of vital cells on tissue culture surfaces (a), P(MMA) (b), hydrolyzed P(MMA) (c) and functionalized P(MMA) (d). No negrotic cells (red color) were detected. The fibroblasts' phenotype changed after the hydrolysis (c) as well as the modification of hydrolyzed P(MMA) surfaces with the functional amines propargylamine, cysteamine and 11-azido-3,6,9-trioxaundecan-1-amine (d).

The cell viability was quantified by the conversion of resazurin into the fluorescent moiety resorufin (AlamarBlue assay). Fibroblasts showed a similar viability on the untreated P(MMA) surface of 105 % ± 12 % in comparison to the reference material (tissue culture dishes, 100 % ± 13 %). Interestingly, the cells exhibited an increased viability of up to 178 % ± 22 % on modified P(MMA) surfaces (Figure 5). The surface modification allowed for the carbodiimide-mediated coupling of different functional amines, i. e. 11-azido-3,6,9-trioxaundecan-1-amine, cysteamine and propargylamine and almost all carboxylic groups were converted after 48 h as determined by titration of the remaining carboxylic groups. The hydrolytic surface modification of methacrylic microparticles enabled the introduction of chemically quite valuable functionalities, i. e. azide, thiol and alkyne moieties. These groups allow for versatile further modifications via copper-catalyzed huisgen cycloaddition as well as UV-initiated thiol-ene reaction.

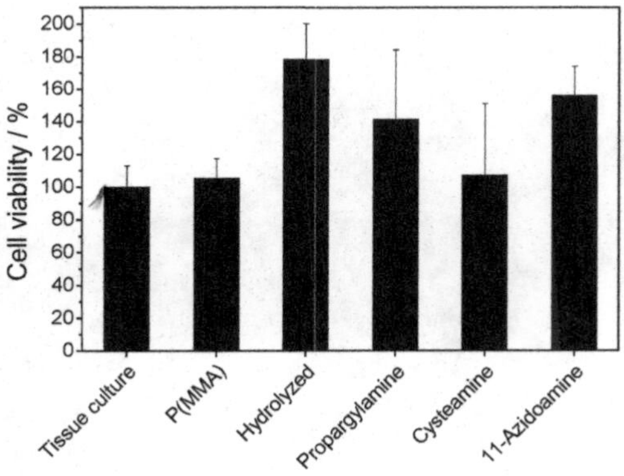


**Figure 5.** Quantitative analysis of the fibroblasts' viability by AlamarBlue® assay. The cells showed a similar viability on the untreated P(MMA) surface in comparison to the reference material (tissue culture dishes). Interestingly, the cells exhibited an increased viability of up to 178 % ± 22 % on hydrolyzed and functionalized P(MMA) surfaces.

## CONCLUSIONS

In this article, we have investigated the surface hydrolysis of poly(methacrylic) microparticles. The hydrolysis yielded a highly effective surface functionalization resulting in functional densities of up to 1.29 μeq $g^{-1}$ under basic conditions, which is equivalent to a surface density of 0.77 carboxylic groups per $nm^2$ as determined by potentiometric titration. The surface modification allowed for the carbodiimide-mediated coupling of different functional amines, i. e. 11-azido-3,6,9-trioxaundecan-1-amine, cysteamine and propargylamine, enabling the introduction of chemically quite valuable functionalities, i. e. azide, thiol and alkyne moieties. These groups allow for versatile further modifications via copper-catalyzed huisgen cycloaddition as well as UV-initiated thiol-ene reaction.

## ACKNOWLEDGMENTS

The authors would like to express their gratitude to Nadine Klechowitz for the cytocompatibility assays (Endress + Hauser Conducta, Gerlingen, Germany). We thank Stefan Güttler (Stuttgart Media University, Stuttgart, Germany) and Christian Seifarth (Fraunhofer Institute for Manufacturing Engineering and Automation IPA, Stuttgart, Germany) for the electro photographic printing. Furthermore, we appreciate the financial support by the Volkswagen Foundation, Hannover, Germany (funding initiative "Innovative methods for the manufacturing of multifunctional surfaces") and the German Chemical Industry Fund (FCI e. V.), Frankfurt.

## REFERENCES

1. C. Speyerer, S. Güttler, K. Borchers, G. Tovar, T. Hirth and A. Weber, MRS Proc. 1340, mrss11-1340-t05-091340 (2011).
2. A.S. Goldmann, A. Walther, L. Nebhani, R. Joso, D. Ernst, K. Loos, C. Barner-Kowollik, L. Barner and A.H.E. Mueller, Macromolecules 42, 3707 (2009).
3. C. Speyerer, K. Borchers, T. Hirth, G. Tovar and A. Weber, Journal of Colloid and Interface Science 397, 185–191 (2013).

## CONCLUSIONS

## ACKNOWLEDGMENTS

## REFERENCES

Mater. Res. Soc. Symp. Proc. Vol. 1569 © 2013 Materials Research Society
DOI: 10.1557/opl.2013.815

# Morphology of crosslinked poly(ε-caprolactone) microparticles

Fabian Friess[1,2], Andreas Lendlein[1,2,3], Christian Wischke[1,3]
[1]Institute of Biomaterial Science, Helmholtz-Zentrum Geesthacht, Teltow, Germany.
[2]Institute of Chemistry, University of Potsdam, Potsdam, Germany
[3]Berlin-Brandenburg Centre for Regenerative Therapies, Berlin and Teltow, Germany.

## ABSTRACT

In order to explore the feasibility for preparing defined crosslinked particulate structures, oligo(ε-caprolactone) [oCL] derived microparticles (MPs) were crosslinked in non-molten, non-dissolved, i.e. solid state in aqueous suspension by applying a controlled regime with well-defined polymer network precursors either with or without photoinitiator. The MPs (diameter ~ 40 µm) were prepared by an oil-in-water emulsion process from linear $^{2}$oCL or 4-arm star-shaped $^{4}$oCL with methacrylate end groups. Crosslinking was initiated by UV-laser irradiation (308 nm) at room temperature. Conversion of methacrylate was monitored by ATR-FTIR spectroscopy and crosslinking was confirmed by a lack of MP dissolution in dichloromethane. In a quantitative evaluation of swelling by dynamic light scattering, higher swelling ratios were detected for particles synthesized with photoinitiator. Wrinkled particle surfaces and distorted particle shapes were observed by light microscopy in the solvent-swollen state and by scanning electron microscopy after deswelling. This work indicated some limitations due to internal inhomogeneity of the MP, but particle crosslinking in solid state was generally possible and may be further improved by higher chain mobility during crosslinking.

## INTRODUCTION

Poly(ε-caprolactone) [PCL] and its copolymers are hydrolytically degradable and are therefore extensively explored for temporary implants and drug carriers [1], in which the degradation is a material function required for excretion from the body. By photo-initiated crosslinking of methacrylate functionalized precursors, PCL networks can be obtained with tailored degradation behavior [2]. Additionally, a suitable network architecture and morphology may facilitate well controllable mechanical properties of the materials.

Typically, the crosslinking of macroscopic samples is performed in the molten state or in solution [3]. When aiming to prepare smaller network matrices like microparticles [MP] as relevant for drug delivery, crosslinking is more challenging. In principle, crosslinked MP may be prepared by emulsion-based techniques, in which droplets could serve as microtemplates [4-6]. After solvent evaporation and polymer precipitation into individual spherical and solid MP, the scattering of light at interfaces and the mobility of the functionalized chains are two important parameters that may affect network formation by polymerization of methacrylate end groups.

In this study it was hypothesized that microparticles from functionalized oligomers can be crosslinked due to their small size in a non-dissolved, non-molten state, which herein is refered to as solid state. Still, certain preconditions need to be fullfilled. Particularly, crosslinking should be performed above the glass transition temperature $T_g$ for enabling destinct chain mobility and below the melting temperature $T_m$ to preserve the particle structure, which is given for semi-crystalline PCL at ambient conditions ($T_g \sim$ -60 °C, $T_m \sim$ 60 °C). If the sample dimension is small as for MPs, then the light scattering effects of crystallites should be negligible. Dispersion of oCL MP in an aqueous phase and performing the crosslinking in this suspension should further improve the accessibility of UV irradiation by separation of particle aggregates and reducing the chance of inter-particulate bonds. Despite photo-initiators are not necessary for this polymerization if high intensity laser light is applied [3, 7], they are often employed for radical formation and to alter the rate of initiation and termination [8]. A bifunctional initiator as used here at high concentration to identify maximum effects should increase initiation events and reduce early termination in the diffusion controlled regime of polymerization [9]. A specific focus is applied on the obtained microparticle topography and semi-crystalline polymer morphologies.

## EXPERIMENT

Polyvinyl alcohol (PVA; Mowiol® 4-88, Kururay, Frankfurt, Germany), ethyl acetate (EA), and dichloromethane (DCM) were purchased from Sigma-Aldrich and used without further purification. A bifunctional 2-hydroxy-2-methyl-1-phenyl propanone [HMPP] type of photoinitiator (Esacure One , Lamberti, Albizzate, Italy) was used in same cases. Linear $^{2}$oCL-diols and four arm star-shaped $^{4}$oCL-tetraols (CAPA®, Perstorp, UK) with a number average molecular weight ($M_n$) of 8 kDa as confirmed by multidetector gel permeation chromatography were functionalized with 2-isocyanatoethyl methacrylate (IEMA) [10] with a degree of functionalization of 98 mol.%. The precursor solution (0.5 ml, 23 wt.% in EA), in some cases supplemented with 3 wt.% HMPP, were emulsified in aqueous PVA (2 ml, 1 wt.%, EA saturated) in a 25 mL screw cap vial by vortexing (2500 rpm, 30 s). The emulsions were transferred in stirred (800 rpm, 3 h) aq. PVA (20 ml, 0.5 wt.%) for solvent evaporation, followed by MP washing with water on 0.45 µm nylon filters and lyophilization. MP resuspended in aq. PVA (1 wt.%) were exposed to laser irradiation at r.t. (308 nm, 26 mW·cm$^{2}$, distance of 4 cm, 2x 10 min) between silanized quartz glass plates ($4 \cdot 6.5 \cdot 0.1$ cm$^{3}$) separated by 1 mm teflon spacers. The following terminology was applied: N-(x)-$^{y}$oCL-IEMA; where "N" indicates UV-irradiation for network formation, "x" determines 0 or 3 wt.% photoinitiator, "y": 2 = linear, 4 = star-shaped precursor structure.

The dry samples were analyzed by ATR-FTIR (Nicolet 6700, Thermo Scientific, Germany). After redispersion in water or DCM (swelling), light microscopy (Axio Imager.A1m, Carl Zeiss Microimaging, Göttingen, Germany) was used to study the MP structure. Additionally, scanning electron microscopy (SEM; Gemini Supra™ 40 VP, Carl Zeiss NTS, Oberkochen, Germany) at 1 kV without sputtering was applied. Average MP diameters from volume weighted distributions were determined by static light scattering (SLS; Mastersizer 2000, Malvern, Herrenberg, Germany) in solid state in water ($d_{aq}$) or in the DCM-swollen state ($d_{DCM}$). The swelling ratio of MP $Q_{MP}$ was determined by $Q_{MP} = (d_{DCM}^{3} \cdot d_{aq}^{-3}) \cdot 100\%$.

Differential scanning calorimetry (DSC; DSC 204 F1; Netzsch, Selb,Germany) was performed in $N_2$ atmosphere from -100 to 150 °C with a heating and cooling rate of 10 K·min$^{-1}$. Wide-Angle X-ray Scattering (WAXS) pattern were accumulated for 300 s on a Bruker D8 Discover X-ray diffractometer, with determination of the degree of crystallinity (DOC) by Pearson VII function fitting and of the crystallite size ($l_c$) by the Sherrer equation ($k = 0.9$). Small-Angle X-Ray scattering (SAXS) analysis was performed for 1 h on a Bruker Nanostar diffractometer and the long periode $L = s^{-1}$ was calculated from Kratky-plots after integration over 360°.

## RESULTS

The synthesis of crosslinked MP using preformed solid MP from linear or star-shaped oCL precursors was achieved by photopolymerization of their IEMA end groups, which was confirmed by disappearance of $=CH_2$ out-of-plane vibration signal of methacrylate groups at 815 $cm^{-1}$ in FTIR analysis (data not shown). Furthermore, the crosslinking to non-soluble materials was investigated by light microscopy (Figure 1), which showed spherical MP shapes with in some cases irregularities and MP debris. While the gel content of individual MP could not be determined, microscopic swelling experiments in DCM illustrated increasing sizes without matrix dissolution as another indicator of successful crosslinking. In the swollen state, patterned, less spherical structures were observed. The MP size, swelling behaviour, and the semi-crystalline morphology were quantitatively analyzed by SLS, DSC, WAXS, and SAXS as summarized in Table I. The addition of photoinitiator (N-(3) vs. N-(0) samples) led to higher swellability as shown by higher $Q_{MP}$. Additionally, a reduced $T_m$, melting enthalpy, DOC, and crystallite size were observed, while the long period increased upon using photoinitiator for both, MPs from linear and four arm star-shaped precursors.

It should be noted, that each individual oCL segment is shorter for $^4$oCL compared to $^2$oCL for the used precursors of similar $M_n$, resulting in lower $T_m$, DOC, and $l_c$ for $^4$oCL materials For discussion of crosslinking effects on thermal properties such as $T_m$ and $\Delta H_m$, attention should be payed to the width of the DSC melting peaks, which showed a broadening especially after crosslinking of $^4$oCL-IEMA (Figure 2).

Finally, the surface structure of dry MPs was analyzed by SEM, which showed smooth MP surfaces for MPs from non-crosslinkable $^2$oCL-diol (Figure 3A-B), which were UV irradiated as all other samples. Crosslinked MP from linear and star-shaped precursors changed from smooth to slightly wrinkled surfaces only when photoinitiator was supplemented (Figure 3C-F). After swelling and deswelling the crosslinked MPs in DCM, all samples exhibited rough surface patterns and deformed shapes (Figure 3G-J). In case of crosslinked MP matrices of linear $^2$oCL, the changes to surface morphology appear stronger with some precipitated material detected around the MPs.

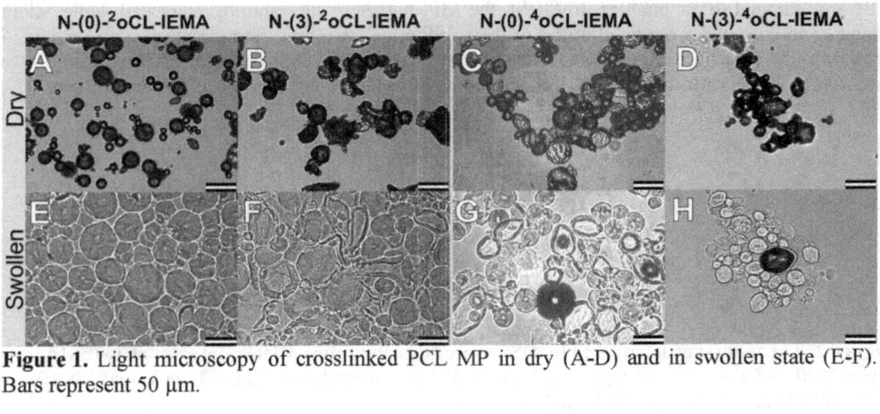


**Figure 1.** Light microscopy of crosslinked PCL MP in dry (A-D) and in swollen state (E-F). Bars represent 50 µm.

**Table I.** Precursor and MP properties.

| MP sample ID | SLS [c] | | | DSC | | WAXS | | SAXS |
|---|---|---|---|---|---|---|---|---|
| | $d_{aq}$ [a] (µm) | $d_{DCM}$ [b] (µm) | $Q_{MP}$ (%) | $T_m$ (°C) | $\Delta H_m$ (J/g) | DOC (%) | $l_c$ (nm) | $L$ (nm) |
| $^2$oCL-IEMA | n.a. | n.a. | n.a. | 56 | 74 | n.d. | n.d. | n.d. |
| N-(0)-$^2$oCL-IEMA | 34 | 49 | 300 | 53 | 67 | 49.1±1.6 | 18.6 | 13.3 |
| N-(3)-$^2$oCL-IEMA | 39 | 60 | 360 | 51 | 59 | 37.7±2.2 | 16.8 | 15.1 |
| $^4$oCL-IEMA | n.a. | n.a. | n.a. | 48 | 55 | n.d. | n.d. | n.d. |
| N-(0)-$^4$oCL-IEMA | 37 | 57 | 370 | 45 | 44 | 37.1±1.6 | 17.2 | 14.5 |
| N-(3)-$^4$oCL-IEMA | 41 | 69 | 480 | 37 | 38 | 29.9±2.1 | 13.5 | 14.9 |

Volume weighted mean diameter determined in $H_2O$ [a]. and in DCM [b]. [c] Uniformity was between 0.36 and 0.78. $Q_{MP}$ = swelling ratio, $T_m$ = melting temperature, $\Delta H_m$ = melting enthalpy, DOC = degree of crystallinity, $l_c$ = crystallite size, $L$ = long period (distance between crystallites). n.a. = not applicable. n.d. = not determined

## DISCUSSION

Crosslinking of MPs from functionalized precursors, as confirmed by FTIR and swelling studies (Figure 1 and 3), would not necessarily be expected considering the slow chain mobility in semi-crystalline materials and the laser light scattering induced by both the polymer crystallites and the numerous interfaces of materials with different refractive indices in the aqueous MP suspension. However, suitable photo-initiation apparently took place and MP with and without photo-initiators converted into matrices that no longer could be dissolved by organic solvents. The extent of precursor incorporation in the crosslinked material as determined for macroscopic samples by the gel content is difficult to measure for single MPs. However, indirect conclusions could be drawn by microscopy-based swelling experiments, where at the resolution of light microscopy no extracted and precipitated material was detected. It should be noted that besides the large number of spherical solid MPs, some ruptured and hollow MPs could be observed possibly due to the sensitive two-step process with intermediate freezing and lyophilization of brittle MPs from the oligomeric precursor. Dark spots in light microscopy images of swollen MPs are assigned to air bubbles. In DCM, the MP diameters substantially increased while MP remained mobile and were not limited in their individual swelling. Some pattern with straight lines from the particle center to its surface become repeatedly visible, what might be caused by extensive swelling of former crystallized structures in an outer, less flexible, crosslinked frame.

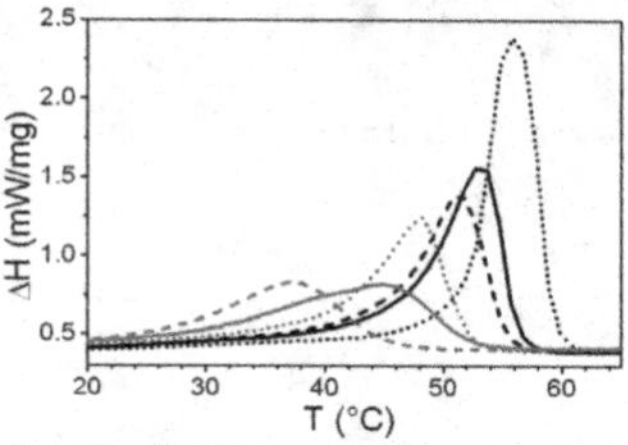


**Figure 2.** Thermograms determined by DSC (second heating run). Black: doted = $^2$oCL-IEMA, line = N-(0)-$^2$oCL-IEMA, dashed = N-(3)-$^2$oCL-IEMA; Grey: doted = $^4$oCL-IEMA, line = N-(0)-$^4$oCL-IEMA, dashed = N-(3)-$^4$oCL-IEMA.

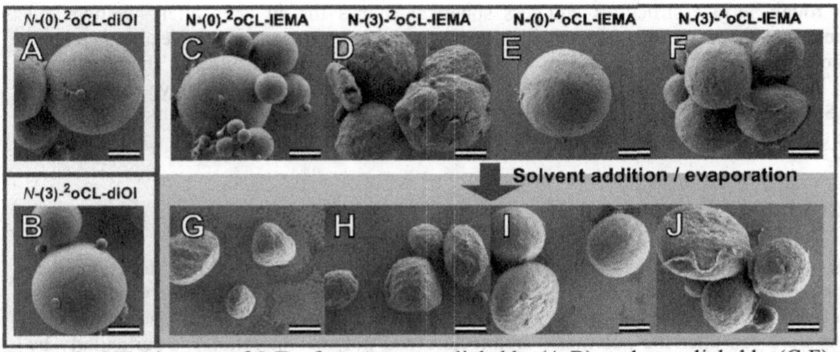


**Figure 3.** SEM-images of MPs from non-crosslinkable (A-B) and crosslinkable (C-F) oCL-precursors after UV-treatment. [G-J] show MPs after swelling and deswelling in DCM. Bars = 10 μm.

When comparing the thermal properties of MPs derived from linear and star-shaped precursor (Figure 2), it is obvious that the shorter individual segment length of $^4$oCL translated into diminished crystallization. Accordingly, also the polymer morphology of the crosslinked $^4$oCL-derived particles was characterized by a lower $T_m$, DOC, $l_c$ and a higher $L$. The higher swellability and lower crosslinking density of $^4$oCL MP, despite their higher initial number of reactive methacrylate groups, might be assigned to sterical restrictions in finding binding partners in the low-mobility solid state thus causing more reacted dangling chain ends or increased intramolecular reactions.

For photo-initiator containing MPs, i) more radicals should be present to start and terminate radical reaction, leading to lower netpoint functionality and lower crosslink density by formation of shorter polymethacrylate chains, and ii) elevated local temperatures might occur when the reaction heat is not fully dissipating to the aqueous environment, thus possibly leading to partial melting of particles despite external cooling. Increased swellability was shown for photo-initiator based samples, which suggested lower crosslink density, less restrictions in chain orientation, effective crystallization, and a high $T_m$ and DOC. Contradictory, the opposite effect on $T_m$ and DOC were observed (Table I). A decrease of DOC and $T_m$ might originate from entrapment of photoinitiator in the densely packed amorphous phase.

Additional to information gained from the peak value of the $T_m$, the width of melting transitions indicates the uniformity of crystallite sizes. The extensive peak broadening as determined for $^4$oCL-derived MP with and without initiator indicated differences in crystallization restrictions throughout the polymer network, possibly because longer polymethacrylate chains (higher functionality of netpoints) were formed during the initial phase of crosslinking than compared to later stages with a further restricted chain mobility.

The SEM analysis of MPs with non-crosslinkable OH-terminated oCL visualized that the smooth surface structure of MPs did not change upon UV-irradiation in presence or absence of the photo-initiator (Figure 3A-B). Topographical differences upon crosslinking of methacrylate-functionalized precursors occurred only when photoinitiator was supplemented with rough surfaces being formed (Figure 3 D,E), possibly caused by volatile reaction products or shrinkage of the methacrylate-rich amorphous phase. However, compared to light microscopy, SEM revealed rough surface structures and less spherical shapes for all MPs after swelling and deswelling. It is appearant that the formed networks are inhomogeneous on a molecular level and it was considered that the presence of crystallites during crosslinking and/or an insufficient covalent incorporation of all precursor molecules (reduced gel content) might have caused these effects. Still, at least for $^4$oCL-derived MP, topographical changes and precipitation of soluble network content around the MPs remained small after DCM treatment.

## CONCLUSIONS

It was shown in this study that photocrosslinking of methacrylate functionalized precursors can be performed in the non-molten, solvent-free state for solid semi-crystalline oCL microparticles above their $T_g$. Switching from linear to star-shaped precursors and addition of photo-initiators can alter the polymeric architecture and polymer morphology, particularly by reducing the extent of crosslinking as apparent by quantitative swelling data and the capability for crystallization. Furthermore, despite generally successful, MP crosslinking in the solid state may be associated to inhomogeneities of crosslink density throughout the MP matrix due to different local light intensities during irradiation. Higher chain mobility during crosslinking or photoinitiators might contribute to improve network homogeneity if combined with a suitable precursor architecture such as star-shaped, IEMA functionalized $^4$oCL.

## ACKNOWLEDGMENTS

Funding of F. Friess by the DFG (Grant No. WI 3637/1-1) and performance of WAXS and SAXS measurements by Dr. Ulrich Nöchel are gratefully acknowledged.

## REFERENCES

1. T.K. Dash and V.B. Konkimalla, *Journal of Controlled Release*, **158**, 15 (2012).
2. A.T. Neffe, G. Tronci, A. Alteheld, and A. Lendlein, *Macromolecular Chemistry and Physics*, **211**, 182 (2010).
3. F. Friess, C. Wischke, M. Behl, and A. Lendlein, *Journal of applied biomaterials & functional materials*, **10**, 273 (2013).
4. C. Vaida, P. Mela, K. Kunna, K. Sternberg, H. Keul, and M. Möller, *Macromolecular Bioscience*, **10**, 925.
5. C.-H. Choi, J.-H. Jung, T.-S. Hwang, and C.-S. Lee, *Macromolecular Research*, **17**, 163 (2009).
6. S. Schachschal, H.-J. Adler, A. Pich, S. Wetzel, A. Matura, and K.-H. Pee, *Colloid and Polymer Science*, **289**, 693.
7. L. Huang, Y. Li, J. Yang, Z. Zeng, and Y. Chen, *Polymer*, **50**, 4325 (2009).
8. T. Scherzer and U. Decker, *Radiation Physics and Chemistry*, **55**, 615 (1999).
9. C. Dietlin, J. Lalevee, X. Allonas, J.P. Fouassier, M. Visconti, G.L. Bassi, and G. Norcini, *Journal of Applied Polymer Science*, **107**, 246 (2008).
10. N.K. Uttamchand, K. Kratz, M. Behl, and A. Lendlein, *Mater Res Soc Symp Proc*, **1190**, 55 (2009).

Mater. Res. Soc. Symp. Proc. Vol. 1569 © 2013 Materials Research Society
DOI: 10.1557/opl.2013.816

# Correlating the Uptake and Dendritic Cell Activation by MDP-loaded Microparticles

Stefanie Schmidt, Toralf Roch, Simi Mathew, Nan Ma, Christian Wischke, and Andreas Lendlein
*Institute of Biomaterial Science and Berlin-Brandenburg Center for Regenerative Therapies, Helmholtz-Zentrum Geesthacht, Kantstraße 55, 14513 Teltow, Germany*

## ABSTRACT

Polymer-based, degradable microparticles (MP) are attractive delivery vehicles for vaccines as the polymer properties can be specifically tailored and the carrier can be loaded with adjuvant. For all newly developed carrier systems it is important to analyze cellular uptake efficiency and the specific effects mediated by the encapsulated agent when phagocytosed by the cells, which is barely reported so far. By the encapsulation of N-acetylmuramyl-L-alanyl-D-isoglutamine (MDP) labeled with fluoresceinisothiocyanat (FITC) in poly[(*rac*-lactide)-*co*-glycolide] (PLGA) MP, the MP was fluorescent and used to visualize the phagocytic uptake. Since encapsulated MDP can activate dendritic cells (DC) via the cytosolic nucleotide-binding oligomerization domain receptors (NOD), it can be investigated whether only cells that have phagocytosed the MP are activated or whether bystander effects occur, resulting in activation of cells, which did not take up MDP-FITC loaded MP. Here, it is demonstrated that increasing MP concentrations in the culture medium had no impact on the viability of DC and that the MP uptake efficiency was dose dependent. Interestingly, it could be shown by the CD86 expression, that only DC, which had engulfed MP, were significantly stronger activated than DC, which had not phagocytosed MDP-FITC loaded MP. On the one hand these results indicate that sufficient amounts of MDP were released from the PLGA carriers into the cytosol of the DC. On the other hand, based on the correlation of uptake and activation on the single cell level, minimal MP induced bystander effects may be expected for *in vivo* applications.

## INTRODUCTION

Even with high hygienic standards, infections for example with *Vibrio cholerae, Mycobacterium tuberculosis* or with influenza viruses are still challenging in their treatment and have a big impact on health economics. Effective vaccine strategies could help to improve the protection against these pathogens. Carrier systems are evaluated for vaccine delivery combining different advantages, such as protection of encapsulated antigens or adjuvants from microenvironment, which could lead to instability, degradation or unwanted *in vivo* effects. In this regard, polymeric PLGA MP are of interest as the degradation rate and hydrophobicity can be influenced by the lactide/glycolide ratio. PLGA has been shown to be an immunologically inert carrier as it was not activating different immune cells [1, 2]. To achieve a long-term protective immunological memory by vaccination, it is of high importance that the antigen/adjuvants are delivered to professional antigen-presenting cells like DC, efficiently processed and subsequently presented to T cells to initiate adaptive immune responses. Depending on the adjuvants used for vaccination, the immune reaction can be modulated and distinct pathways can be triggered. Ligands of cytosolic NOD receptors, such as MDP, have the capacity to activate DC [3]. As the NOD-receptors are expressed in the cytosol rather than on the cell surface, efficient MP uptake by antigen presenting cells (APC) is mandatory. Thus, it is of high interest to analyze the influence of MP concentrations on the cellular uptake and its direct effect on cell activation. While previous studies tried to correlate overall uptake and overall immune activation in separate experiments [4], the MP system described here, allows to specifically analyze the activation status of DC, which had taken up MP compared to DC, which had not. We hypothesize, that the uptake efficiency of DC increases with MP concentration leading to specific activation only of DC, which had engulfed the MP, while DC that had not phagocytosed the MP remained in their immature state. Therefore, MP loaded with FITC labeled MDP were used to track MP uptake by DC. Furthermore cell viability and cell activation were evaluated by surface marker expression on single cell level and correlated with MP uptake. Additionally the cytokines released by the DC after incubation with the MP were determined.

## EXPERIMENT

### Preparation of microparticles and size determination

MP were prepared by water-in-oil-in-water double emulsion/solvent evaporation method as reported before [2, 5]. Briefly, 75 µl of 0.7 % (w/v) MDP-FITC (Invivogen, San Diego, USA) in water were added to a 19 wt. % PLGA solution in dichloromethane and emulsified by sonication (PLGA: 50 mol. % glycolide, carboxyl end groups, number average molecular weight $M_n$ = 5 kDa and polydispersity PD = 3.2 as determined by multidetector GPC; Resomer® RG 503H, Boehringer Ingelheim, Germany). The water-in-oil emulsion was then added to 2 % (w/v) poly(vinyl alcohol) (PVA) solution (Mowiol 4-88; Kuraray Europe GmbH, Frankfurt, Germany) and homogenized by water-water homogenization (24.000 rpm, Ultra Turrax, IKA, Germany). After solvent evaporation, washing, and lyophilization, the volume-weighted size distribution was determined by laser diffraction (Mastersizer 2000, Malvern, Herrenberg, Germany). The obtained MP size distribution was d (0.1) = 1.2 µm, d (0.5) = 4.2 µm, d (0.9) = 8.5 µm.

### Monocyte isolation, dendritic cell culture, assessment of cell viability, MP uptake and activation status of DC

Peripheral blood mononuclear cells were purified from buffy coats of individual healthy donors by density gradient centrifugation using Biocoll™ (Biochrom, Berlin, Germany). To obtain CD14 positive monocytes, CD14 negative cells were depleted, using the Monocyte Isolation Kit II (Miltenyi Biotec, Bergisch Gladbach, Germany) as previously described [2]. Routinely, more than 94 % of the purified cells were CD14 positive, determined by flow cytometry using CD14-PC7 (Beckman-Coulter, Krefeld, Germany), CD19-FITC, and CD3-PE (Miltenyi Biotec) antibodies. To induce differentiation into DC, isolated monocytes ($2 \cdot 10^6$) were cultured for six days in RPMI medium (Biochrom) supplemented with 10 vol % FBS, IL-4 (Miltenyi Biotec), and GM-CSF (R&D systems, Wiesbaden, Germany) (P92006, Biochrom). The differentiation was confirmed by the expression of CD209 (DCsign) as a DC marker.

For MP uptake studies, harvested DC were seeded at $1 \cdot 10^6$ cells per 1 mL in 24 well plates and incubated for 24 hours with increasing MP concentrations varying from 1.56 $\mu g \cdot mL^{-1}$ to 800 $\mu g \cdot mL^{-1}$. As controls untreated DC or DC treated with 1 $\mu g \cdot mL^{-1}$ LPS (*E.coli* 0111:B4, Axxora, Lörrach, Germany) were used. After cultivation, supernatants were collected and stored at -20°C until further analysis. Cells were harvested and labeled for flow cytometry analysis (MACSQuant flow cytometer, Miltenyi Biotec). Frequencies of dead cells were determined by identification of 4′,6-diamidino-2-phenylindole (DAPI) (Roth, Karlsruhe, Germany) positive cells. Frequencies of DC, which had taken up the MP were identified as $CD209^+/FITC^+$ double positive cells (anti-CD209-APC, Miltenyi Biotech). The DC activation status was determined by the mean fluorescence intensity (MFI) of surface marker expression of CD86 (anti-CD86-PC7,

BD Pharmingen, Heidelberg, Germany). The flow cytometric data were analyzed using FlowJo (Tree Star, Inc, Ashland, Oregon, USA). Cytokine secretion was assessed by Bio-Plex® assay (Bio-Rad®, München, Germany).

**Statistics**

Data were statistically analyzed by GraphPad Prism (La Jolla, CA, USA) and the mean and standard error of the mean (SEM) were calculated for all samples. A two-way analysis of variance (ANOVA) was used to determine whether differences in CD86 expression between DC, which had taken up MP and DC, which had not, are significant.

## RESULTS

**Flow cytometric analysis of DC – gating strategy**

To discriminate cells, which phagocytosed the MP from cells, which did not and to compare their activation status, the following gating strategy was applied (Figure 1). Cells were discriminated from debris by FSC/SSC parameters, followed by aggregate exclusion and identification of viable DC (Figure 1 A-C). DC, which had engulfed MDP-FITC MP were identified according to untreated DC by higher FITC signals compared to cells, which had not taken up MP (Figure 1D). The expression of CD86 as an activation marker was analyzed and compared between the two subsets and a representative histogram is shown in Figure 1E.

This gating strategy allows on single cell level the comparison between DC, which had taken up the MPs and which had not, regarding their activation status.

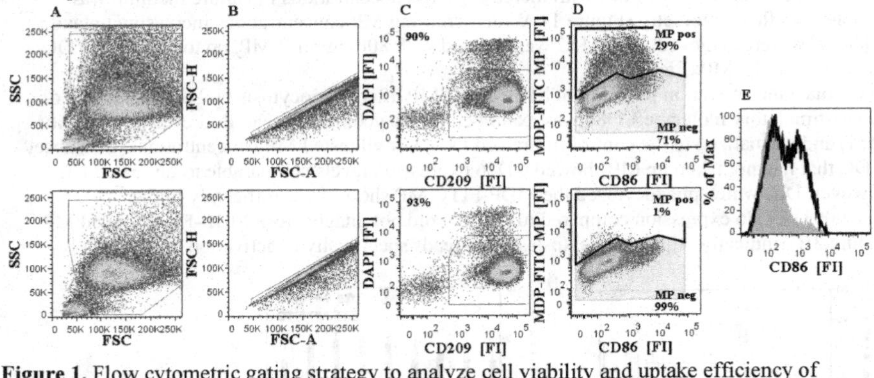


**Figure 1.** Flow cytometric gating strategy to analyze cell viability and uptake efficiency of MDP-FITC MP by DC. The top row demonstrates DC incubated with 100 $\mu g \cdot mL^{-1}$ MP. DC incubated without MP are shown in the lower row. **(A)** Scatter plot discriminating cells from debris. **(B)** Doublet exclusion by FSC-A versus FSC-H. **(C)** CD209 versus DAPI was used to determine viable DCs. **(D)** Discrimination of DC, which phagocytosed the MDP-FITC MP (MP pos.), thus being FITC positive, and FITC negative DC, which did not phagocytose MP (MP neg). **(E)** Histogram of CD86 expression of FITC positive DC (black line) compared to FITC negative DC (filled grey). FI indicates for fluorescence intensity.

**DC viability**

The DC viability depending on increasing MP concentrations varying from 1.56 μg·mL$^{-1}$ up to 800 μg·mL$^{-1}$ was evaluated by DAPI exclusion (Figure 1C). The cell survival of the DC was not influenced by the MP with concentrations up to 400 μg·mL$^{-1}$ and was comparable to untreated DC. The highest concentration of 800 μg·mL$^{-1}$ showed decreased cell viability, which was comparable to LPS-activated DC.

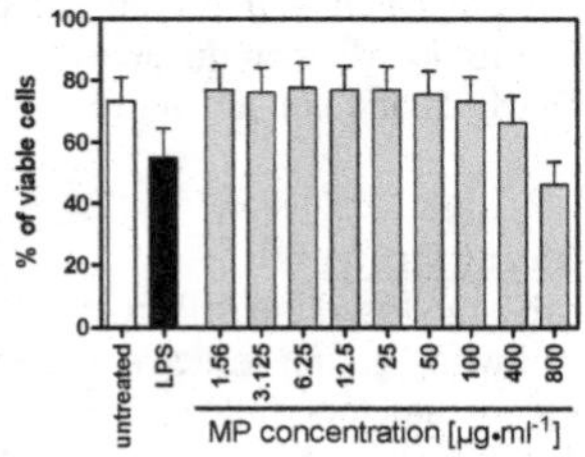


**Figure 2.** Cell viability of DC depending on increasing MP concentrations. Average of three different experiments with in total seven individual donors is shown (mean ± SEM).

**Uptake efficiency and activation status of DC**

The uptake efficiency of DC with increasing MP concentrations in culture medium was evaluated by flow cytometry (Figure 1). With increasing MP concentrations increasing uptake efficiencies were achieved. When DC were treated with 800 μg·mL$^{-1}$ MP, up to 45% of the DC phagocytosed the MP (Figure 3A).

To evaluate the activation potential of MDP-FITC MP after phagocytosis by DC, expression of the co-stimulatory molecule CD86 was examined by the geometric mean fluorescence intensity (MFI) and normalized to the control (untreated DC). For all tested MP concentrations, the subset of DC that had not taken up MP showed CD86 expression levels comparable to untreated DC. However, DC, which phagocytosed the MDP-FITC MP showed a significantly higher ($p<0.0001$) CD86 expression compared to DC that did not phagocytose MDP-FITC loaded MP (Figure 3B), indicating that phagocytosis of MP leads specifically to activation of DC.

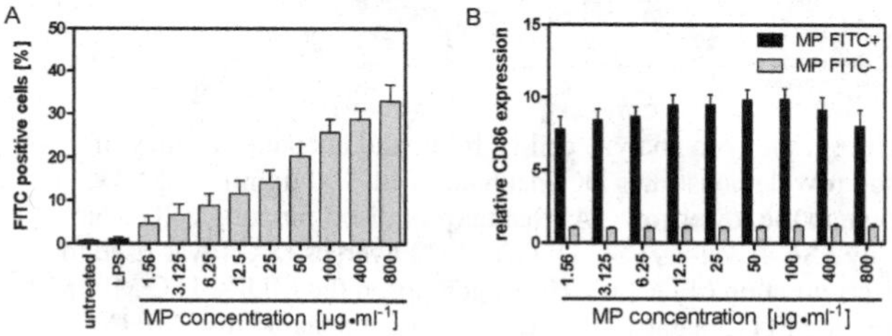


**Figure 3.** Uptake of MDP-FITC MP and subsequent DC activation. **(A)** Frequency of MDP-FITC positive DC was identified according to the gating shown in Figure 1D. **(B)** Relative CD86 expression of DC, which had taken up MP (black bars) compared to DC, which had not

taken up MP (grey bars). Gating strategy was applied according to Figure 1E. Average of three different experiments with in total seven individual donors is shown (mean ± SEM).

**Cytokine expression of DC in response to increasing MP concentrations**

T cell response can be directed by the secretion of cytokines produced by DC. Thus, analyzing cytokine secretion after MP uptake is of high interest when improving vaccine strategies. Here, the cytokine secretion of DC in response to increasing concentrations of MP with MDP as NOD receptor agonist was investigated. The cell culture supernatants were collected after 24 hours incubation with MP and pro-inflammatory cytokines were analyzed using a multiplex suspension array. The secretion of the pro-inflammatory cytokines IL-1β, IL-6, IFN-γ, and TNF-α was observed when DCs were stimulated with LPS. When DC were incubated with different concentrations of MP, a slight increase of IFN-γ and TNF-α secretion could be observed at higher MP concentrations (Figure 4). The IL-1β and IL-6 secretion was comparable to the control for all MP concentrations (data not shown).

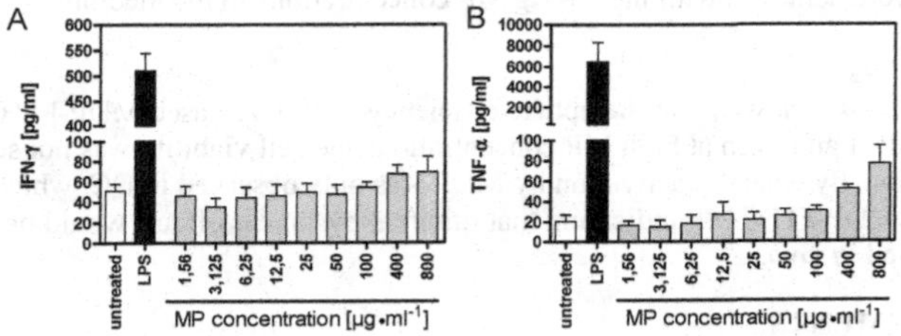


**Figure 4.** Cytokine secretion by DC after cultivation with increasing MDP-FITC MP concentrations. DC stimulated with 1 $\mu g{\cdot}mL^{-1}$ LPS (black bar) served as a positive control. Cytokine concentrations of **(A)** IFN-γ and **(B)** TNF-α were quantified using the multiplex suspension array.

## DISCUSSION

The fluorescent MDP-FITC MP were used to track DC, which engulfed the MP determining the uptake efficiency. We showed that the frequency of DC, which had taken up the MP increased in a concentration dependent manner. With the aim to use these MP for antigen delivery to antigen-presenting cells, it is important that cell viability is not influenced by increasing MP concentrations in the cell culture. The viability testings performed in this study showed that DC survival was not influenced by MDP-FITC MP from PLGA, with exception of the highest concentration of 800 $\mu g{\cdot}mL^{-1}$. At this concentration, cell toxic levels might either be attributed to polymer or to a response towards the MDP payload leading to reduced DC survival. Since cell viability rates were comparable to LPS-treated DC, physiological apoptosis due to DC activation could be an explanation. It is described that LPS-activated DC undergo apoptosis to maintain self-tolerance preventing from autoimmune diseases [6].

A specific activation of DC could be shown by surface expression of the co-stimulatory molecule CD86, which was only observed in the DC that had engulfed the MDP-FITC loaded MP. Although the activating effect of MP loaded with MDP on DC was already demonstrated [2], here, by using encapsulated FITC labeled MDP, we were able to track the particles on single cell basis and could show, that only those DC, which had phagocytosed the MP, were activated specifically, whereas DC, which did not phagocytose the MP remained immature. These data indicated the importance that only a clear discrimination of cells that phagocytosed MP from cells, which did not, allows a comprehensive interpretation of MP uptake studies.

High concentrations of pro-inflammatory cytokines were not observed. However, slight dose dependent increase of IFN-$\gamma$ and TNF-$\alpha$ was detected. Considering that the amount of cytokines detectable in the supernatant is dependent on the amount of activated cells, which were at maximum 45% of all DC, the low cytokine levels may not be surprising. Stronger activation of cells indicated by increasing secretion of pro-inflammatory cytokines could possibly be induced by enhanced particle uptake. Here, we showed that increasing MP concentration leads to higher uptake efficiency. However, since the surface charge and hydrophobicity of MP is described to influence cellular uptake, higher uptake efficiency could be achieved by varying these parameters by altering the polymer composition [7].

Since the MDP only acts on NOD receptors, which are located intracellularly, phagocytosis and high uptake efficiencies of the MDP-FITC MPs are mandatory in order to activate DC and were achieved with increasing MP concentrations in the medium.

## CONCLUSION

In this study, we showed that the uptake efficiency of PLGA-based MDP-FITC MP by DC is dose dependent and even at high MP concentrations the cell viability was not substantially influenced. Additionally, specific activation by MDP was only observed in DC, which have phagocytosed the MDP-FITC MP indicating that minimal bystander effects would occur when such MP are applied *in vivo*.

## ACKNOWLEDGMENTS

We thank Anja Müller-Heyn for the technical support and the Bundesministerium für Bildung und Forschung (BMBF) for partial funding within the project 0315696A „Poly4bio BB“.

## REFERENCES

1. S. Mathew, T. Roch, M. Frentsch, A. Lendlein, and C. Wischke, *Journal of Applied Biomaterials & Functional Materials*, **10**, 229 (2013).
2. C. Wischke, S. Mathew, T. Roch, M. Frentsch, and A. Lendlein, *J Control Release*, **164**, 299 (2012).
3. I. Tattoli, L.H. Travassos, L.A. Carneiro, J.G. Magalhaes, and S.E. Girardin, *Seminars in immunopathology*, **29**, 289 (2007).
4. A.M. Hafner, B. Corthesy, M. Textor, and H.P. Merkle, *Biomaterials*, **32**, 2651 (2011).
5. S. Mathew, A. Lendlein, and C. Wischke, *Macromolecular Symposia*, **309-310**, 123 (2011).
6. I. Zanoni, R. Ostuni, G. Capuano, M. Collini, M. Caccia, A.E. Ronchi, M. Rocchetti, F. Mingozzi, M. Foti, G. Chirico, B. Costa, A. Zaza, P. Ricciardi-Castagnoli, and F. Granucci, *Nature*, **460**, 264 (2009).
7. S. Lorenz, C.P. Hauser, B. Autenrieth, C.K. Weiss, K. Landfester, and V. Mailänder, *Macromolecular Bioscience*, **10**, 1034 (2010).

**Mater. Res. Soc. Symp. Proc. Vol. 1569 © 2013 Materials Research Society**
**DOI: 10.1557/opl.2013.798**

# Polycaprolactone/α-Alumina and Hydroxyapatite-based Micro- and Nano- Structured Hybrid Fibers

Simón Y. Reyes-Lopéz[1], Bonifacio Alvarado-Tenorio[1*], Ángel Romo-Uribe[2]

[1]Departamento de Ciencias Químico-Biológicas, Instituto de Ciencias Biomédicas, Universidad Autónoma de Ciudad Juárez, Chihuahua 32310, MEXICO.
[2]Lab. de Nanopolímeros y Coloides, Instituto de Ciencias Físicas, Universidad Nacional Autónoma de México, Cuernavaca Mor. 62210, MEXICO.
* To whom correspondence should be addressed: bonifacio.alvarado@uacj.mx

## ABSTRACT

The micro- and nanostructured phase behavior of non-woven hybrid electrospun fibers of Polycaprolactone (PCL-f), Polycaprolactone/α-Alumina (PCL/AA) and Polycaprolactone/Hydroxyapatite (PCL/HA) were elucidated. Nanoparticles of Hydroxyapatite featured 7-20 nm while α-Alumina was 500 nm in average size. Analysis by Transmission Electron Microscopy (TEM), Small-Angle Light Scattering (SALS), Polarized Optical Microscopy (POM) and Differential Scanning Calorimetry (DSC) enabled us to suggest a simple structural model for the series of hybrid fibers. PCL with a molecular weight of 70,000 g/mol was dissolved in chloroform. POM exhibited qualitative birefringent regions for all samples at the microscale level. Furthermore, polarized light distinguished between amorphous and micro-ordered structures along the fibers. On the other hand, SALS patterns suggested a needle-like morphology for all fibers. The influence of both, Hydroxyapatite and α-Alumina, was apparent in the SALS patterns of the PCL semicrystalline phase. Thermal analysis by DSC showed that the crystalline phase of PCL was disrupted by the presence of the inorganic nanoparticles. TEM micrographs showed that Hydroxyapatite and α-Alumina nanoparticles were embedded along the structure of PCL microfibers.

## INTRODUCTION

One of the primary challenges for reconstruction or regeneration of bone tissue by bone scaffoldings materials is the basic understanding of the structure-property relationship of those biomaterials. Design and fabrication of functional biodegradable scaffolds is a complex task. Recently, Woodruff et al [1] summarized that morphology and other properties (physico-chemical, degradation kinetics, surface) as pore size and pore interconnectivity of the bone scaffoldings materials plays an important role for cell/tissue growth and maturation. Megelski et al [2] found a correlation ship between the surface area of non biodegradable polymeric fibers and the solvent used in the electrospinning process. The results from optical microscopy, transmission electron microscopy and atomic force microscopy showed that different polymer architectures (semicrystalline and amorphous) can yield a variety of micro and nanopores on electrospun fiber surfaces. On the other hand, few reports are related to micro and nanostructure analysis for biodegradable spun fibers. Particularly, some researchers have focused on nanofibers study of poly caprolactone (PCL), poly lactic acid (PLA), polyhydroxy alkanoates (PHA), poly vinyl alcohol (PVA), poly ethylene oxide (PEO), and poly urethane (PU) [3-6]. Nanofibers electrospun from PCL with calcium carbonate and PLA with Hydroxyapatite (HA) also have demonstrated recently [7, 8]. Alumina is another known tolerable material with minimum tissue reaction after implantation [9, 10]. In this context, we propose in the present research the preparation of Polycaprolactone/Hydroxyapatite (PCL/HA), Polycaprolactone/α-Alumina (PCL/AA) and fibers of polycaprolactone (PCL-f) for their potential medical applications. Hydroxyapatite and α-Alumina nanoparticles were used for the preparation of the hybrid fibers, respectively. Micro- and nanostructure phases were elucidated into these electrospun fibers by POM, SALS and TEM techniques.

## EXPERIMENTAL

PCL was purchased from Aldrich (Milwaukee, USA). The number average molecule weight ($\overline{M}_n$) of PCL was 70,000. PCL solutions were prepared with concentration of ranging from 10 to 15 wt % in chloroform. Hybrid fibers of PCL/AA and PCL/HA were prepared by varying the voltage in an electrospinning process. Hydroxyapatite and α-Alumina nanoparticles were dispersed in the solutions. Biodegradable PCL and nanoparticles solutions were mixed and then electrospun at room temperature. Decomposition temperatures of the mats were examined by Thermogravimetric Analysis (TGA) using a TGA-DTA-DSC analyzer SDT2960 model (manufactured by TA Instruments). All fibers were carried out at 5°C/min. Polarized Optical Microscopy was performed by using a Leitz Laborlux 12. The polarization direction of the incident beam was set by a polarizer (P) with the polarization axis set vertically. The 0°-90° orientation of the polarizer and analyzer sets the so-called $H_V$ polarization condition. The thermal behavior were characterized by differential scanning calorimetry (DSC) using the DSC6000® DSC (Perkin Elmer Inc., CT, USA). 10 mg were loaded into standard aluminum pans. The heating and cooling scans were carried out at 10 °C/min under nitrogen atmosphere. Small Angle Light Scattering patterns were obtained for all fibers. An in-house equipment and He-Ne Laser was used [11]. Morphology of electrospun PCL nonwoven mats was studied by transmission electron microscopy TEM (Phillips TECNAI F20 super Twin, at 200 kV) and SEM (JEOL JSM-6400 SEM, at 15 kV coupled with EDS detector Bruker AXS Inc. XFlash 4010).

## RESULTS and DISCUSSION

Figure 1 exhibit the plot of weight loss as a function of temperature for the fibers. The onset of decomposition temperatures for the hybrid fibers PCL/HA (trace 1-i) and PCL/AA (trace 1-ii) were found to be 308°C. On the other hand PCL fiber showed (trace 1-iii) an onset temperature at 332°C. According to the first derivative of weight loss respect to the temperature (not shown in figure 1), the major weight loss region for the hybrid fibers PCL/HA and PCL/AA (60 wt% in both cases) occurred around 382°C, and was due to the degradation of PCL phase. PCL fiber showed its major weight loss region at 392°C and 67 wt%. The next peaks of the first derivative of weight loss for PCL/AA, PCL/HA and PCL-f fibers occurred at 466°C, 458°C and 425°C respectively, and are attributed to the cleavage backbone of PCL phase. The total weight loss for PCL/AA (99.98 wt%), PCL/HA (92.61 wt%) and PCL-f (99.99%) at 600°C indicates that hydroxyapatite nanoparticles embedded into the PCL/HA fibers significantly improve the thermal stability of the PCL phase.

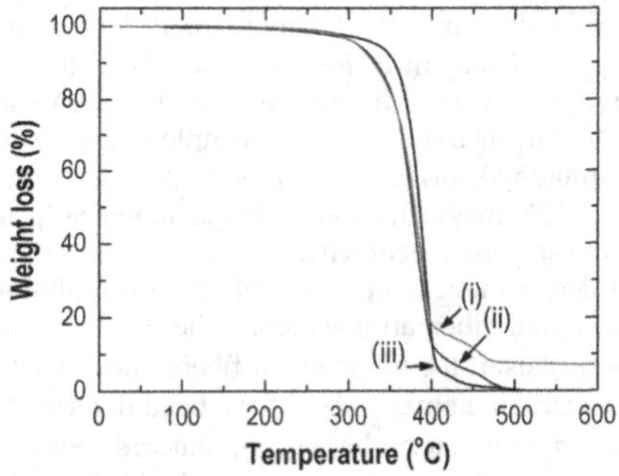


Figure 1. Traces of weight loss in function of temperature for (i) PCL/HA, (ii) PCL/AA and (iii) PCL-f mats.

All Hybrid fibers for POM characterization had a thickness of few tens of microns (30- 40 μm). On the other hand, porosity will be analyzed in our future research to check its use in biomedical applications. Figure 2 shows polarized optical micrographs for the hybrid fibers PCL/HA, PCL/AA, and PCL fiber.

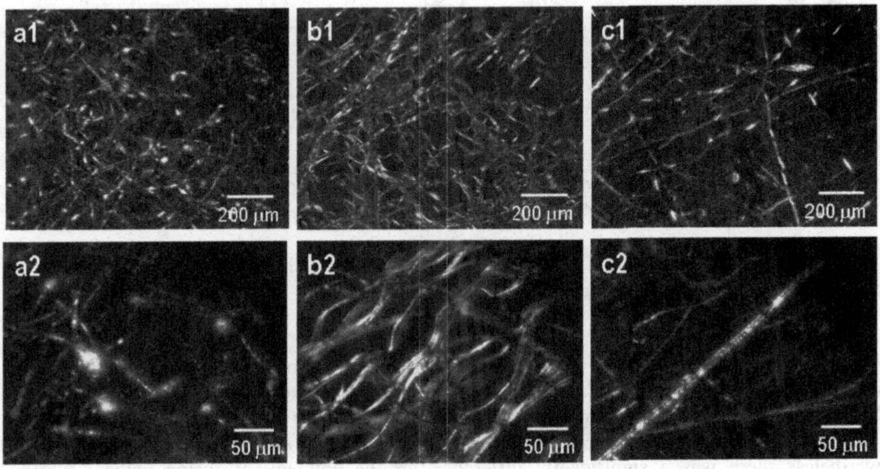


Figure 2. Polarized optical micrographs of the fibers a1 and a2 for the PCL/HA sample, b1 and b2 for the PCL/AA sample, and c1 and c2 for the PCL sample.

Under cross-polarized light conditions, all spun fibers exhibited birefringence (anisotropic and crystalline arrays), i.e mostly of the regions of the fibers were not dark under the crossed polarizers. Crystalline morphology in all samples is attributed to the ordered phase of polycaprolactone. PCL/HA, PCL/AA, and PCL samples that correspond to figures 2-a1, 2-b1 and 2-c1 were observed with a 10X magnification view (scale bar 200 μm). For this case, some defects such as beads occurred in PCL/HA (figure 2-a1) and PCL/AA samples (figure 2-c1), but mostly populated in PCL/HA. Surprisingly, PCL/AA sample (figure 2-b1) did not exhibit bead defects but rather showed a smoothed surface. It is important to note that bead defects exhibited also birefringent properties. A 32X magnification view (scale bar 50 μm) of the fibers (Figures 2-a2, 2-b2 and 2-c2) exhibited different geometrical arrangements. For example, the PCL/HA sample showed a pearl-necklace arrangement and a fiberlike morphology. On the other hand, PCL/AA sample showed a rippled fiber arrangement. The 32X magnification view let us to magnify the fiber cross section sizes of the birefringent fibers, the diameters ranging from ~2 μm to ~10 μm. Furthermore, the average diameter size of the bead defects also was viewed, ranging from ~25 μm to ~35 μm. In summary, the cross-polarized light condition showed how Hydroxyapatite and α-Alumina nanoparticles were embedded in PCL fibers, and these nanoparticles can modify the morphology of the PCL fibers without compromising the birefringent properties.

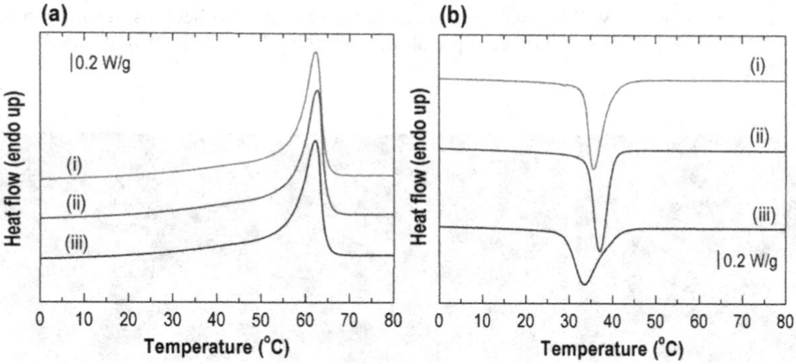


Figure 3. DSC thermograms of (i) PCL/HA, (ii) PCL/AA and (iii) PCL-f samples. (a) Second heating, 10°C/min and (b) First cooling at 10°C/min.

Thermal properties of the electrospun microfibers are shown in figure 3 (3a, 3b) and summarized in table 1. A first heating was applied to the fibers to erase their thermal history. DSC results showed that the melting temperature of the PCL phase does not significantly change with electrospinning conditions, for all electrospun fibers. All melting temperatures were ~62°C; nevertheless there was a significantly difference for the enthalpy of fusion values. While PCL/HA and PCL/AA exhibited values of 35.49 J/g and 35.62 J/g respectively, PCL fiber reached a value of 43.08 J/g. Similar tendency was obtained for the enthalpy of crystallization,

PCL/HA and PCL/AA had values of 47.15 J/g and 48.96 J/g respectively, and PCL was 54.65 J/g. Respect to crystallization behavior, a lower value was calculated at 33.66°C for the PCL sample. PCL/HA and PCL/AA showed crystallization temperatures of 35.67°C and 37.06°C respectively. The degree of crystallinity ($\chi$) of the fibers was calculated by the following formula [13]:

$$\chi(\%) = \Delta H_m \frac{1-x}{\Delta H_m^o} \times 100$$

Where $\Delta H_m$ are the experimental heat of fusion values, $\Delta H^o_m$ is the theoretical heat of fusion [14] for 100% crystalline of PCL (139 J/g), and a correction (1- $x$) for the weight percent of the HA and AA nanoparticles embedded, $x$. 10 wt% of each nanoparticle was used to prepare the respective hybrid fiber. A lower degree of crystallinity was obtained for the fibers containing HA and AA nanoparticles (table 1) which let us to assume a disruption of the crystallinity for the PCL phase in both hybrid fibers.

Table 1. Thermal properties of PCL fibers: PCL/HA, PCL/AA and PCL-f.

| Fiber | $T_{m,PCL}$ (°C) | $\Delta H_m$ (J/g) | $T_{c,PCL}$ (°C) | $\Delta H_c$ (J/g) | $\chi$(%) |
|---|---|---|---|---|---|
| PCL/HA | 62.38 | 35.49 | 35.67 | 47.15 | 22.97 |
| PCL/AA | 62.72 | 35.62 | 37.06 | 48.96 | 23.06 |
| PCL-f | 62.35 | 43.08 | 33.66 | 54.65 | 27.89 |

Figure 4 shows the $H_v$ SALS patterns of PCL/HA (4-a), PCL/AA (4-b), and PCL-f (4-c) fibers. Patterns of PCL/HA and PCL fibers show clearly a diamond-shaped patterns, and this is typical of a needle-like morphology [12]. Moreover, PCL/AA hybrid fiber shows its lobes nearest to the beam stop which indicates a larger size of the crystalline structure compared to PCL/HA and PCL-f fibers.

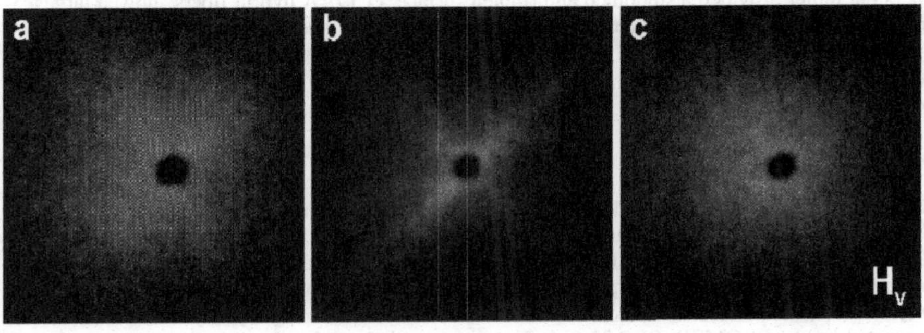


Figure 4. SALS patterns of (a) PCL/HA, (b) PCL/AA and (c) PCL-f electrospun fibers. $H_v$ polarization conditions.

TEM photographs of electrospun non-woven fibers and hybrid fibers were performed as shown in figure 5. Scale bar for all fibers is 2 µm. In figure 5a, PCL/HA fibers showed a narrow distribution with an average cross section about 250 nm and featuring a smoothed surface. Nano-aggregates of hydroxyapatite appear in dark contrast. α-Alumina nanoparticles embedded in the PCL matrix are less populated along the fiber (figure 5b). PCL fiber in figure 5c features a non uniform cross section fiber. TEM characterization was useful to elucidate nano-aggregates of hydroxyapatite and α-Alumina embedded along the PCL microfibers.

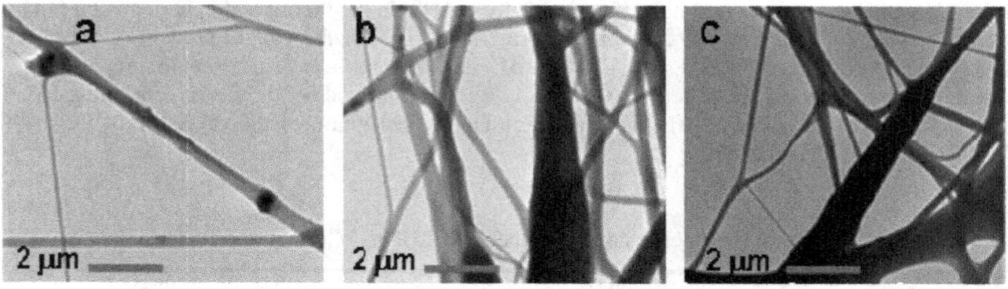


Figure 5. TEM micrographs of (a) PCL/HA, (b) PCL/AA and (c) PCL-f electrospun fibers.

## CONCLUSIONS

Non-woven PCL hybrid fibers containing Hydroxyapatite and α-Alumina nanoparticles were fabricated by electrospinning process. Thermogravimetric analysis exhibited that HA and AA nanoparticles decrease the onset of decomposition temperature of the PCL phase, furthermore HA nanoparticles improved the thermal stability of PCL/HA fibers. HA nanoparticles were embedded in PCL fibers and induced bead defects at microscale and formed HA nano-aggregates at nanoscale. AA nanoparticles did not induce bead defects along the PCL fibers but rather provided a smoothed and rippled surface. PCL/AA hybrid fibers showed largest ordered structures in agreement to SALS patterns. All fibers showed birefringent properties. Thermal analysis showed that HA and AA nanoparticles significantly decreased the $\Delta H_m$ melting values of PCL phase in the fibers, indicating that the crystallinity of PCL was disrupted by the nano-aggregates.

## ACKNOWLEDGMENTS

Angel Romo-Uribe gratefully acknowledges the financial support of SEP-CONACyT, under Ciencia Básica 2011 program, grant 168095. Bonifacio Alvarado-Tenorio is grateful to the Consejo Nacional de Ciencia y Tecnología (CONACYT) for the financial support under the program Repatriación de Investigadores. S.Y. Reyes-López acknowledges the funding of the proyect "Sintesis de Nanocompositos de Alumina e Hidroxiapatita para Uso Biomédico" supported by the PROMEB, SEP-23-005. They are also indebted to Sayil López and Francisco Solorio for the electron microscopy, UMSNH.

## REFERENCES


1. M.A. Woodruff, C. Lange, J. Reichert, *et al*, *Materials Today* **15**, 430 (2012).
2. S. Megelski, J.S. Stephens, D.B. Chase and J.F. Rabolt, *Macromolecules* **35**, 8456 (2002).
3. M. Bognitzki, W. Czado, T. Frese, *et al*, *Adv. Mater.* **13**, 70 (2001).
4. G.H. Kim, H. Han, J.H. Park and W.D. Kim, *Polym. Eng. Sci.* **47**, 707 (2007).
5. H. Vargas-Villagrán, E. Terán-Salgado, M. Domínguez-Díaz, O. Flores, B. Campillo, A. Flores, and A. Romo-Uribe, *Mater. Res. Soc. Symp. Proc.* Vol. **1376**. (2012). DOI: 10.1557/opl.2012.285
6. M. Dominguez-Diaz, A. Flores, R. Cruz-Silva, and A. Romo-Uribe, *Mater. Res. Soc. Symp. Proc.* **1466** (2012), DOI: 10.1557/opl.2012.1256
7. K. Fujihara, M. Kotaki and S. Ramakrishna, *Biomaterials* **26**, 4139 (2005).
8. S.I. Jeong, E.K. Ko, J. Yum, C.H. Jung, Y.M. Lee and H. Shin, *Macromol. Biosci.* **8**, 328 (2008).
9. L. Sedel and A. Raould, *Proc. Inst. Mech. Eng. [H]*, **221**, 21 (2007).
10. D. Rokusek, C. Davitt, A. Bandyopadhyay, S. Bose and H.L. Hosick, *J. Biomed. Mater. Res. Part A* **75**, 588 (2005).
11. A. Romo-Uribe, B. Alvarado-Tenorio and M.E. Romero-Guzmán, *Rev. LatinAm. Metal. Mat.*, **30** (1), 190 (2010).
12. Stein, R. S. *Newer Methods of Polymer Characterization*. New York: Interscience Publishers, (1964) p. 155.
13. Y. Katoh and M. Okamoto, *Polymer* **50**, 4718 (2009).
14. V. Crescenzi, G. Manzini, G. Calzolari and C. Borri, *Eur. Polym. J.* **8**, 449 (1972).

**Multifunctional Biomaterials in Sensors and Applications**

**Mater. Res. Soc. Symp. Proc. Vol. 1569 © 2013 Materials Research Society**
**DOI: 10.1557/opl.2013.806**

# Biosensor based on DNA directed immobilization of enzymes onto optically sensitive porous Si

Giorgi Shtenberg[1], Naama Massad-Ivanir[2], Oren Moscovitz[3], Sinem Engin[4], Michal Sharon[3], Ljiljana Fruk[4] and Ester Segal[2,5]
[1]The Inter-Departmental Program of Biotechnology, [2]Department of Biotechnology and Food Engineering, [5]The Russell Berrie Nanotechnology Institute, Technion – Israel Institute of Technology, Haifa 32000, Israel, [3]Department of Biological Chemistry, Weizmann Institute of Science, Rehovot 76100, Israel, [4]Karlsruhe Institute of Technology, DFG – Center for Functional Nanostructures, Karlsruhe 76131, Germany

## ABSTRACT

Optical biosensor for monitoring proteolytic activity is constructed by DNA-directed immobilization of enzymes onto porous Silicon nanostructures. This sensor configuration allows both protease recycling and easy surface regeneration for subsequent biosensing analysis by means of mild dehybridization conditions. We demonstrate real-time analysis of minute quantities of proteases paving the way for substrate profiling and the identification of cleavage sites. The biosensor is compatible with common proteomic methods and allows for a successful downstream mass spectrometry analysis of the reaction products.

## INTRODUCTION

The catalytic activity of proteases is crucial for life. These enzymes regulate a multitude of biological processes through the modification of protein activity or by controlling protein turnover [1]. They have the unique ability to irreversibly hydrolyze peptide bonds, which results not only in protein degradation, but also in the introduction of new levels of information content into the signaling pathways [2]. Identifying protease substrate repertoire is important for understanding the functions of these enzymes, revealing their biological roles, and generating lead compounds for new therapeutic strategies [3]. Most common methods to assess protease activity suffer from low efficiencies and extended pretreatments that can easily affect the native activity of the desired enzyme. Even the rapid technologies e.g., chromatography and mass-spectrometry, that possess high selectivity, reproducibility and sensitivity, have drawbacks associated with the meticulous sample preparation prior to analysis [4]. Thus, the need for rapid information screening requires the development of new and improved bioanalytical tools, allowing for fast and high-throughput analysis of minute samples with minimal losses.

In recent years, porous Si (PSi) has emerged as a promising nanomaterial for optical biosensing applications due to its large internal surface area and tunable optical properties [5-9]. It has been demonstrated that PSi-based interferometers, which operate by measuring a change in refractive index in a volume of solution contained within the porous nanostructure, allows for label-free detection of a variety of biomolecular interactions, including enzymes, DNA fragments, proteins, and antibodies [5]. In this work, we report on a generic approach for reversible enzyme conjugation to porous $SiO_2$ ($PSiO_2$) support by applying DNA-directed immobilization (DDI). The DNA serves as handle for specific immobilization through Watson-Crick base pairing with complementary DNA sequence. Successful immobilization of different classes of enzymes onto a range of solid surfaces has already been demonstrated [10], keeping in mind that the most crucial step in this approach is to preserve the functionality of both conjugate components [11]. Moreover, given that DNA hybridization is reversible, by using relatively mild procedures the surface and the enzyme, can be regenerated and reused for multiple reaction cycles [12].

## EXPERIMENT

### Preparation of Porous $SiO_2$ Nanostructures

Single side polished on the <100> face oriented and heavily doped p-type Si wafers (1.0 mΩ-cm resistivity, B-doped, from Siltronix Corp, France) are electrochemically etched in a 3 : 1 (v/v) solution of aqueous HF (48%, Merck) and ethanol (99.9%, Merck) at a constant current density of 385 mA $cm^{-2}$ for 30 s. After etching, the surface of the wafer is rinsed with ethanol several times and dried under a dry nitrogen gas. The freshly-etched PSi samples are thermally oxidized in a tube furnace (Thermolyne) at 800°C for 1 h in ambient air, resulting in a porous $SiO_2$ ($PSiO_2$) layer. Note: All materials were purchased from Sigma Aldrich Chemicals unless mentioned otherwise.

### Preparation of Enzyme-DNA Conjugates

To prepare the enzyme-DNA conjugate, 100 μL of a 100 μM solution of 5'-thiol modified single stranded oligonucleotide (cD sequence) in TE Buffer (10 mM Tris and 1 mM EDTA) is mixed with 60 μL 1,4-dithiothreitol (1M) and incubated overnight at 37°C. 1 mg of Horseradish peroxidase (HRP) type VI, is dissolved in 200 μL 50 mM Phosphate buffered saline (PBS, pH 7.4) and incubated for 1 h at 37°C with Sulfo-succinimidyl-4-(N-maleimido-methyl)cyclohex-an-1-carboxylate (sulfo-SMCC) obtained from Pierce Products (2 mg in 60 μL of N,N-Dimethylformamide). Detailed purification data by two consecutive gel-filtration and anion exchange chromatography is as previously described [13]. Trypsin-DNA conjugates are prepared using the same conditions. The final concentration of the conjugates is determined spectrophotometrically.

### Biofunctionalization of $PSiO_2$

The $PSiO_2$ films are first immersed in a solution of 3-Glycidoxypropyl(trimethoxy)silane (GPTS) dissolved in toluene (90 mM) for 2 h. The samples are then rinsed with toluene, ethanol, acetone and dried under a nitrogen gas. The sequence of the capture strand is 5'-amino TCCTGTGTGAAATTGTTATACGCC-3' (aD). A solution of the amino-modified probe (2μL, 100 μM) is applied onto the silanized $PSiO_2$ surface and incubated for 2 h in a humidity chamber, followed by a post-cleaning process. Afterwards, 2μL of the complimentary strand modified-enzyme conjugates (cD), 2.6 μM DNA-HRP and 4.6 μM DNA-trypsin conjugates, are incubated onto the $PSiO_2$ for 30 min. Finally, the $PSiO_2$ samples are rinsed with 50 mM PBS, soaked in the buffer for 30 min, and rinsed again with PBS to remove any un-bounded species from surface. All processes are carried out at room temperature; bellow the melting temperature of the DNA. Mild basic conditions are used for dissociation of the hybridized complex (0.1 M NaOH).

**Enzymatic Activity**

HRP and trypsin activity test were previously described [12]. Note: All primary data are normalized to its activity in a solution by using the equivalent volumes and concentrations. Proteolytic activity: The trypsin-modified $PSiO_2$ is washed with 0.1 M HEPES buffer solution pH 8.2 for 20 min. Then, 2 mg $mL^{-1}$ of myoglobin in 0.1 M HEPES buffer is introduced and continually cycled through the flow cell. Finally, the surface is washed with HEPES buffer for the removal of the entrapped proteins and fragments. The experiments are conducted at 37°C, which is the optimal temperature for trypsin activity.

**Scanning Electron Microscopy**

High-resolution scanning electron microscopy (HRSEM) of the neat $PSiO_2$ is performed using a Carl Zeiss Ultra Plus HRSEM, at an accelerating voltage of 1 keV.

**Optical Measurements**

Interferometric reflectance spectra of $PSiO_2$ samples are collected using an Ocean Optics CCD USB 4000 spectrometer. Experimental setup and spectra analysis were previously described [6].

**Mass Spectrometry (MS) Analysis**

Mass spectrometry analysis of the retrieved fragments (aliquots of 2 µL) is performed by using nano-electrospray ionization (ESI) quadrupole time-of-flight instrument (Applied Biosystems, Foster City, CA). Proteomic analysis of the separated peptides is performed by ESI-MS/MS. Detailed characterization data is described elsewhere [12].

## DISCUSSION

**Biofunctionalization of $PSiO_2$ with DNA-enzyme conjugates**

The $PSiO_2$ surfaces are prepared from a highly doped p-type single-crystal Si wafer using electrochemical anodization [6]. The resulting PSi nanostructure is thermally oxidized to generate a stable and a more hydrophilic scaffold. The structural properties, i.e., thickness, porosity and pore diameter, are as previously described by Massad-Ivanir et al. [14]. Figure 1 left shows HRSEM images of a typical $PSiO_2$ scaffold. Top-view micrograph of the film (inset Fig. 1 left) reveals its highly porous nature (>80%) and clearly shows that the pores entrance is large enough to allow an easy access of biomolecules to the internal void space of the porous layer. The semi-synthetic approach for anchoring DNA-enzyme conjugates onto the $PSiO_2$ surface is based on a well-established methodology [11], as schematically outlined in Figure 1 right. First, the oxidized nanostructure is epoxy-silanized with GPTS to obtain an activated surface for grafting of amine-modified single stranded capture DNA (aD). The complimentary thiolated oligonucleotides strand is cross-linked with sulfo-SMCC to the Lysine residues of the enzymes [10]. In the final step, the resulting DNA-enzyme conjugates (HRP-cD or trypsin-cD) are hybridized with aD to enable their immobilization onto the $PSiO_2$ surface. Detailed characterization is as previously described [12-13].

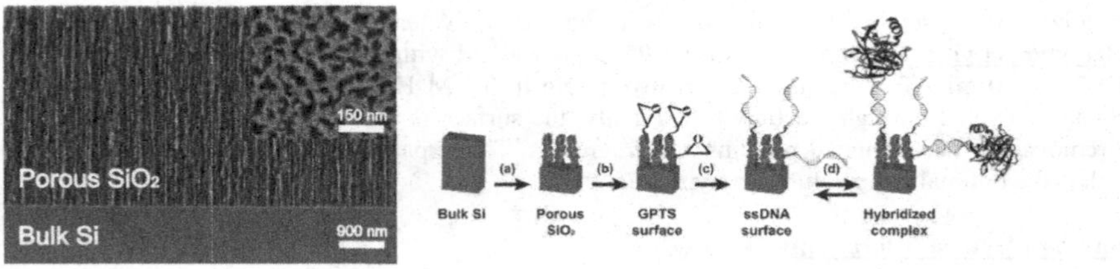


**Figure 1.** (left) HRSEM micrographs of a typical $PSiO_2$ film; (right) A schematic representation of the synthetic steps required for enzyme immobilization onto $PSiO_2$. (a) Si wafer is electrochemically etched at a constant current density followed by thermal oxidation to yield a $SiO_2$ surface; (b) The $PSiO_2$ modified with GPTS; (c) The epoxy-terminated surface reacts with amine-modified single stranded capture DNA (aD); (d) DNA-enzyme conjugate hybridization to the anchored aD onto ssDNA surface (reversible process). Note: the schematics are for illustration purposes only and are not drawn in scale, also all modifications occur inside the porous layer.

__Enzymatic activity tests__

The enzymatic activity of the anchored enzymes onto the porous scaffold is studied by monitoring the oxidized and the proteolytic products of HRP and trypsin, respectively. Figure 2 shows the results of the colorimetric assays for the immobilized conjugates. Both, trypsin and HRP, exhibit high relative activity when anchored to the $PSiO_2$. Values of 59% and 74% were obtained for trypsin and HRP, respectively, compared to the native enzyme activity in solution.

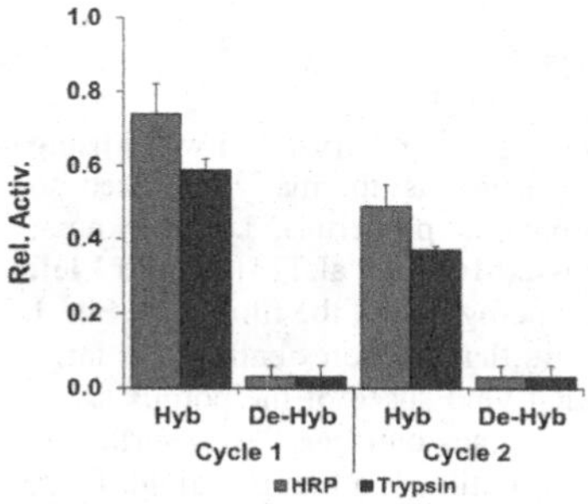


**Figure 2.** Relative activity of HRP and trypsin, immobilized via DDI onto $PSiO_2$. Mild dehybridization (0.1 M NaOH) allows fast removal of the DNA-enzyme conjugates from the surface. We show two consecutive cycles of DNA-enzyme hybridization/dehybridization to demonstrate the high reversibility and reproducibility of this immobilization strategy. Data are reported as mean ± standard deviation ($n \geq 4$).

The important advantage of the DDI anchoring approach is the ability to regenerate the surface or the DNA-enzyme conjugate by applying mild dehybridization conditions (base wash). Indeed, applying these conditions allows for complete surface regeneration, as insignificant enzymatic activity of less than 3% is detected (Fig. 2, cycle 1). To further demonstrate the applicability of the DDI strategy in the characterization and analysis of minute enzyme samples grafted onto optical transducers, a consecutive cycle of hybridization/dehybridization is carried out. These experiments clearly show the high reversibility and reproducibility of this immobilization strategy (as >40% and <3% of relative activity, before and after dehybridization, respectively, are measured for both HRP and trypsin), as shown in Figure 2, cycle 2.

**Optical detection of enzymatic activity**

The potential of this platform to act as an optical biosensor for monitoring proteolytic activity is investigating through Reflective Interferometric Fourier Transform Spectroscopy (RIFTS) experiments [5,7-9]. This method presents high sensitivity to small changes in the average refractive index of the porous thin-film, allowing for direct and real-time monitoring the binding of different species to the pore walls [5-6,12-13]. Thus, trypsin-modified $PSiO_2$ are fixed in a flow cell setup and are exposed to various buffer/substrate combinations. The reactions are monitored in real-time by acquisition of the reflectivity spectra from the porous thin film. The recorded effective optical thickness (EOT) values provide a direct measure of the amount of digestion products infiltrating the pores. Figure 3 shows the normalized EOT change of trypsin-modified $PSiO_2$ following the introduction of myoglobin. A gradual increase of approximately 0.35% in the EOT value is observed (Fig. 3b), which is attributed to infiltration and accumulation of the myoglobin in the nanostructure [13]. Additionally, proteolytic reaction occurring within the porous nanostructure influences the change in the EOT (Fig. 3b). When HEPES buffer is introduced a rapid decrease in the EOT value occurs (Fig. 3c), which is attributed to the removal of myoglobin and its digestion fragments from the nanostructure.

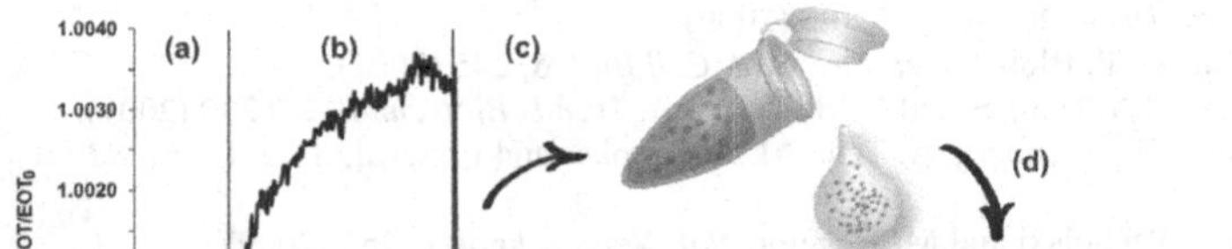

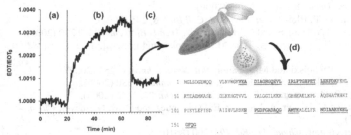


**Figure 3.** Optical response, expressed as relative EOT, of trypsin-modified $PSiO_2$ nanostructures upon myoglobin introduction. The trypsin-modified $PSiO_2$ nanostructure is pre-treated with 0.1 M PBS, pH 7.4 to minimize nonspecific adsorption of proteins. (a) Wash with 0.1 M HEPES buffer; (b) Addition of 2 mg $mL^{-1}$ of myoglobin in HEPES buffer; (c) HEPES buffer wash step. The $PSiO_2$ biosensor is fixed in a custom-made flow-cell and the reflectivity spectra are recorded every 15 s; (d) Amino acid sequence of myoglobin (sequence in bold reflect the identified peptides in the LC-MS/MS analysis).

Next, the generated peptides are sequence identified by proteomic analysis (Fig. 3d). Such examination does not only allow to identify the substrate specificity of the protease but it also enables identification of the cleave site. As expected, the retrieved peptides have lysines or arginines in their N-terminal. These results demonstrate the ability to anchor a protease onto the porous Si nanostructures, maintain its activity, generate and retrieve peptides, and use them for sequence identification.

## CONCLUSIONS

An optical biosensor, based on a $PSiO_2$ Fabry-Pérot thin Film, is synthesized and characterized, and its potential applicability as a biosensor for proteolytic activity of proteases is investigated. The major advantages of the system the specific and reversible nature of the enzyme immobilization (DDI technique) that permits the preservation of material and the simplicity of the optical readout procedure enables real-time tracking of the reaction process. Given that to date the substrate degradome of hundreds of proteases remains unknown, we trust that such a platform will be of practical importance for substrate profiling.

## ACKNOWLEDGMENTS

E.S acknowledges the financial support of the NEVET program administered by the Russell Berrie Nanotechnology Institute and by the Marie Curie Reintegration Grant (NANOPACK) administered by the European Community. This work was also supported by Weizmann-Technion-KIT Start up grant and DFG-CFN Excellence Initiative Project A5.7.

## REFERENCES

1. O. Schilling and C. M. Overall, *Curr. Opin. Chem. Biol.* ***11***, 36 (2007).
2. H. Neurath, *Science* ***224***, 350 (1984).
3. B. Turk, *Nat. Rev. Drug Discovery* ***5***, 785 (2006).
4. C. M. Overall and C. P. Blobel, *Nat. Rev. Mol. Cell Biol.* ***8***, 245 (2007).
5. A. Jane, R. Dronov, A. Hodges and N. H. Voelcker, *Trends Biotechnol.* ***27***, 230 (2009).
6. N. Massad-Ivanir, G. Shtenberg, A. Tzur, M. A. Krepker and E. Segal, *Anal. Chem.* ***83***, 3282 (2011).
7. M. M. Orosco, C. Pacholski and M. J. Sailor, *Nat. Nanotechnol.* ***4***, 255 (2009).
8. K. A. Kilian, T. Boecking and J. J. Gooding, *Chem. Commun.*, 630 (2009).
9. L. A. DeLouise, P. M. Kou and B. L. Miller, *Anal. Chem.* *77*, 3222 (2005).
10. L. Fruk, J. Mueller, G. Weber, A. Narvaez, E. Dominguez and C. M. Niemeyer, *Chem.--Eur. J.* ***13***, 5223 (2007).
11. C. M. Niemeyer, *Angew. Chem., Int. Ed.* ***49***, 1200 (2010).
12. G. Shtenberg, N. Massad-Ivanir, O. Moscovitz, S. Engin, M. Sharon, L. Fruk and E. Segal, *Anal. Chem.* ***85***, 1951 (2013).
13. G. Shtenberg, N. Massad-Ivanir, S. Engin, M. Sharon, L. Fruk and E. Segal, *Nanoscale Res. Lett.* *7*, (2012).
14. N. Massad-Ivanir, G. Shtenberg, T. Zeidman and E. Segal, *Adv. Funct. Mater.* ***20***, 2269 (2010).

Mater. Res. Soc. Symp. Proc. Vol. 1569 © 2013 Materials Research Society
DOI: 10.1557/opl.2013.1099

# Efficient Production of Fluorescent Polydiacetylene-Containing Liposomes for Pathogen Detection and Identification

Cassandra J. Wright-Walker, Caroline E. Hansen, Michael A. Evans, Emily S. Nyers, and Timothy W. Hanks
Department of Chemistry, Furman University, 3300 Poinsett Highway, Greenville, SC 29613, U.S.A.

## ABSTRACT

Amphiphilic diacetylenes (DAs) can self-assemble into photopolymerizable liposomes that can be used to construct effective pathogen sensors. Here, modified commercial inkjet printers are used to disperse DAs into water, facilitating self-assembly. The liposomes are of similar size, but are significantly less polydisperse than liposomes formed using conventional sonication methods. The process is efficient, readily scalable and tolerant of structural modification. The derivitization of approximately 5% of the DA head groups and the incorporation of fluorophores into the hydrophobic bilayer allows for the preparation of novel multifluorophore PDA sensing systems that can provide enhanced bacterial discrimination in a single experiment by way of a fluorescent fingerprint.

## INTRODUCTION

Liposomes are bilayer assemblies of long chain amphiphiles that were first characterized in the 1960s [1]. Much of the interest in these structures is due to the synthetic ease with which the hydrophobic layer membrane and the surface can be covalently derivatized, as well as the facile trapping of various functional species within the aqueous interior. In particular, we are interested in systems containing highly conjugated polydiacetylene polymer chains in the hydrophobic layer. This not only stabilizes the assembly, but adds a chromatic functionality that is sensitive to mechanical disturbances of the vesicle. Polydiacetylene (PDA) liposomes, once polymerized, exhibit a deep blue color ($\lambda_{max}$=650nm). Upon interaction with stimuli such as temperature, pH, mechanical stress, or biological species, the PDAs will change from blue to bright red ($\lambda_{max}$=540nm) due to a disruption of the backbone planarity [1].

In combination with other modifications, these liposomes can be made into environmental sensors (pH, temperature, mechanical stress, biological species, etc.) [2] as well as nanocapsules for the delivery and controlled release of drugs, nanoparticles and other species. The result is a "smart" device capable of detecting and responding to its environment.

PDA liposomes have previously been produced by a variety of methods, with the most prevalent being via sonication. Sonication, however, often results in a wide distribution of liposome sizes and may degrade sensitive components [3]. Recently, a modification of the "ethanol injection method" was shown to efficiently produce liposomes from simple phospholipids [4]. We have extended this method to DA-containing lipids, optimized preparation conditions and explored the effects of liposome surface derivitization and bilayer incorporation of fluorophores.

## EXPERIMENTAL DETAILS

**Materials** 10,12-pentacosadynoic acid (PCDA) acquired from GFS Chemicals (Powell, OH). Fluorophores, specifically 1-pyrenedecanoic acid (1-pyrene), 4,4-difluoro-5-(2-thienyl)-4-bora-3a,4a-diaza-*s*-indacene-3-dodecanoic acid (BODIPY 558/568 $C_{12}$ or BO 558), and 4,4-difluoro-5,7-dimethyl-4-bora-3a,4a-diaza-*s*-indacene-3-undecanoic acid (BODIPY FL C11; BO FL) ) were acquired from Sigma Aldrich (1-pyrene) or Molecular Probes (BODIPYs). Amino acid derivatives of PCDA were synthesized in-house. Lipolyzed polysaccharide (LPS) from E. Coli serotype 026:B6, Salmonella enterica serotype enteritidis, and Pseudomonas aeruginosa serotype 10 were purchased from Sigma-Aldrich. Commercially available inkjet printers and cartridges were acquired (HP D1530 and HP21, respectively). Other reagents and solvents were purchased from commercial sources and used as received.

**Printer Modification** HP printers were remodeled such that a stir plate and small beaker could be placed beneath the print head. Initial models were similar to those found in other labs [5] in that each 'page' required manual signaling of the print sensor. However, careful deconstruction bypassed this, removing the need for regular input from the user and permitting the continuous printing of an entire cartridge of amphiphiles solution in a single step. During modification, it was found to be paramount to only remove non-essential components such that paper will still trigger the media sensors within the printer. It was found that 2 3/4" wide strips of paper placed in the printer was effective and convenient for this purpose.

The grey 'lids' were removed from the HP21 cartridges with care to ensure that the entirety of the 'lid' remained intact such that solvent could not evaporate through cracks in it. Ink and sponge from the cartridges were removed followed by careful rinsing and sonication of the cartridge using deionized water. To ensure all ink had been effaced from the nozzles, solutions of 2-propanol were printed through the cartridge.

**PCDA Liposome Synthesis** "Ink" solutions were prepared by dissolving 5mg of PCDA in 3.3 mL of 2-propanol (Figure 1). These were dispersed via the modified inkjet printer into small beakers containing 10mL of DI water maintained at 60 °C under stir. After dispersal, liposome solutions were kept at 4 °C for a minimum of 12 hours to allow self-assembly. Solutions were polymerized (254nm light) for 10 minutes and filtered using Whatman 0.45μM PES syringe filters.

**Printing Optimization** Several water-miscible solvents (ethanol, isopropanol, acetone, and tetrahydrofuran) were investigated for their suitability as the base for the DA "ink". The selection criteria included the ability to solvate the DA, to completely disperse the PCDA into water, and the ease with which it could be removed from the aqueous dispersion so as not to trigger the blue-red sensor transition over a two-week window. The PCDA:solvent ratio was also varied (1:1, 3:2 or 2:1) to maximize the concentration of liposomes in the resulting solution and to minimize the polydispersity. Finally, the temperature of the liposome solution during printing was optimized by printing in two different conditions: room temperature and 60 °C; optimal conditions were defined as production of the highest concentration (UV-VIS) without increasing polydispersity (DLS).

**Bacterial Sensor Synthesis** Partially derivatized amino acid liposomes (AA-PDA) incorporating a fluorophore were prepared similarly to control liposomes described above. Each amino acid (arginine, tryptophan, or tyrosine) corresponded to a specific concentration of fluorophore in tetrahydrofuran: 1-pyrene (50uL of 2.0µM), BO 558 (100uL of 0.4µM), or BO FL (50uL 0.4µM), respectively. PCDA, derivatized PCDA (10uL of 1M AA-DA), corresponding fluorophore, and 2-propanol were mixed and then printed using the modified ink cartridge; the solutions were then printed, polymerized and filtered as described above.

**Bacterial Challenge** LPS from E. Coli serotype 026:B6, Salmonella enterica serotype enteritidis, and Pseudomonas aeruginosa serotype 10 were dissolved separately in phosphate buffer (pH 7.4; 3mg/mL). AA-PDA & PDA liposome solutions were exposed to 10 µL LPS solutions for 1hr (T=25°C) followed by analysis in an emission spectrometer ($\lambda_{ex}$=348nm).

Samples were analyzed using dynamic light scattering (DLS), UV-VIS, and emission spectroscopy ($\lambda_{ex}$=348nm). The chromatic transformation of the liposomes solution was quantified using the Colorimetric Ratio Equation (CR), where A refers to the absorbance of either the blue ($\lambda$=650nm) or the red ($\lambda$=540nm) in the UV-Vis spectrum; CR provides an estimate of form conversion percentage attributed to stimulus exposure (Equation 1) [6]. Fluorophore emission peak intensities were determined by normalizing the fluorescence from the LPS-exposed solutions to the maximum fluorescence output, which was considered to be fluorescence after heating the sample to 90 ˚C. Peaks 1 & 2 corresponded to 1-pyrene (376 & 395nm). Peak 3 corresponded with BO 558 (510nm) and Peak 4 with BO FL (570nm). Results were analyzed using Student's t-test or ANOVA with Tukey post-hoc tests ($\alpha$=0.05). PDA fluorescence from the red form was small enough to ignore under these conditions.

$$PB_0 = \frac{A_{blue}}{(A_{blue}+A_{red})} \qquad PB_1 = \frac{A_{red}}{(A_{blue}+A_{red})} \qquad CR = \frac{(PB_0 - PB_1)}{PB_0} \qquad \text{(1) Colorimetric Ratio}$$

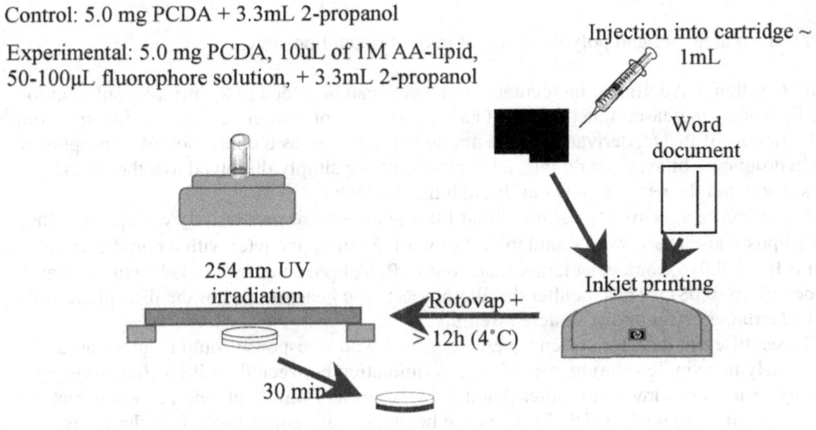


**Figure 1:** Schematic of optimized liposome printing procedure

## DISCUSSION

**Printing Optimization** Inkjet printing generates small, reproducible droplets that provide a constant level of supersaturation of the DA lipids. This allows for the highly reproducible formation of unilamellar vesicles, with very little aggregation. Printed PDA liposomes were found to be of similar size (130.2 ± 21.9 (Ave ± SD) nm; $p=0.209$) with a significantly decreased polydispersity index (0.091±0.020; $p=0.022$) as compared to liposomes formed via the conventional probe sonication method (141.8 ± 18.6 nm and 0.145 ± 0.046, respectively). The size of these lipids is within the range of 50-200 nm reported by others for PDA liposomes [4].

Only ethanol and isopropanol were found to be acceptable solvents for liposome printing, as other solvents damaged the ink cartridge. Liposomes prepared from ethanol solutions showed lower long-term stability than those from isopropanol. The two-week stability test using isopropanol showed the greatest liposome stability was produced from an ink with a 3:2 PCDA to isopropanol ratio (14.6% shift towards red phase; $p=0.292$) compared to 1:1 (22.4%) and 2:1 (20.3%). The 3:2 ratio of PCDA to solvent also resulted in the most efficient production of liposomes ($F=171.3$; $p<0.001$) as determined by UV-VIS. The results showed a significant increase in liposome concentration compared to both the 1:1 ($p<0.001$) and 2:1 ($p<0.001$) ratios. Additionally, increasing temperature to 60 °C for printing resulted in further increased liposome concentration while maintaining size and polydispersity (Figure 2).

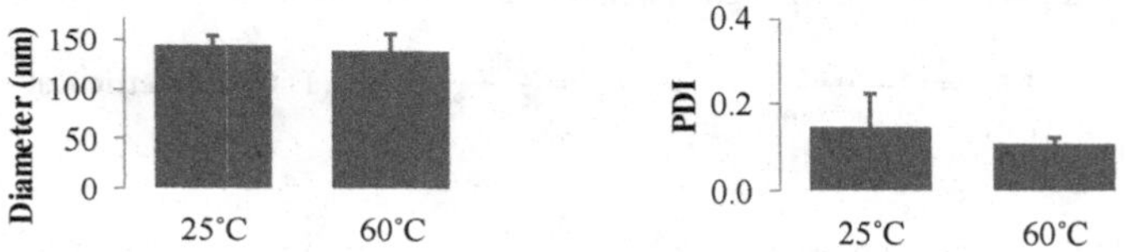


**Figure 2:** Mean diameter and polydispersity of inkjet printed liposomes.

**Bacterial Challenge** Additional molecular components can be added to the printing solution to give the liposomes increased functionality. Our bacterial sensor design calls for the incorporation of small amounts of PCDA derivatized with amino acids as well as the insertion of fluorophores into the hydrophobic bilayer. To do this, all components are simply dissolved together in the alcohol solvent and dispersed into water through the printhead.

Fluorescence spectroscopy showed that fluorophores were successfully entrapped in the AA-PDA liposomes. These were found to be 142.0 ± 0.5 nm in diameter with a polydispersity index of 0.105 ± 0.015, somewhat larger than control PDA liposomes ($p = 0.034$) with similar polydispersity ($p=0.081$). Thus, neither the alterations to the headgroups nor the disruption of the bilayer had a major effect on the vesicle assembly.

Three different detector systems were generated whose response could be monitored simultaneously in a single solution, providing discrimination between the LPS tested. Some preliminary results are shown in Figures 3 and 4. Fingerprints from E.coli and Pseudomonas were very similar, with only the BO FL response being slightly reduced with Pseudomonas.

Ongoing work includes is focused on the exploration of other fluorophores and headgroups as well as the optimization of fluorophore concentrations. The baseline fluorescence is currently rather high, thus optimization of the system to improve the quenching in the initial blue phase is ongoing. Baseline reduction will allow for greater differentiation within emission intensities and thus production of unique fingerprints for E. coli and Pseudomonas. We expect to be able to improve discrimination by the simultaneous detection four or more fluorophores simultaneously.

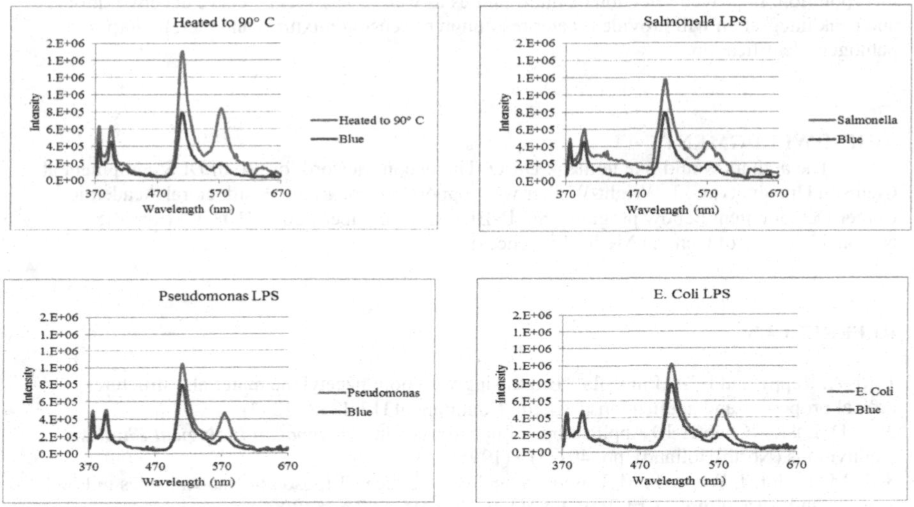


**Figure 3:** A comparison of the heated control solution and the solutions exposed to LPS solutions at $\lambda_{ex}$=348nm.

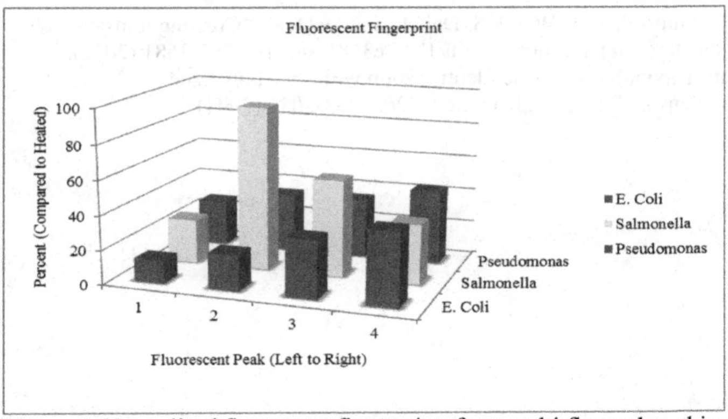


**Figure 4:** Normalized fluorescent fingerprints from multi-fluorophore biosensing PDA liposome vesicles exposed to LPS solutions.

## CONCLUSION

Inkjet printing has been found to give high quality, narrowly dispersed PDA liposomes and is easily amenable to forming liposomes with mixed amphiphiles and with fluorophore incorporation. Improved selection of fluorophores as well as analysis by curve deconvolution and peak integration will provide greater resolution of sensing mixtures and thereby improve pathogen identification.

## ACKNOWLEDGMENTS

The authors would like to thank Dr. O. Thomspon Mefford for use of DLS equipment at Clemson University. C. J. Wright-Walker was supported through the Postdoctoral Academic Career Development Fellow program (SC-INBRE grant number P20GM103499 from the National Institute of General Medical Sciences).

## REFERENCES

1. M. A. Reppy and B. A. Pindzola, "Biosensing with polydiacetylene materials: structures, optical properties and applications," Chem. Commun., 4317-4338 (2007).
2. D. D. Lasic, "Chapter 10 Applications of liposomes," in *Handbook of Biological Physics,* Anonymous (North-Holland), pp. 491-519 (1995).
3. T. M. Taylor, J. Weiss, P. M. Davidson, and B. D. Bruce, "Liposomal Nanocapsules in Food Science and Agriculture," Crit. Rev. Food Sci. Nutr. 45, 587-605 (2005).
4. S. Hauschild, U. Lipprandt, A. Rumplecker, U. Borchert, A. Rank, R. Schubert, and S. Förster, "Direct Preparation and Loading of Lipid and Polymer Vesicles Using Inkjets," Small 1, 1177-1180 (2005).
5. A. B. Owczarczak, S. O. Shuford, S. T. Wood, S. Deitch, and D. Dean, "Creating transient cell membrane pores using a standard inkjet printer." J Vis Exp. e3681, doi:10.3791/3681 (2012).
6. M. Rangin and A. Basu, "Lipopolysaccharide Identification with Functionalized Polydiacetylene Liposome Sensors." J. Am. Chem. Soc. 126, 5038-5039 (2004).

**Mater. Res. Soc. Symp. Proc. Vol. 1569 © 2013 Materials Research Society**
**DOI: 10.1557/opl.2013.1100**

# Distyrylbenzene oligoelectrolyte (DSBN+) for human cervical carcinoma cells imaging

*Roza Pawlowska,[1] Paulina Gwozdzinska[1], Logan Garner[2] and Arkadiusz Chworos[1,*]*

[1]Centre of Molecular and Macromolecular Studies, Polish Academy of Sciences, Sienkiewicza 112, 90363 Lodz, Poland

[2]Department of Chemistry and Biochemistry, University of California, Santa Barbara, California 93106, United States

achworos@cbmm.lodz.pl

## ABSTRACT

In the presented study, a new application for distyrylbenzene oligoelectrolyte, named DSBN$^+$, as a marker for bioimaging is presented. DSBN$^+$ is a water-soluble, conjugated oligoelectrolyte (COE) with novel photophysical and solvatochromatic properties. Previous studies have shown that this compound spontaneously inserts into bilayer membranes in both synthetic and microbial living systems and can facilitate visualization of cell membranes through fluorescence imaging. In the presented research, we seek to further study and exploit the multifunctional nature of DSBN$^+$ in terms of membrane interactions and photophysical properties for visualization of membranous structures of more complex mammalian cells, namely a human cervical carcinoma (HeLa) cell line. Obtained results confirm the possibility of applying DSBN$^+$ as a fluorescent dye for bioimaging of membranes in human cell cultures systems, both in live-cell imaging and in the studies required formaldehyde fixation. Due to the defined structure of this conjugated oligoelectrolyte we suspect that it will display organelle membrane selectivity, but this has to be further investigated.

## INTRODUCTION

The molecular structure of DSBN$^+$ consists of a phenylenevinylene framework flanked on each end with an alkylamino donor group bearing a tethered charged functionality (Figure 1) that results in a unique combination of solubility in polar media and photophysical features common to donor-pi-donor chromophores. [1] This two-photon absorption (TPA) chromophore was originally developed to study the effects of polar media on the TPA cross-section [2,3]. In later studies it was found that DSBN$^+$ was able to incorporate into a model membrane system *in vitro*. Thus, it is useful for characterizing membrane insertion in phospholipid vesicle model [4]. Inside membranes, the hydrophobic long molecular axis positions within the inner lipid bilayer perpendicular to the plane of the membrane with the polar terminal groups oriented towards the polar lipid head groups near the membrane surface. This observation has been confirmed by confocal imaging and cyclic voltammetry experiments, which show, that incorporation of DSBN$^+$ facilitates transmembrane electron transport across lipid bilayers [4]. The first studies in this area have already been made in Baker's yeast cells and have shown that this conjugated electrolyte is able to incorporate into yeast membranes, giving a strong fluorescent signal [4]. Furthermore, $DSBN^+$ exhibits a notable increase in quantum efficiency when incorporated within membranes as compared to its performance in aqueous surroundings. This contributes to optimal visualization of membranes coupled with little-to-no background fluorescence from internal and external aqueous phases [5]. The previously reported results have provided precedent for incorporation of $DSBN^+$ into the membranes in eukaryotic living cells and possible further applications as a new class of bioimaging dyes.

In the present study, we have verified the ability of $DSBN^+$ to penetrate into the membranes of human cells in both living and fixed cell imaging experiments. As expected from previous results, our data suggest that DSBN spontaneously accumulates within cell membranes and shows increased emission. Such concentration gradient enhanced by quantum efficiency makes this class of molecules very efficient for cells staining.

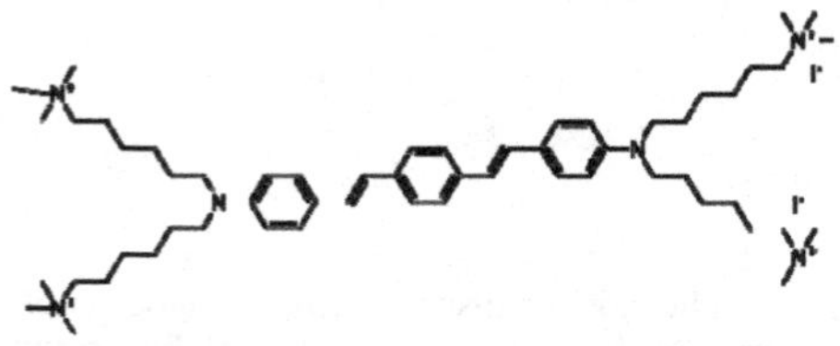

**Fig 1** Structure of 1,4-bis(4′-(N,N-bis(6″-(N,N,N- trimethylammonium)hexyl)amino)-styryl) benzene tetraiodide ($DSBN^+$)

## EXPERIMENTAL DETAILS

The 1,4-bis(4'-(N,N-bis(6"-(N,N,N-trimethylammonium)hexyl)amino)-styryl)benzene tetraiodide (DSBN+) was synthesized as previously reported [2,4].

The HeLa (human cervical carcinoma) cell line was cultured in RPMI 1640 supplemented with 10% FBS, 100 U/ml penicillin and 100 μg/ml streptomycin.

For staining experiment, cells were trypsinized with 0.05% trypsine in PBS and seeded into 24-well plate at a density of $1x10^5$ cells per well. After 24 hours incubation, the culture medium was replaced with 10 μM $DSBN^+$ in RPMI. The cells were then cultured at 37°C in an atmosphere of 5% $CO_2$ for 2 hours.

For $DSBN^+$ visualization in fixed HeLa cells, the culture was washed by phosphate saline buffer (PBS) and fixed at room temperature with 3.8% paraformaldehyde for 10 min. Afterwards, cells were washed three times with PBS. The analysis was conducted using Nikon Eclipse fluorescent microscope.

For live-cell imaging, the fluorescence microscopy visualization was done immediately after 2 hours incubation with 10 μM $DSBN^+$ on both microscope glass slide and in the 24 well plate.

## RESULTS

The main goal of presented research was to investigate the applicability of $DSBN^+$ as a new membrane stain for bioimaging studies of human cell cultures. For this purpose, we have conducted a series of experiments using a human cervical carcinoma (HeLa) cell line as a model of human cells. It has been shown previously, that this type of conjugated oligoelectrolyte has the ability to incorporate within membranes, in phospholipid vesicle models, and in microbial cells [4], although the application of $DSBN^+$ for bioimaging of human cells has yet to be investigated. Additionally we seek to ascertain how formaldehyde cell fixing, preceded with extensive sample wash, will influence the stability of the conjugated oligoelectrolyte in the cell membrane. We wanted to confirm that the fluorescence dye would not be washed out of the sample, and additionally that formaldehyde treatment would not alters the performance of stain. Both techniques (live and fixed cell imaging) offer advantages, such as comparison of visualization of life processes vs. captured state.

Our results to date have shown that $DSBN^+$ can be used for staining living HeLa cells in cell culture conditions without the need for washing excess dye from the culture medium. After 2 hours of incubation a strong intracellular fluorescent signal was observed from membranous structures. Due to the photophysical and solvatochromatic properties of $DSBN^+$, namely the difference in quantum yield in water vs. hydrophobic environment [4], the fluorescent emission was observed only in the cell membranes without background fluorescence (Figure 2). This advantageous property makes $DSBN^+$ a good candidate to be used as a fluorescent marker, especially for tests requiring live cell staining.

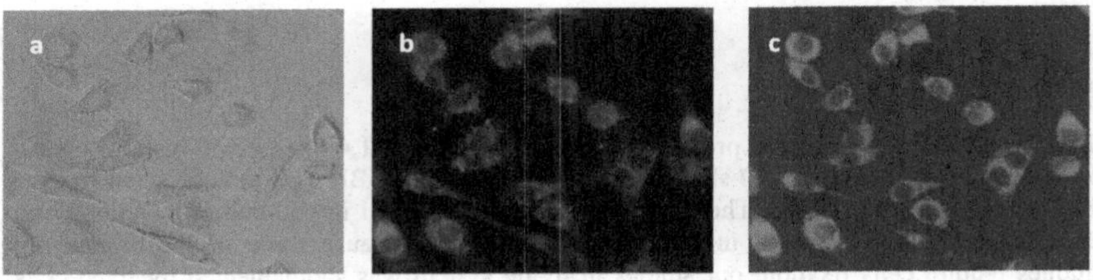


**Fig 2** HeLa cells after 2 hours incubation with 10 μM $DSBN^+$ in RPMI. Live-cells visualization. (a) phase contrast, (b) FITC filter (ex= 465-495, em=515-555) (c) UV2 filter (ex= 330-380, em=420)

Although living cell visualization is essential for kinetic studies, other techniques, such as immunestaining using specific antibodies, requires cell fixation [6,7]. Therefore we sought to determine whether it is possible to observe the fluorescent signal from the $DSBN^+$ inside HeLa cells following cell fixation procedures.

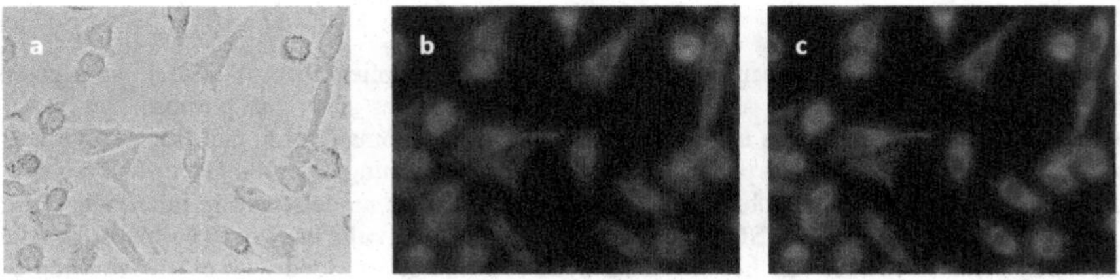


**Fig 3** The fluorescence microscopy visualization of HeLa cells after 2 hours incubation with 10 μM $DSBN^+$ and followed cells fixation with formaldehyde. (a) phase contrast, (b) FITC filter (ex= 465-495, em=515-555) (c) UV2 filter (ex= 330-380, em=420)

The results obtained corresponding to our cell fixation experiments confirm that a fixation procedure using 3.8% paraformaldehyde does not result in elution of $DSBN^+$ from the cell membranes or render the dye inactive in terms of fluorescence emission. The fluorescent signal observed after formaldehyde treatment was comparable to the signal from living cells (Figure 3).

These observations suggest that $DSBN^+$ can be employed as a useful bioimaging stain for both live-cell imaging and visualization of fixed cells. As presented here $DSBN^+$ not only gives fast staining of cellular structures which, due to low background emission, can be performed without media exchange. Additionally while not expected, $DSBN^+$ could potentially react to the cell media environment (aggregation, phase separation) but this was not observed.

## CONCLUSIONS

In this communication we have presented the applicability of 1,4-bis(4'-(N,N-bis(6"-(N,N,N-trimethylammonium)hexyl)amino)-styryl)benzene tetraiodide ($DSBN^+$) as a fluorescent marker for bioimaiging of human cells. The experiments were performed in cell culture condition and fluorescent signal was observed in living cells. Moreover, because some of the bioimaging techniques require cells fixation, our studies in living system was supplemented by fixed cells analysis. In all tested conditions, $DSBN^+$ treated human cervical carcinoma cells gave an intracellular fluorescent signal from membranous cellular structures with little-to-no undesired background emission. Obtained results indicate that the tested distyrylbenzene oligoelectrolyte may serve as a novel new type of marker for human cell staining and visualization.

## ACKNOWLEDGMENTS

This work was supported by the National Science Centre (NCN) grant, based on the decision DEC-2012/05/B/ST5/00364. The authors thank Guillermo C. Bazan for all inspiration and support.


## REFERENCES


1. Albota M, Beljonne D, Brédas JL, Ehrlich JE, Fu JY, Heikal AA, Hess SE, Kogej T, Levin MD, Marder SR, McCord-Maughon D, Perry JW, Röckel H, Rumi M, Subramaniam G, Webb WW, Wu XL, Xu C, *Science*, 281, 1653-1656 (1998)

2. Woo HY, Liu B, Kohler B, Korystov D, Mikhailovsky A, Bazan GC, *J. Am. Chem. Soc.* 127, 14721-14729 (2005)

3. Woo HY, Korystov D, Mikhailovsky A, Nguyen TQ, Bazan GC, *J. Am. Chem. Soc.* 127, 13794–13795 (2005)

4. Garner LE, Park J, Dyar SM, Chworos A, Sumner JJ, Bazan GC, *J. Am. Chem. Soc.* 132, 10042-10052 (2010)

5. Ortony JH, Chatterjee T, Garner LE, Chworos A, Mikhailovsky A, Kramer EJ and Bazan GC, *J. Am. Chem. Soc.* 133, 8380-8387 (2011)

6. Schnell U, Dijk F, Sjollema, Giepmans BNG, *Nat Methods.* 9, 152-158 (2012)

7. Massoud TF, Gambhir SS, *Genes Dev.* 17, 545-580 (2003)

Mater. Res. Soc. Symp. Proc. Vol. 1569 © 2013 Materials Research Society
DOI: 10.1557/opl.2013.681

# Top-down Fabricated Polysilicon Nanoribbon Biosensor Chips for Cancer Diagnosis

Hsiao-Kang Chang[1], Xiaoli Wang[1], Noppadol Aroonyadet[1], Rui Zhang[2], Yan Song[2], Ram Datar[3], Richard Cote[3], Mark Thompson[2] and Chongwu Zhou[1]

[1]Department of Electrical Engineering and [2]Chemistry, University of Southern California, Los Angeles, CA 90089, U.S.A.
[3]Department of Pathology, University of Miami, Miami, Florida 33136, U.S.A.

## ABSTRACT

Nanobiosensors have drawn significant research interest in recent years owing to the advantages of label-free, electrical detection. However, nanobiosensors fabricated by bottom-up process are limited in terms of yield and device uniformity due to the challenges in assembly. Nanobiosensors fabricated by top-down process, on the other hand, exhibit better uniformity but require time and costly processes and materials to achieve the critical dimensions required for high sensitivity. In this report, we introduce a top-down nanobiosensor based on polysilicon nanoribbon. The polysilicon nanoribbon devices can be fabricated by conventional photolithography with only materials and equipments used in the standard CMOS process, thus resulting in great time and cost efficiency, as well as scalability. The devices show great response to pH changes with a wide dynamic range and high sensitivity. Biomarker detection is also demonstrated with clinically relevant sensitivity. Such results suggest that polysilicon nanoribbon devices exhibit great potential toward a highly efficient, reliable and sensitive biosensing platform.

## INTRODUCTION

In recent years label-free, electrical nanobiosensors have drawn lots of research interests due to the potential of achieving superior time and cost efficiency to current state-of-the-art biosensing platform such as ELISA. Among nanobiosensors studied by various research teams, most of them are fabricated by "bottom-up" technique [1-4], that is, nanostructures are assembled to make devices. One of the major challenges for nanosensors fabricated by bottom-up technique is assembly, which can significantly limit the yield and uniformity of such nanosensors. The yield is highly related to cost and throughput, and uniformity is essential to the reliability of nanobiosensors. Although intensive research efforts have been made toward assembly of nanostructures, most of the techniques still lack controllability, reproducibility and scalability. The other school of process in nanotechnology is "top-down" fabrication, which seeks to create nanoscale devices by using larger, externally-controlled ones to direct their assembly. Top-down fabrication is much more controllable than their bottom-up counterpart, thus leads to more uniform device performance. As a result, nanobiosensors fabricated using top-down approaches can yield more reliable diagnosis. One of the major challenges for top-down nanobiosensors, however, is to achieve large surface-to-volume ratio, as surface-to-volume ratio is directly linked to sensitivity. Research efforts have been made toward reducing the critical dimension of nanostructures (hence increase surface-to-volume ratio) by applying techniques such as electron beam lithography and directional etching [5, 6]. However, such techniques are extremely time-consuming, cost-inefficient, and have poor scalability. Those drawbacks significantly limit the commercial impact the aforementioned nanobiosensors can potentially generate. In this regard, a recent study demonstrates that highly sensitive nanobiosensors can be achieved without pushing critical dimensions to a limit [7, 8]. By carefully limiting the active layer thickness, silicon nanoribbon based biosensors can be fabricated by conventional photography (~μm critical dimension) with clinically relevant sensitivity. Despite the promising result, single-crystalline silicon on insulator (SOI) wafers with extremely thin (~50nm) active layer are required to produce the nanobiosensors. SOI wafers with such thin active layers are not easily available, and thus can be very expensive. Also, precise oxidation and wet etching are needed to achieve the desired thickness, and thus further limit the time and cost efficiency of such platform.

In this paper, we describe a nanobiosensor platform based on polysilicon nanoribbons. The devices are fabricated with ~100% yield and great uniformity due to top-down process. The polysilicon is deposited using a highly scalable, precise and cost-efficient low-pressure chemical vapor deposition (LPCVD). Unlike SOI wafers, the thickness of polysilicon active layer and dielectric layer can be well controlled and easily customized for different applications. The whole process is compatible with conventional photolithography with only two masks required, thus are time and cost efficient. Moreover, the fabrication can be performed on full wafers, which results in great scalability. By performing pH sensing experiment with a wide dynamic range and high sensitivity, we demonstrate that the polysilicon nanoribbon sensors are highly sensitive to ions. Finally biomarker detection is performed with clinically relevant sensitivity, and thus confirms the practical value of polysilicon nanoribbon biosensors. Multiple devices are monitored simultaneously during all sensing experiments, and the response shows great uniformity. We demonstrate that polysilicon nanoribbon can act as a highly efficient, reliable and scalability platform for nanobiosensors. Such a platform exhibits great potential toward label-free, electrical biosensor with enormous practical impact.

## EXPERIMENT

The device fabrication process is shown in Figure 1. The fabrication starts with a simple silicon wafer with a layer of thermal oxide as in the inset of Figure 1(a). The oxide thickness is determined by the desired dielectric thickness. A thin layer (typically 50nm) of polysilicon is then deposited via low-pressure chemical vapor deposition (LPCVD). (shown in Figure 1(a)) Spin-on dopant (purchased from Emulsitone cooperation) of desired doping concentration is applied to the polysilicon via spin coating. The diffusion is performed at 1100°C for 15 minutes in nitrogen environment. This process is shown in Figure 1(b). A buffered HF solution is used to remove spin-on dopant upon the completion of diffusion process. Photolithography and a $CF_4$ dry etch are then performed to define the contact lead and nanoribbon area. (Figure 1(c)) A second photolithography is then performed to define the metal contact. Finally, 5nm Ti and 45nm Au are deposited as electrode via thermal evaporation. (Figure 1 (d)) A HF dipping is required before the metal evaporation to remove native oxide. A nitride passivation layer can be deposited via PECVD if necessary. We note that the device fabrication is highly efficient and scalable with only two masks required and all the process involved can be performed on wafer scale. Shown in Figures 1(e) and 1(f) are a SEM image of polysilicon nanoribbons and a photographic image of hundreds of polysilicon nanoribbon sensor arrays fabricated on a whole 3" wafer, respectively. The yield is ~100%, and the devices exhibit very little device-to-device variation due to the top-down process.

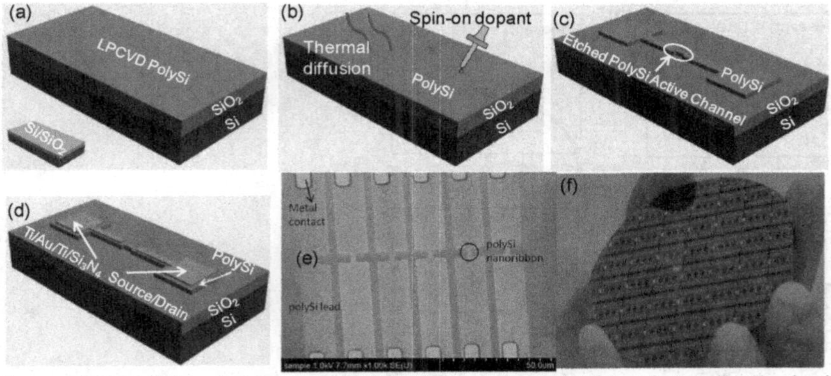


**Figure 1** (a) Polysilicon is deposited via LPCVD. Inset shows Si/$SiO_2$ substrate (b) Conduction is aided by spin-on dopants that are thermally diffused into the polysilicon. (c) Active mesas defined by photolithography and dry etching. (d) Metal contact and passivation layer defined by photography and thermal evaporation. (e) SEM image of an array of polysilicon nanoribbon devices. (f) Photographic image of nanosesnor arrays on a 3" wafer.

During the fabrication, spin-on boron dopant solutions of three different concentrations, ($C_0 = 1\times10^{17}$, $5\times10^{17}$ and $1\times10^{18}$) are applied to the polysilicon. We characterized the electrical properties for devices of all three doping concentrations with an Agilent semiconductor analyzer. During the measurement, back gate voltage is applied to the silicon substrate with 500nm silicon oxide as dielectric, and the result is shown in Figure 2. Plotted in Figure 2 (a) and (b) is the source-drain current ($I_{ds}$) versus source-drain voltage ($V_{ds}$) under various back gate voltage ($V_g$) and $I_{ds}$ versus $V_g$ under various $V_{ds}$ for a device with $C_0 = 1\times10^{17}$ doping, respectively. The device exhibits ohmic contact behavior with strong gate dependence. On/off ratio reaches 250 and such value is high enough for nanobiosensor application. We note that the on/off ratio can be further improved by using dielectrics with higher dielectric constant or by applying the gate voltage via liquid gate. The $I_{ds}$ versus $V_{ds}$ under various $V_g$ and $I_{ds}$ versus $V_g$ under various $V_{ds}$ for a device with $C_0 = 5\times10^{17}$ doping concentration are shown in Figure 2 (c) and (d), respectively. The device is 10 times more conductive compared to the $1\times10^{17}$ device. However, the on/off ratio is reduced to 62. The $I_{ds}$ versus $V_{ds}$ under various $V_g$ and $I_{ds}$ versus $V_g$ under various $V_{ds}$ for a device with $C_0 = 1\times10^{18}$ doping concentration are shown in Figure 2 (e) and (f), respectively. The device exhibits higher on-state current and lower on/off ratio compared to devices with lower doping concentrations. Further study indicates that devices with low doping concentration yield better sensitivity as nanobiosensors. Thus the $1\times10^{17}$ doped devices are selected for further sensing experiments in this chapter unless otherwise stated.

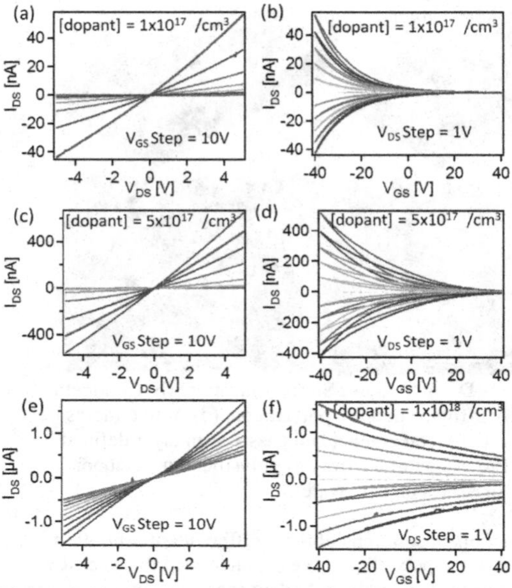


**Figure 2** (a) $I_{ds}$ versus $V_{ds}$ under various $V_g$ for a device doped with spin-on boron solution of $C_0 = 1\times10^{17}$ (step: 10V). (b) $I_{ds}$ versus $V_g$ under various $V_{ds}$ for a device with $1\times10^{17}$ doping (step: 1V). (c) $I_{ds}$ versus $V_{ds}$ under various $V_g$ for a device with $5\times10^{17}$ doping (step: 10V). (d) $I_{ds}$ versus $V_g$ under various $V_{ds}$ for a device with $5\times10^{17}$ doping (step: 1V). (e) $I_{ds}$ versus $V_{ds}$ under various $V_g$ for a device with $1\times10^{18}$ doping (step: 10V). (f) $I_{ds}$ versus $V_g$ under various $V_{ds}$ for a device with $1\times10^{18}$ doping (step: 1V).

## DISCUSSION

### pH sensing

To demonstrate the polysilicon FETs are ion-sensitive, we preformed pH sensing with three devices simultaneously. During the sensing experiment, a 200mV source-drain voltage and a 200mV liquid gate voltage is applied to the devices. The source-drain current is monitored by a semiconductor analyzer. The devices were first soaked in a buffer of pH 9, and then exchanged to a buffer of pH8. All three devices showed significant decrease in conductance after the buffer exchange, and such decrease in conductance can be explained by the p-type transistor behavior of the polysilicon devices. Further buffer exchanges were performed to change the pH to 7, 6, 5, 4, 5, 6, 7, 8, and 9, respectively. The normalized current versus time for all 3 devices is plotted in Figure 3(a). All three devices showed decreased conductivity under lower pH. The variation in conductivity from pH4 to pH9 is around 200%. Further pH sensing experiment is carried out with a pH step of 0.2, and the result is shown in Figure 3(b). The device shows a response of

10% with high signal-to-noise ratio even to a 0.2 variation in pH, suggesting the polysilicon nanosensor is highly sensitive to ions.

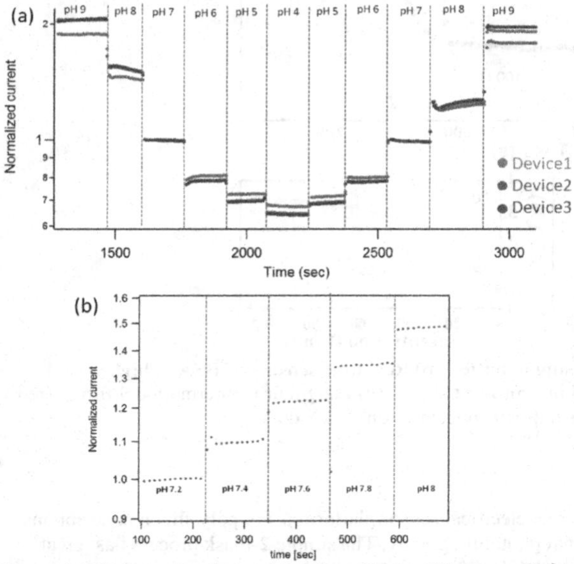


**Figure 3** (a) Normalized current under pH4 to pH9 (step= 1) for three devices measured simultaneously. (b) Normalized current versus time with a pH step of 0.2.

## CA125 biomarker sensing

We also performed detection for an ovarian cancer biomarker, antigen 125 (CA-125), in buffer. The polysilicon devices are first conjugated with CA-125 for selectivity. During the sensing experiment, a 200mV source-drain voltage and liquid gate voltage are applied to the device. CA-125 was progressively added to the sensor surface.

The normalized current versus time for a polysilicon device is plotted in Figure 4 (a). The current started to show response at a CA-125 concentration of 1U/ml (5pM). Such limit of detection is two orders of magnitude lower than the clinically relevant level for diagnosis. The device showed larger responses to CA-125 at higher concentrations. In comparison, the polysilicon sensor did not show significant response to 250nM bovine serum albumin (BSA), as shown in Figure 4 (b), even the concentration is several hundred times higher than the target analyte. The response to BSA can be further suppressed using passivation methods. The response versus CA-125 concentration is plotted in Figure 4 (c) along with the simulated Langmuir isotherm model fitting. From this model, we obtain a dissociation constant of 22 U/ml. The sensing data shows that the polysilicon nanosensor can detect biomarkers with clinically relevant sensitivity and great selectivity. The results suggest that polysilicon nanoribbon biosensors have great potential as a platform for clinical diagnosis of diseases such as cancers.

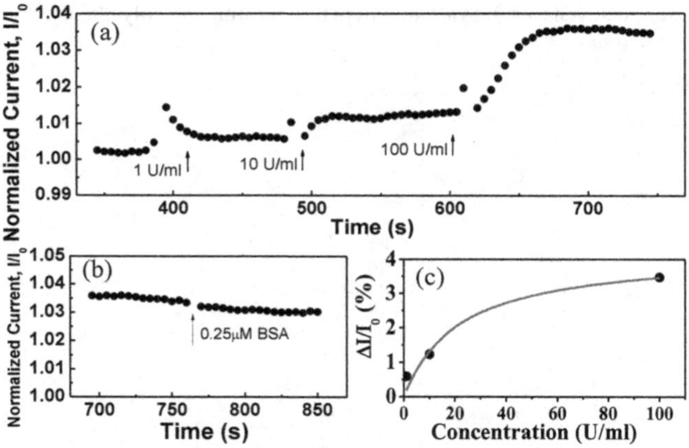


**Figure 4** (a) Real time CA-125 sensing in buffer. (b) Real time sensing of BSA at higher concentration than CA-125 showed insignificant signal. (c) Langmuir isotherm model fitting (red line) of the CA-125 response versus antigen concentration (black dot).

## CONCLUSIONS

We have developed a label-free, electrical sensor platform using polysilicon nanoribbons that can be fabricated by conventional photolithography. The simple 2-mask process has great time and cost efficiency, as well as scalability. The devices are very responsive to pH changes with a wide dynamic range and high sensitivity. Biomarker detection is demonstrated using CA-125, an ovarian cancer biomarker, as model with clinically relevant sensitivity. Conductivity of multiple devices simultaneously measured during pH sensing confirms the uniformity of device performance. Such results suggest that polysilicon nanoribbon sensors have great potential as a highly efficient, reliable and sensitive platform for cancer diagnosis.

## REFERENCES

[1] G. Zheng, F. Patolsky, Y. Cui, W. U. Wang, and C. M. Lieber, *Nat Biotechnol,* vol. 23, pp. 1294-301, Oct 2005.
[2] G. Gruner, *Anal Bioanal Chem,* vol. 384, pp. 322-35, Jan 2006.
[3] B. L. Allen, P. D. Kichambare, and A. Star.
[4] I. Heller, A. M. Janssens, J. Mannik, E. D. Minot, S. G. Lemay, and C. Dekker, *Nano Lett,* vol. 8, pp. 591-5, Feb 2008.
[5] Y. L. Bunimovich, Y. S. Shin, W. S. Yeo, M. Amori, G. Kwong, and J. R. Heath, *J Am Chem Soc,* vol. 128, pp. 16323-31, Dec 20 2006.
[6] R. Juhasz, N. Elfström, and J. Linnros, *Nano Lett,* vol. 5, pp. 275-280, 2005/02/01 2004.
[7] N. Elfström, A. E. Karlström, and J. Linnros, *Nano Lett,* vol. 8, pp. 945-949, 2008/03/01 2008.
[8] E. Stern, J. F. Klemic, D. A. Routenberg, P. N. Wyrembak, D. B. Turner-Evans, A. D. Hamilton, *et al.*, *Nature,* vol. 445, pp. 519-22, Feb 1 2007.

**Mater. Res. Soc. Symp. Proc. Vol. 1569 © 2013 Materials Research Society**
**DOI: 10.1557/opl.2013.1101**

# Electrically Conducting PEDOT:PSS – Gellan Gum Hydrogels

Holly Warren[1], Marc in het Panhuis[1,2]

[1]Soft Materials Group, School of Chemistry, University of Wollongong, Wollongong NSW 2522, Australia.
[2] Intelligent Polymer Research Institute, ARC Centre of Excellence for Electromaterials Science, AIIM Facility, University of Wollongong, Wollongong, NSW, 2522, Australia.

## ABSTRACT

Hydrogels consisting of the conducting polymer PEDOT:PSS and the biopolymer gellan gum (GG) were characterized using electrical, mechanical and rheological methods. Compression testing and rheological analysis showed that the gels weakened with increasing PEDOT:PSS content. In contrast, the increasing PEDOT:PSS content resulted in an increasing electrical conductivity.

## INTRODUCTION

Bionic and soft-robotic applications require soft materials which are both electrically conducting and mechanically compliant[1]. Gels based on biopolymers are suitable candidates for such materials as they are classified as soft materials[2]. Conducting polymers are well-known materials, which can be used to fabricate conducting composites. For example, it has been found that conducting polymers have enhanced the lifetime of neural implants[3] by reducing localized tissue damage. Gellan gum is a linear anionic polysaccharide, derived from the bacterium *Sphingomonas Elodea*[4]. It is a gel-forming biopolymer with a tetrasaccharide repeating unit structure; commonly used as a food additive with Food and Drug Administration and European Union approval (E418). It is cross-linked with cations at high temperatures (~80°C) to form rigid hydrogels upon cooling. It has been found that divalent cations efficiently promote the gelation of gellan gum due to the formation of ionic bonds between the carboxyl groups on adjacent chains[5].

Poly(3,4-ethylenedioxythiophene) poly(styrenesulfonate) (PEDOT:PSS) is a block copolymer consisting of anionic PSS and cationic PEDOT. It is a conducting polymer which has been used in applications such as humidity and temperature sensors[6], anti-static coatings[7] and actuators[8].

In this study, gellan gum is combined with the conducting polymer PEDOT:PSS, to assemble electrically conducting gels. These hydrogels are characterized using mechanical compression testing, rheological analysis and electrical impedance analysis.

## EXPERIMENTAL

Low acyl gellan gum (GG) was received as a gift from CP Kelco (Gelzan, Lot # 1I1443A). All solutions and subsequent gels were prepared with Milli-Q water (resistivity ~ 18.2

MΩ·cm). Gellan solutions were prepared by adding dry GG powder to heated (~80°C) water under rapid stirring (800 r.p.m) using an overhead stirrer (IKA RW 20 digital). PEDOT:PSS, (conductive grade, 1.3 % solution in water, batch no. MKBJ4242V) was purchased from Sigma Aldrich, Australia.

PEDOT:PSS-GG hydrogels were prepared as follows: PEDOT:PSS dispersions were heated to ~80°C, appropriate quantities of hot GG solutions were added while stirring to result in gels with GG concentration 0.5 % w/v. $CaCl_2$ solution was added such that the composite solution contained 5 mM $Ca^{2+}$ ions. For example, a PEDOT:PSS-GG gel with PEDOT:PSS content of 1.05 % w/v was prepared by combining 15 mL PEDOT:PSS dispersion with 5 mL GG solution (2% w/v) and 100 μL $CaCl_2$ solution (1M). Hot liquids were poured into plastic molds (diameter 16 mm, height 10 mm), forming gels upon cooling. GG hydrogels were prepared by cross-linking a GG solution (0.5% w/v) with $CaCl_2$ as described above.

Electrical impedance analysis was carried out using a custom-built instrument. Electrical conductivity ($\sigma$) was evaluated from impedance measurements (1 Hz – 100 kHz). Mechanical testing was carried out using a Shimadzu Universal Tester at a rate of 2 mm/min using a 50N load cell. Oscillatory rheology was performed using an Anton Paar Physica MCR 301 Digital Rheometer (parallel plate tool, diameter 15 mm) at a frequency of 10 Hz while performing a shear strain sweep.

## DISCUSSION

Hydrogels were prepared by combining PEDOT:PSS dispersions with GG solutions and cross-linked with $Ca^{2+}$. GG hydrogels were prepared by cross-linking GG solutions. Electrical impedance analysis was used to evaluate the electrical conductivity of these GG and PEDOT:PSS-GG hydrogels (Figure 1). The GG gels are conducting ($\sigma = 0.138 \pm 0.007$ S/m), even in the absence of PEDOT:PSS, due to the presence of ionic charge carriers. PEDOT:PSS volume fractions were calculated from the PEDOT:PSS (0.8 $g/cm^3$[9]), GG (1.3 ± 0.03 $g/cm^3$[10]) and water densities. The conductivity increased with increasing PEDOT:PSS volume fraction. For example, the conductivity of PEDOT:PSS-GG gel (1.05 % w/v PEDOT:PSS) increased by 128 % compared to the corresponding value for a GG hydrogel, i.e. from 0.138 ± 0.007 S/m to 0.315 ± 0.014 S/m.

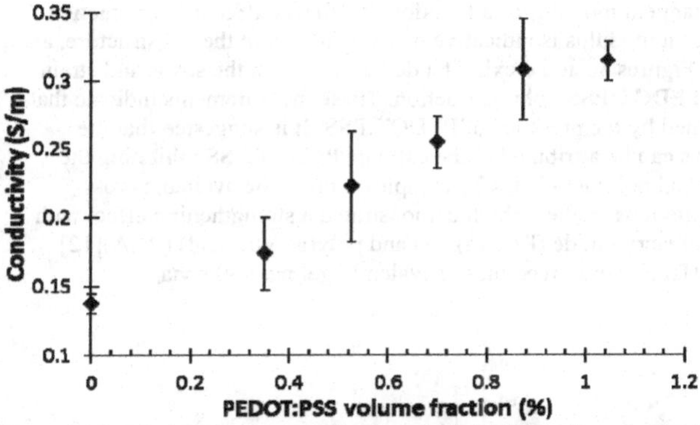


**Figure 1.** Electrical conductivity as a function of PEDOT:PSS volume fraction for rigid PEDOT:PSS-GG samples, measured using a custom-built apparatus for electrically characterizing liquids and gels.

Mechanical and rheological analysis was used to probe the effect of PEDOT:PSS on the mechanical characteristics of GG gels. GG hydrogels (0.5 % w/v) had an average storage modulus of 24.3 ± 0.7 kPa #in the linear viscoelastic (LVE) region, prior to the addition of PEDOT:PSS. Figure 2 shows the storage (elastic) modulus in the LVE region as a function of shear strain. For all PEDOT:PSS concentrations, the loss modulus was always lower than the storage modulus in the LVE region, indicating that these materials were true gels. The storage modulus decreases with increasing PEDOT:PSS content. For example, it was found to decrease by one order of magnitude when the PEDOT:PSS volume fraction was increased to 1.05 %. This suggests that addition of PEDOT:PSS weakens the GG network structure.

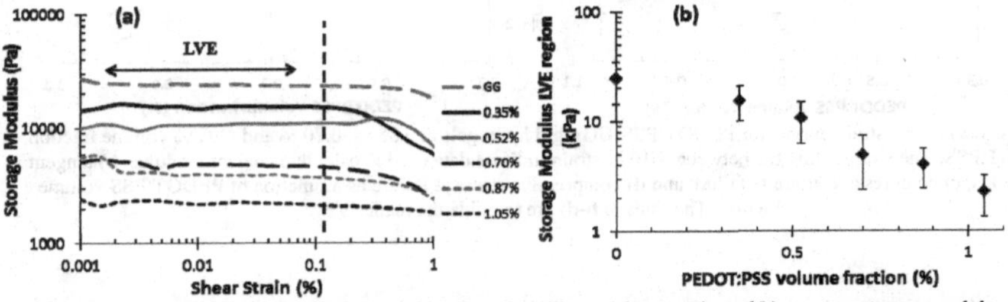


**Figure 2. a)** Rheological storage modulus as a function of oscillatory shear strain and **b)** average storage modulus in the LVE region for rigid PEDOT:PSS-GG hydrogels with varying PEDOT:PSS volume fractions between 0 and 1.05 % v/v.

These observations were further supported by the results from mechanical compression testing.Typical stress-strain curves of PEDOT:PSS gels are shown in Figure 3a. This data is used to calculate tangent modulus (5 – 10 % strain), stress at failure and strain at failure values.

Figure 3b is a plot of the tangent modulus as a function of PEDOT:PSS volume fraction. The depicted decrease in the tangent modulus is indicative of a weakening in the gel structure, as observed in the rheological data. Figures 3c and d exhibit a decrease in both the stress and strain values at failure as a function of PEDOT:PSS volume fraction. These measurements indicate that GG gels are mechanically weakened by the presence of PEDOT:PSS. It is suggested that the reduction in mechanical properties can be attributed to the cationic PEDOT:PSS inhibiting the $Ca^{2+}$ cross-linking of the anionic GG polymer chains by complexation of the available cross-linking sites. This is contrary to previous studies which demonstrated a strengthening effect with the addition of PEDOT:PSS to polyacrylamide (PAAm)[11] and poly(acrylic acid) (PAA)[12] gels. It was suggested that PEDOT:PSS reinforced these covalently gel networks via entanglement.

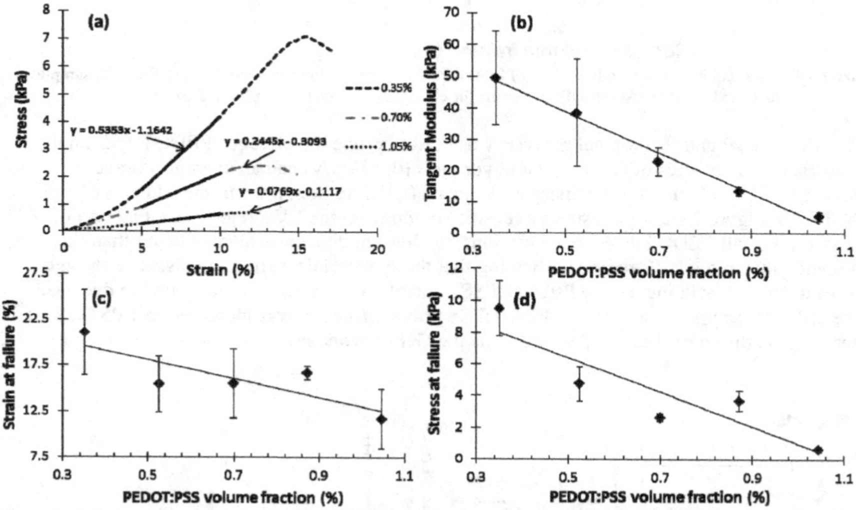


**Figure 3. a)** Stress-strain curves for PEDOT:PSS-GG rigid hydrogels at 0.35 %, 0.70 % and 1.05 % volume fraction PEDOT:PSS. The straight line fits between 5-10 % strain indicate data used to calculate tangent modulus. **b)** tangent modulus, **c)** compressive strain at failure and **d)** compressive stress at failure as a function of PEDOT:PSS volume fraction. The lines in **b-d)** are to guide the reader's eye.

## CONCLUSIONS

The preparation and characterization of PEDOT:PSS-gellan gum hydrogels is presented. It was shown that the electrical conductivity increased proportionally with PEDOT:PSS volume fraction between 0.35 % and 1.05 % to a value of $0.315 \pm 0.014$ S/m. However, this increase in conductivity with PEDOT:PSS volume fraction is coupled with a decrease in storage modulus, tangent modulus, stress at failure and strain at failure. This work has shown that conducting hydrogels can be prepared by combining a biopolymer with a conducting polymer.

## ACKNOWLEDGMENTS

This work was supported by the University of Wollongong and the Australian Research Council Centre of Excellence and Future Fellowship programs. Dr. R. Clark (CP Kelco) is thanked for the provision of gellan gum.

## REFERENCES

1. Ilievski, F., Mazzeo, A. D., Shepherd, R. F., Chen, X. & Whitesides, G. M. *Angewandte Chemie* **123**, 1930–1935 (2011).
2. Ferris, C. J., Gilmore, K. J., Wallace, G. G. & in het Panhuis, M. *Soft Matter* **9**, 3705–3711 (2013).
3. Green, R. A., Lovell, N. H., Wallace, G. G. & Poole-Warren, L. A. *Biomaterials* **29**, 3393–3399 (2008).
4. Morris, E. R., Nishinari, K. & Rinaudo, M. *Food Hydrocolloids* **28**, 373–411 (2012).
5. Nickerson, M. T., Paulson, A. T. & Speers, R. A. *Food Hydrocolloids* **17**, 577–583 (2003).
6. Daoud, W. A., Xin, J. H. & Szeto, Y. S. *Sensors and Actuators B* **109**, 329–333 (2005).
7. Jonas, F. & Morrison, J. T. *Synthetic Metals* **85**, 1397–1398 (1997).
8. Okuzaki, H., Suzuki, H. & Ito, T. *Synthetic Metals* **159**, 2233–2236 (2009).
9. Patra, P. K., Calvert, P. D., Warner, S. B., Kim, Y. K. & Chen, C. H. *National Textile Centre Annual Report* (2007).
10. Pidcock, G. C. & in het Panhuis, M. *Advanced Functional Materials* **22**, 4789-4799 (2012).
11. Aouada, F., Guilherme, M., Campese, G., Girotto, E., Rubira, A. & Muniz, E. *Polymer Testing* **25**, 158–165 (2006).
12. Dai, T., Qing, X., Lu, Y. & Xia, Y. *Polymer* **50**, 5236–5241 (2009).

**Mater. Res. Soc. Symp. Proc. Vol. 1569 © 2013 Materials Research Society**
**DOI: 10.1557/opl.2013.1090**

# Incorporating Biodopants into PEDOT Conducting Polymers: Impact of Biodopant on polymer properties and biocompatibility

Paul J. Molino[1], Anthony Tibbens[1], Robert M.I. Kapsa[1], Gordon G. Wallace[1]

[1]ARC Centre of Excellence for Electromaterials Science, Intelligent Polymer Research Institute, University of Wollongong, Wollongong, NSW 2522, Australia.

## ABSTRACT

Poly(3,4-ethylenedioxythiophene) (PEDOT) was polymerized with the biological dopants dextran sulphate and chondroitin sulphate. Polymer physical and mechanical properties were investigated using quartz crystal microgravimetry with dissipation monitoring and atomic force microscopy, revealing polymer shear modulus and interfacial roughness to be significantly altered as a function of the dopant species. The adsorption of fibronectin, an important extracellular protein that is critical for a range of cellular functions and processes, was investigated using QCM-D, revealing protein adsorption to be increased on the DS doped PEDOT film relative to the CS doped film. PEDOT films have traditionally been doped with synthetic counterions such as polystyrene sulphonate (PSS), however the incorporation of biological molecules as the counterion, which has been shown to improve polymer biofunctionality, has received far less attention. In particular, there has been little detailed study on the impact of incorporating polyelectrolyte biomolecules into the PEDOT polymer matrix on fundamental polymer properties which are critical for biomedical applications. This investigation provides a detailed characterization of the interfacial and mechanical properties of biologically doped PEDOT films, as well as the efficacy of the composite films to bind and retain extracellular proteins of the type that are critical to the biocompatibility of the polymeric material.

## INTRODUCTION

Inherently conducting polymers (ICPs) have attracted significant interest in the area of biomaterials research due to their biocompatibility and ability to perform a variety of biologically relevant functions such as the controlled release of growth factors and targeted drugs[1] and mechanical and electrical stimulation[2,3]. Of the ICPs, Poly(3,4-ethylenedioxythiophene) (PEDOT) has attracted particular attention due to its favorable mechanical properties and chemical stability. During ICP synthesis, the positive charge on the polymer backbone is neutralized by the incorporation of an anionic species, a process known as doping. While PEDOT has typically been doped with a range of synthetic dopants, little attention has been directed to understanding the influence of the incorporation of large biological dopants on biologically relevant polymer properties, including polymer modulus and roughness, as well as the ability of the polymer surface to attract and retain proteins.

It is widely accepted that artificial implanted surfaces may never directly interact with the cellular environment. Rather, proteins adsorbed to the material surface will interact with the surrounding environment and thus mediate subsequent cellular interactions [4]. Numerous proteins such as fibronectin (FN) and vitronectin are critical for cellular anchorage and also in triggering a cascade of intracellular reactions that ultimately direct cellular behavior on the material surface. It is therefore important when designing new biomaterials that the nature of protein interactions at the material interface are considered and well understood.

Herein we employ QCM-D and AFM to investigate the influence of the large polyelectrolyte biological dopants dextran sulphate and chondroitin sulphate on shear modulus and interfacial surface roughness. QCM-D is a highly sensitive surface sensing technique that can be applied to characterise the physical and mechanical properties of mass adsorbed to either the native electrode sensor surface, or a sensor modified with a range of materials, including polymeric films. In addition to characterizing the PEDOT films themselves, we employ this technique to evaluate the ability for the PEDOT films to adsorb fibronectin; an important extracellular protein that is critical for a range of cellular processes, including cellular adhesion, mobility and differentiation.

## EXPERIMENT

3,4-Ethylenedioxythiophene (EDOT) was purified by distillation and stored at -18 °C. Electrochemical Polymerisation of 3,4-Ethylenedioxythiophene (EDOT) with the biological dopants Dextran Sulphate (DS) and Chondroitin Sulphate (CS) was performed in a three electrode Q-Sense Electrochemistry module (QEM 401) connected to a Q-Sense E4 Quartz Crystal Microbalance (Q-Sense AB, Västra, Frölunda, Sweden) and an eDAQ e-corder 410 recorder and EA163 potentiostat. PEDOT films were polymerized on Q-Sense gold coated sensor crystals. An aqueous solution of 0.01M EDOT and 2mg/ml of the dopant was polymerized at a current density of 0.25mA/cm$^2$ for 2 mins. Modelling of the polymerized polymer film properties was performed using the Q-tools software package employing the Voigt viscoelastic model.

Atomic Force Microscopy (AFM) was employed to investigate the polymer interfacial surface properties. AFM was used to investigate the morphology of the polymer films using a MFP-3D AFM (Asylum Research, CA) with a Mikromash NSC15 cantilever (spring constant ~ 37 N/m). Images of the polymer surfaces were obtained AC mode (5 x 5μm) at a 0.5 Hz scan rate in air. The Asylum Research Analysis software with the Igor Pro Software Package (WaveMetrics, OR) was used to determine the RMS roughness and surface area values as reported. Force measurements were performed using a JPK Nanowizard II BioAFM system. Force distance curves were first converted to force-indentation curves, and thereafter fitted with the Hertz model to calculate Elastic Modulus. Elastic modulus measurements were further converted to Shear Modulus using equation (1).

The mass of the protein fibronectin (FN) adsorbed to the polymer film surface was investigated using QCM-D. A 50μg/ml solution of FN in PBS was introduced to the QCM-D sensor chamber containing a polymer coated sensor at 10μl/min for 60 mins, and thereafter rinsed with PBS. Protein adsorption experiments were performed in triplicate.

## DISCUSSION

### Polymer morphology and surface roughness

The interfacial properties of the PEDOT films were investigated by AFM (Figure 1) with the critical interfacial polymer parameters $R_{RMS}$ roughness and surface areas outlined in Table 1. The surface properties of both biodoped PEDOT films varied according to the biodopant incorporated during electrochemical polymerisation. The PEDOT films demonstrated a nodular morphology, with PEDOT-CS exhibiting small, finer nodules compared to the DS doped films. Accordingly, PEDOT-DS demonstrated a higher average $R_{RMS}$ roughness value (23.5 ± 2.9 nm) compared to PEDOT-CS (21.9 ± 1.1 nm). However the specific interfacial surface area (represented as the percentage increase above the geometric AFM scan area, i.e., 25μm/cm$^2$) was inversely related to $R_{RMS}$ roughness, with PEDOT-CS demonstrating an increase of 24.4 ± 2.9% compared to 10.69 ± 3.43% for PEDOT-DS.

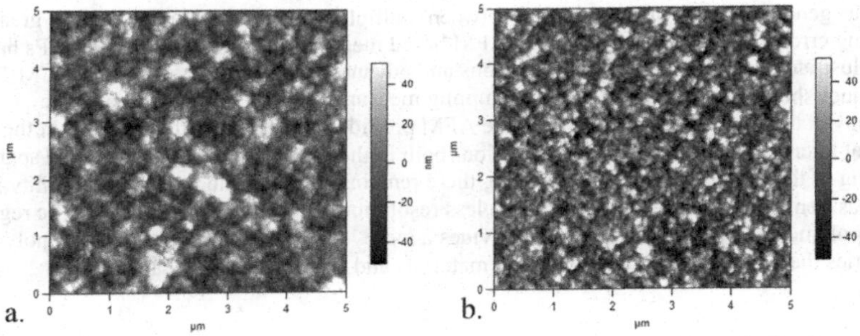

**Figure 1.** AFM topographic images (5μm x 5μm) of PEDOT-DS (a) and PEDOT-CS (b).

**Table 1.** Polymer Interfacial Surface Properties

| | PEDOT-CS | PEDOT-DS |
|---|---|---|
| Roughness $_{RMS}$ (nm) | 21.91 ± 1.14 | 23.59 ± 2.9 |
| Specific Surface Area (%) | 24.4 ± 2.93 | 10.69 ± 3.43 |

### Polymer modulus

Figure 2 illustrates the QCM-D modeling results of the shear modulus of the PEDOT biocomposite polymer films grown for 2 mins at 0.25mA/cm$^2$. The highest shear modulus was exhibited by PEDOT-DS (29.1 ± 4.9 MPa), with PEDOT-CS demonstrating a significantly lower modulus of 5.4 ± 0.4 MPa. To test the general efficacy of the QCM-D model used to resolve the polymer modulus, AFM force measurements were undertaken to quantify the polymer elastic modulus using the Hertz model, with the results from the two different techniques compared by converting the elastic modulus to shear modulus using the equation:

$$G = \frac{E}{2(1+\nu)} \qquad (1)$$

where $G$ is the shear modulus, $E$ the Young's modulus, and $\nu$ the Poisson's ratio (0.33). The converted shear modulus values were higher than those recorded using the QCM-D technique, with mean values of 58 MPa and 27 MPa for PEDOT-DS and PEDOT-CS, respectively; however the general trend between the two polymers remained the same (Figure 2). This variation is likely representative of the specific technique, and the area of the polymer probed: QCM-D provides a bulk measurement of the entire thickness of the polymer, yielding an average measurement over this area. In comparison, AFM only probes the top few nanometers of the polymer film, and this measurement may be confounded by the specific geometry/porosity of the polymer at the indentation site (~10 nm in size:- the size of the AFM tip), thus generating significant variation between multiple measurements and therefore greater sampling error. Considerable variation in AFM based measurement of the moduli of ICPs has been illustrated previously [4]. The limitations and advantages of both the AFM and QCM-D techniques should be considered when attempting measurements on porous, viscoelastic materials of the type reported herein. While AFM provides a measure of the modulus at the material interface at extremely high resolution (both in the forces applied/measured and spatial mapping of the modulus across the surface), there remains considerable inherent variability in the measurements, whilst QCM-D provides less resolution in terms of probing a specific region of the polymer (i.e. interface), however provides a more 'uniform' measure of the bulk polymer properties that is far better suited to porous materials and presents far less variation.

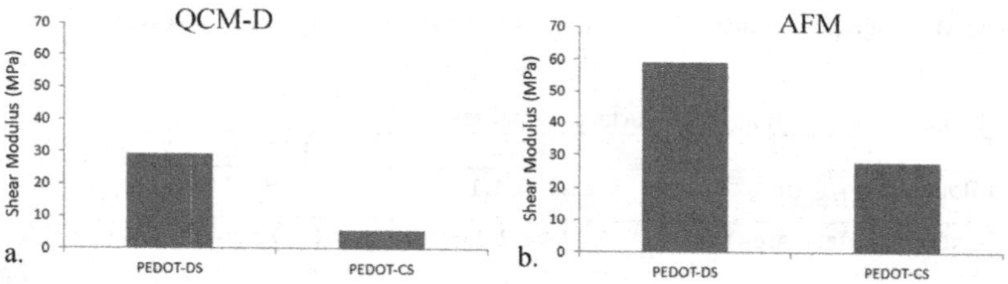


**Figure 2.** QCM-D (a) and AFM (b) based measurements of the shear modulus of PEDOT-DS and PEDOT-CS polymer films grown on QCM-D sensor crystals.

**Protein adsorption**

QCM-D is a powerful technique that provides the ability to resolve the adsorption of proteins to a substratum surface in their native, hydrated condition and in real-time, without the need for laborious staining or sample preparation techniques. Figure 3 illustrates the raw frequency and dissipation shifts resulting from the adsorption of FN to the biodoped PEDOT films, with FN injected into the sample chamber at time zero. PEDOT-DS illustrates significantly greater *f* and *D* parameter shifts compared to PEDOT-CS (Figure 3). When the raw *f* parameter shifts are normalized against the variable surface area exhibited by the two polymer films, direct comparisons of the ability of each polymer to adsorb protein per unit surface area can be resolved. PEDOT-DS demonstrated significantly greater FN adsorption, demonstrating a $\Delta f$ of -65.9 ± 2.8Hz.cm$^2$ compared to PEDOT-CS which demonstrated a frequency shift of -14.1 ± 2.2Hz.cm$^2$.

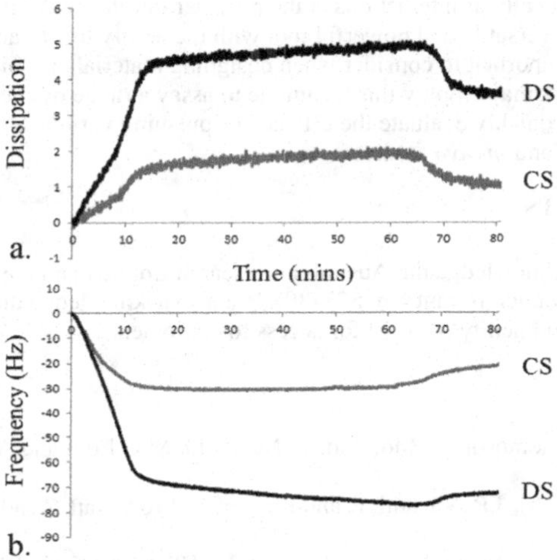


**Figure 3.** Representative QCM-D frequency and dissipation responses from the adsorption of FN to PEDOT-DS (black) and PEDOT-CS (grey).

A range of polymer properties contribute to the overall biocompatibility of a biomaterial, and its ability to perform the desired function within the body. Mechanical compliance of the biomaterial with surrounding tissue is critical to reducing inflammatory responses and fibrotic tissue encapsulation of the material [5,6]. PEDOT-CS demonstrated a lower modulus compared to the DS doped films. Polypyrrole films doped with DS and CS have previously demonstrated significantly higher moduli than those presented herein, with pyrrole films generally demonstrated to be stiffer or less viscoelastic than PEDOT polymers when possessing similar dopants. However the relationship between the films incorporating each dopant generally remained the same, with the DS doped films demonstrating a higher elastic modulus (706 ± 44 MPa) compared to the CS doped PPy (293 ± 31 MPa) [4].

The development of a proteinaceous film on the material surface has also been demonstrated to be critical in mediating cellular and tissue response to the implant. PEDOT-DS demonstrating a considerably greater ability to bind FN to the polymer surface, compared to

PEDOT-CS, and therefore is likely to provide an enhanced polymer surface conditioning layer with which cells and tissues may adhere and interact.

## CONCLUSIONS

This study has illustrated PEDOT doped with DS to provide a stiffer substratum compared to CS doped PEDOT, and to bind a greater mass of the extracellular matrix protein FN that is important to direct cellular interactions at the polymer interface. Additionally the QCM-D technique proved to be a versatile and powerful tool with the ability to characterise a number of film properties that are important to consider when designing materials for biological applications. Future work may employ this technique to assay a range of chemistries and compositions in order to quickly evaluate the efficacy of pursuing various biomaterial formulations for *in-vitro* and *in-vivo* studies.

## ACKNOWLEDGMENTS

The authors gratefully acknowledge the Australian Research Council and the National Health and Medical Research Council (Grant No. 573430). We also acknowledge the Australian National Nanofabrication Facility (ANFF) for access to equipment.

## REFERENCES

1. R.T Richardson, B. Thompson, S. Moulton, C. Newbold, M.G Lum and A. Cameron, Biomat. **28(3),** 513 – 523 (2007).
2. C.E Schmidt, V.R Shastri, J.P. Vacanti, R and Langer, R. Proc. Natl. Acad. Sci. USA **94(17),** 8948 – 8953 (1997).
3. A. Gelmi, M.J. Higgins and G.G. Wallace, Biomat. **31**, 1974 – 1983 (2010).
4. C.A. Wilson, R.E. Clegg, D.I. Leavesley and M.J. Pearcy, *Tissue Engineering* **11**, 1-18 (2005).
5. A.J. Engler, M.A. Griffin, S. Sen, C.G. Bonnemann, H.L. Sweeney and D.E. Discher. J. Cell Biol. **166(6),** 877-887 (2004).
6. T. Chaudhuri, F. Rehfeldt, H.L. Sweeney and D.E. Discher. Methods Mol. Biol. **621**, 185-202 (2010).

Mater. Res. Soc. Symp. Proc. Vol. 1569 © 2013 Materials Research Society
DOI: 10.1557/opl.2013.828

# Biotemplating Magnetic Nanoparticles on Patterned Surfaces for Potential Use in Data Storage

Johanna M. Galloway*, Scott M. Bird, Jonathan P. Bramble, Kevin Critchley and Sarah S. Staniland
School of Physics and Astronomy, The University of Leeds, Woodhouse Lane, Leeds, LS2 9JT, UK.

## ABSTRACT

Thin-films of magnetic nanoparticles (MNPs) with high coercivities are deposited onto surfaces for use in data storage applications. This usually requires specialist clean-room facilities, sputtering equipment and high temperatures to achieve the correct crystallographic phases. One possible cheaper and more environmentally friendly alternative could be to use biomolecules. Many biomineralization and biotemplating molecules have been identified that are able to template a wide range of technologically relevant materials using mild, aqueous chemistry under physiological reaction conditions. Here, we have designed a dual affinity peptide (DAP) sequence to template MNPs onto a surface. One end of the DAP has a high binding affinity for $SiO_2$ and the other for MNPs of the $L1_0$ phase of CoPt, a high coercivity magnetic material. Images of the biomineralized substrates show that nanoparticles of CoPt are localized onto the areas that were functionalized with the biotemplating DAP. Magnetic force microscopy (MFM) plots of the biotemplated nanoparticles show that there is magnetic contrast on the patterned surface.

## INTRODUCTION

Magnetic nanoparticles (MNPs) are used in a wide range of technological applications, which includes the storage of electronic data on computer hard disks [1-3]. It is essential that the magnetic response of the MNPs is consistent and predictable, so that their behavior when an external field is applied is reliably known [4]. The magnetic properties of a material are not only dependent on the material an MNP is made from, but also on its size, shape and crystal structure [3-5]. As such, the MNPs used in any application must be uniform in all of these respects to ensure the required consistency in magnetic behavior. To synthesize such monodispersed MNPs from a bulk solution usually requires high temperatures and toxic reagents [6-9], which is not environmentally friendly. In the natural world, biomineralization proteins are able to template the formation of at least 60 different biominerals [10]. These proteins precisely control the composition, specific crystallographic phases and hierarchical ordering of mineral-organic hybrid structures, which often have superior properties to their artificially synthesized counterparts [9, 11-14]. Excitingly, some of these biomineralization proteins, as well as smaller biotemplating peptide sequences, are able to direct the formation of consistent, uniform nanoparticles from aqueous solutions at room temperature *in vitro* [7, 8, 14-19].

MNPs must be attached to surfaces and the particles should have a high coercivity (i.e. be magnetically hard) to be useful in data storage. This usually requires clean-room facilities, photolithography and sputtering at high temperatures and under vacuum, which is extremely costly and energy intensive. Soft-lithographic techniques, such as micro-contact printing (μCP), have been used to functionalize surfaces with biomineralization proteins and peptides [7, 8, 16-18]. These biomolecules have been shown to be able to template the formation of nanoparticles and/or thin-films of materials onto the patterned surfaces under mild reaction conditions. In recent work, we showed for the first time that the biomineralization protein Mms6 is able to template a single layer of high quality, uniform magnetite MNPs when selectively immobilized onto a patterned surface [7, 8].

However, magnetite has a low coercivity (i.e. is magnetically soft), so it is unlikely that Mms6 templated magnetite MNPs will be adopted for use in data storage. Current research into the development of materials for use in data storage applications is focusing on the high coercivity $L1_0$ phase of FePt and CoPt alloys [20-24]. Conventionally, high temperature annealing (650°C) [22] or sputtering (450-550°C) [23, 24] are required to form this desired phase. Here we present our work on using a biotemplating peptide sequence for the formation of CoPt MNPs onto patterned surfaces at room temperature. The peptide sequence was designed so that the *N*-terminal section has an affinity for silicon oxide (HPPMNASHPHMH) [25, 26] and the *C*-terminal section for templating $L1_0$ CoPt (KTHEIHSPLLHK) [22], which are joined together by a flexible linker (GSG). This dual affinity peptide (DAP) should bind strongly to silicon oxide at one end, leaving the other end free to interact with aqueous cobalt and platinum salts to template metallic CoPt MNPs onto the patterned substrate under mild reaction conditions. By functionalizing a substrate with a biotemplating molecule, an environmentally friendly route towards patterned high coercivity CoPt MNPs for data storage applications may be found.

## EXPERIMENTAL

A silicon wafer was cut into $\approx 1\ cm^2$ squares and a cured polydimethyl siloxane (PDMS) stamp cut from a photolithographically patterned master. These were then sonicated in ethanol for 15 minutes before being dried with $N_2$ and UV/ozone treated (UVOCS) for 20 minutes. This created a native oxide layer on the silicon substrate and a hydrophilic surface on the PDMS. The DAP (Ac-HPPMNASHPHMH-GSG-KTHEIHSPLLHK-Am, GensScript, >95% purity, *N*-terminal acetylation, *C*-terminal amidation) was dialyzed into phosphate buffered saline (PBS from Invitrogen: 10 mM sodium phosphate, 2.68 mM KCl, 140 mM NaCl, pH 7.4). The stamp was inked with the 1 mg $mL^{-1}$ peptide solution for 1 minute, thoroughly dried with $N_2$, and gently pressed onto the Si substrate for 1 minute. The stamp was removed, and the substrate rinsed with deoxygenated MilliQ water and dried with $N_2$ before being transferred to glass jar for mineralization. Various stocks of $Co^{2+}$ (30 mM $CoCl_2 \cdot 6H_2O$, $CoSO_4 \cdot 7H_2O$ and $Co(NO_3)_2 \cdot 6H_2O$) and $Pt^{2+}$ salts (10 mM $Na_2PtCl_4$ and $K_2PtCl_4$) were prepared in deoxygenated HEPES buffered saline (HBS, 50 mM HEPES, 100 mM NaCl, pH 7.5). 500 μL each of $Co^{2+}$ and $Pt^{2+}$ solutions were added to each sample, and 3.5 mL HBS added before sealing against air under an $N_2$ atmosphere. After ≈10 minutes incubation, 500 μL of 100 mM $NaBH_4$ in HBS was added. The peptide patterned substrates were incubated for 30 minutes, thoroughly rinsed in deoxygenated MilliQ water and dried with $N_2$.

The metalized biotemplated substrates were imaged using scanning electron microscopy (SEM) on a LEO 1530 Gemini FEGSEM and processed using Zeiss SmartSEM software. Energy dispersive X-ray (EDX) spectra were collected using an Oxford Instruments AZtecEnergy EDX system on the SEM at 10 keV. Transmission electron microscope (TEM) was used to image control MNPs on carbon coated copper grids with a Phillips CM200 (FEG)TEM using the digital micrograph software. EDX spectra were recorded using an Oxford Instruments INCA EDX system and a Gatan Imaging Filter. Grainsize was recorded as the length and width of ≈400 MNPs for each sample using ImageJ [27]. Magnetic force microscopy (MFM) plots were recorded using magnetized MFM tips (Cr/Co coated MESP, Veeco) on a Multimode Nanoscope III. The topography was recorded in tapping mode, and the magnetic interactions between the tip and samples recorded at a lift height of between 30-50 nm. The AFM/MFM data were processed using WSxM [28].

## RESULTS AND DISCUSSION

### Images and Elemental Analysis of Biotemplated CoPt Nanoparticles

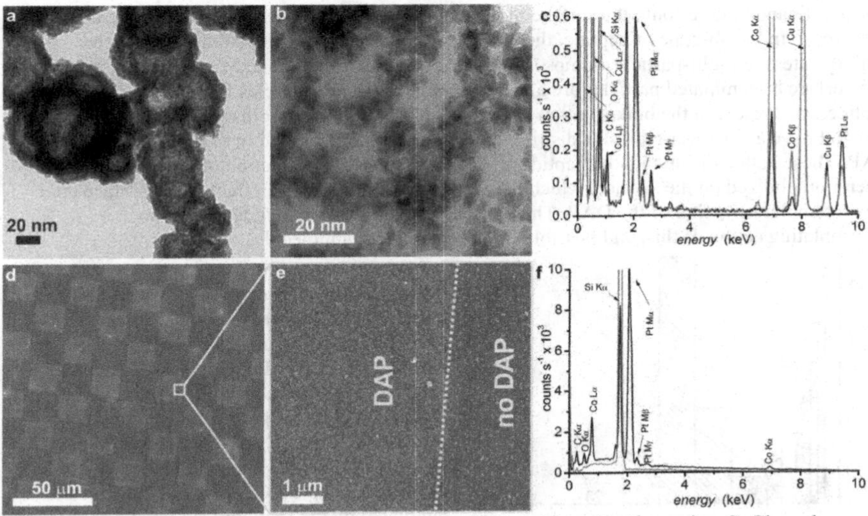

**Figure 1**. SEM images and EDX spectra of µCP DAP after mineralization using $CoCl_2$ and $Na_2PtCl_4$. (a) TEM image of control particles precipitated from a bulk solution and (b) in the presence of the DAP (10 µg $mL^{-1}$) in the bulk solution. (c) EDX spectra of bulk precipitated control (grey line) and DAP in bulk solution control (black line). High energy electrons (200 keV) in the TEM generate peaks at high energies (e.g. Pt Lα peak at 9.44 keV), and the Cu signal is from the TEM grid. (d-e) Chessboard pattern shows clear contrast, as the areas functionalized with the DAP show heavy mineralization of nanoparticles when compared to the unfunctionalized areas. (f) EDX spectra of biotemplated area (black line) and substrate background (grey line) recorded at 10 keV generates more detectible X-rays at lower energies than seen in the TEM.

Patterning of the biotemplating DAP before mineralization using µCP demonstrates how the biofunctionalized surface and unfuntionalized surfaces differ after metallization. Of the six different metal salt combinations investigated, the use of $CoCl_2 \cdot 6H_2O$ and $Na_2PtCl_4$ produced the greatest contrast in nanoparticle density between the biomineralized and non-biotemplating areas. SEM images, TEM images and EDX spectra of biotemplated nanoparticles are shown in **Figure 1**. The EDX spectra of the controls shows the presence of both Co and Pt in the MNPs, but there is a much higher proportion of Co in the bulk MNPs when compared to those formed in the presence of the CoPt DAP. This shows that the presence of the peptide is able to exert some control over the stoichiometry of the CoPt formed. The EDX spectrum recorded on a biomineralized area clearly shows the presence of Co and Pt, in an approximate 50:50 ratio for the peptide surface templated particles, which is required for the high coercivity $L1_0$ phase of CoPt. The spectrum recorded on a background area of the substrate only shows the presence of silicon, further supporting the relative lack of nanoparticle formation onto the non-biotemplating areas of the patterned substrate.

**Figure 2** shows that the areas functionalized with the DAP are able to template a high density of nanoparticles onto the surface in those locations, with a narrow grainsize distribution (23±8 nm). In the absence of peptide, the MNPs precipitated in the bulk solution appear to agglomerate into multi-particle clumps, and have a wider distribution and smaller size (5±5 nm). The surface biotemplated particles are also much larger than the particles formed when the peptide was present in the bulk solution (4±2 nm), which are far better dispersed onto the TEM grid and have a narrower grainsize distribution than the control formed in the absence of the DAP. These data indicate that the peptide is able to template the formation of larger CoPt MNPs when immobilized on the surface, which may be due to a stabilization effect of the surface upon the biotemplating action of the DAP. Thus surface immobilization may facilitate the biotemplating action of this, and possibly other biotemplating molecules.

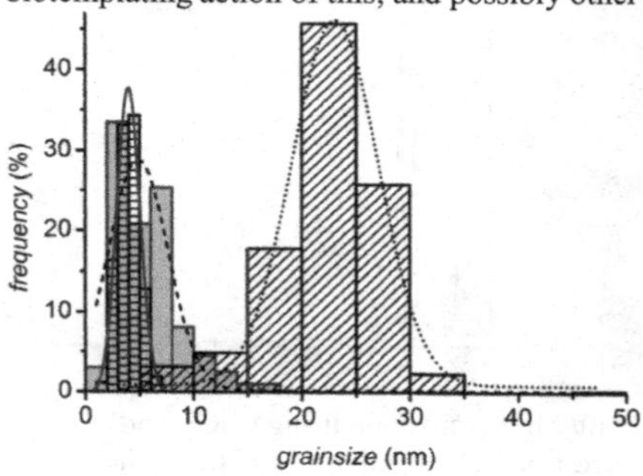


**Figure 2**. Grainsize analysis of CoPt MNPs formed from aqueous solution at room temperature, frequency distributed into 10 bins and peaks fitted in Origin. Controls are MNPs formed from the bulk solution in the absence of peptide (grey fill, solid grey fit) and with 10 μg $mL^{-1}$ DAP in the bulk solution (horizontal fill, solid black fit). These are much smaller than the MNPs templated by the immobilized DAP (diagonal fill, black dotted fit).

## Magnetic Properties of Biotemplated CoPt Nanoparticles

Magnetic characterization of the biotemplated patterns was carried out using MFM. The biotemplated MNPs on the surface have zones of magnetic repulsion and attraction between the surface and the tip (**Figure 3**). As many of these zones occur independently of the topographical height of the particles, these multi-particle magnetic domains are unlikely to be due to imaging artifacts. The multi-particle magnetic zones of attraction and repulsion can be seen most clearly in Figure 3e, and may be due to exchange coupling of the MNPs on the surface, similar to those found in our previous work [7, 8], and that of others [29]. Importantly, this indicates that the biotemplated CoPt MNP patterns are able to maintain their magnetic orientation at room temperature, which is an essential criterion for use in data storage applications.

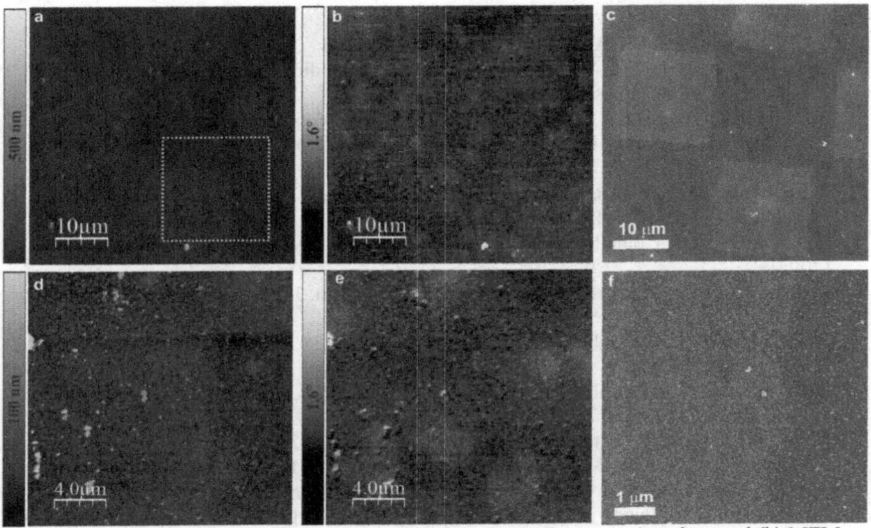


**Figure 3.** (a) AFM topography plot showing the biotemplated patterned surface and (b) MFM phase contrast plot of the same area at 50 nm lift-height. (c) An SEM image of a similar area. (d) AFM topography plot of the area highlighted in (a) and (e) MFM phase contrast at 30 nm lift height. (f) An SEM image of a similar area. Magnetic attraction between the tip and the biotemplated nanoparticles appears dark, with repulsion appearing light.

## CONCLUSIONS

This study shows for the first time that a biomolecule is able to template the formation of CoPt MNPs when immobilized onto a substrate. In the areas functionalized with the DAP, a fairly dense film of nanoparticles containing Co and Pt are biotemplated *in situ*. Probing of the magnetic interactions between biotemplated MNPs on the surface and a magnetized MFM tip show that the particles may be exchange coupled on the surface. In the future, we plan to optimize this biomineralization protocol for the formation of thin-films of high coercivity MNPs, and then further characterize the magnetic and crystallographic properties of the biotemplated nanomagnets. It may then be possible to use such biotemplated MNPs to record data, and allow us to compare our biotemplated surfaces to their high temperature synthesis counterparts.


## ACKNOWLEDGMENTS

We would like to thank our colleagues at Leeds, especially S. Evans, S. Baldwin, R. Bushby, A. Rawlings, J. Bain, V. Micó Egea, M. Booth and K. Curtis, for their invaluable input into the discussion and development of this work, and J. Harrington for his assistance with SEM imaging. Finally, we would like to thank the EPSRC for the Post Doctoral Prize Fellowship (EP/K503071/1) supporting JMG, and a CDT studentship (EP/J500458/1) supporting SMB. The BBSRC (BB/H005412/1) and the EPSRC (EP/I032355/1) for funding SSS and JPB, and the BHRC (University of Leeds) and EPSRC (EP/J01513X/1) for supporting KC.

**REFERENCES**


1. S.N. Piramanayagam, J. Appl. Phys. **102**, 011301 (2007).
2. O. Hellwig, A. Berger, T. Thomson, E. Dobisz, Z.Z. Bandic, H. Yang, D.S. Kercher and E.E. Fullerton, Appl. Phys. Lett. **90**, 162516 (2007).
3. S.J. Lister, T. Thomson, J. Kohlbrecher, K. Takano, V. Venkataramana, S.J. Ray, M.P. Wismayer, M.A.d. Vries, H. Do, Y. Ikeda and S.L. Lee, Appl. Phys. Lett. **97**, 112503 (2010).
4. S. Staniland, W. Williams, N. Telling, G. Van Der Laan, A. Harrison and B. Ward, Nat. Nanotechnol. **3**, 158 (2008).
5. Q.A. Pankhurst, J. Connolly, S.K. Jones and J. Dobson, Phys. D: Appl. Phys. R167 (2003).
6. D. Kim, N. Lee, M. Park, B.H. Kim, K. An and T. Hyeon, J. Am. Chem. Soc. **131**, 454 (2009).
7. J.M. Galloway, J.P. Bramble, A.E. Rawlings, G. Burnell, S.D. Evans and S.S. Staniland, Small **8**, 204 (2012).
8. J.M. Galloway, J.P. Bramble, A.E. Rawlings, G. Burnell, S.D. Evans and S.S. Staniland, J. Nano Res. **17**, 127 (2012).
9. J.M. Galloway and S.S. Staniland, J. Mater. Chem. **22**, 12423 (2012).
10. R.A. Metzler, I.W. Kim, K. Delak, J.S. Evans, D. Zhou, E. Beniash, F. Wilt, M. Abrecht, J.-W. Chiou, J. Guo, S.N. Coppersmith and P.U.P.A. Gilbert, Langmuir **24**, 2680 (2008).
11. L. Addadi and S. Weiner, Angew. Chem. Int. Ed. **31**, 153 (1992).
12. S. Mann, *Biomineralization: Principals and Concepts in Bioinorganic Materials Chemistry*, (Oxford University Press, Oxford, UK, 2001).
13. M.A. Meyers, J. McKittrick and P.-Y. Chen, Science **339**, 773 (2013).
14. J.M. Galloway, J.P. Bramble and S.S. Staniland, Chem. Eur. J. **in press**, 10.1002/chem.201300721 (2013).
15. L.L. Brott, R.R. Naik, D.J. Pikas, S.M. Kirkpatrick, D.W. Tomlin, P.W. Whitlock, S.J. Clarson and M.O. Stone, Nature **413**, 291 (2001).
16. R.R. Naik, S.J. Stringer, G. Agarwal, S.E. Jones and M.O. Stone, Nat. Mater. **1**, 169 (2002).
17. B. Wang, K. Chen, S. Jiang, F. Reincke, W. Tong, D. Wang and C. Gao, Biomacromolecules **7**, 1203 (2006).
18. M. Matmor and N. Ashkenasy, J. Mater. Chem. **21**, 968 (2011).
19. J.M. Galloway, A. Arakaki, F. Masuda, T. Tanaka, T. Matsunaga and S.S. Staniland, J. Mater. Chem. **21**, 15244 (2011).
20. C. Mao, D.J. Solis, B.D. Reiss, S.T. Kottmann, R.Y. Sweeney, A. Hayhurst, G. Georgiou, B. Iverson and A.M. Belcher, Science **303**, 213 (2004).
21. B.D. Reiss, C. Mao, D.J. Solis, K.S. Ryan, T. Thomson and A.M. Belcher, Nano Lett. **4**, 1127 (2004).
22. M.T. Klem, D. Willits, D.J. Solis, A.M. Belcher, M. Young and T. Douglas, Adv. Funct. Mater. **15**, 1489 (2005).
23. L. Zhang, Y.K. Takahashi, K. Hono, B.C. Stipe, J.-Y. Juang and M. Grobis, J. Appl. Phys. **109**, 07B703 (2011).
24. O. Mosendz, S. Pisana, J.W. Reiner, B. Stipe and D. Weller, J. Appl. Phys. **111**, 07B729 (2012).
25. E. Eteshola, L.J. Brillson and S.C. Lee, Biomol. Eng. **22**, 201 (2005).

26. R. Nochomovitz, M. Amit, M. Matmor and N. Ashkenasy, Nanotechnology **21**, 145305 (2010).
27. M.D. Abramoff, P.J. Magalhaes and S.J. Ram, Biophotonics Int. **11**, 36 (2004).
28. I. Horcas, R. Fernandez, J.M. Gomez-Rodriguez, J. Colchero, J. Gomez-Herrero and A.M. Baro, Rev. Sci. Instrum. **78**, 013705 (2007).
29. K. Yamamoto, C.R. Hogg, S. Yamamuro, T. Hirayama and S.A. Majetich, Appl. Phys. Lett. **98**, 072509 (2011).

**Mater. Res. Soc. Symp. Proc. Vol. 1569 © 2013 Materials Research Society**
**DOI: 10.1557/opl.2013.803**

# Feasibility Study of Carbon Nanotube Microneedles for Rapid Transdermal Drug Delivery

Bradley J. Lyon[1], Adrianus I. Aria[1], and Morteza Gharib[1]
[1]Graduate Aerospace Laboratories, California Institute of Technology, Pasadena, CA 91125, U.S.A.

## ABSTRACT

We introduce a new approach for fabricating hollow microneedles using vertically-aligned carbon nanotubes (VA-CNTs) for rapid transdermal drug delivery. Here, we discuss the fabrication of the microneedles emphasizing the overall simplicity and flexibility of the method to allow for potential industrial application. By capitalizing on the nanoporosity of the CNT bundles, uncured polymer can be wicked into the needles ultimately creating a high strength composite of aligned nanotubes and polymer. Flow through the microneedles as well as *in vitro* penetration of the microneedles into swine skin is demonstrated. Furthermore, we present a trade study comparing the difficulty and complexity of the fabrication process of our CNT-polymer microneedles with other standard microneedle fabrication approaches.

## INTRODUCTION

Microneedles are envisioned to provide a painless, self-administered alternative to standard hypodermic injection. Specifically, hollow microneedles allow for a delivery architecture that is more flexible than other microneedle designs such as solid microneedles, drug-coated microneedles, or dissolving microneedles. This is because hollow microneedles allow for variable delivery rate from rapid injection to emulate a hypodermic injection to slow steady delivery to mimic intravenous drug therapy. Additionally, since the hollow microneedle is designed to be inert with respect to the drug, many current drugs that are delivered into the skin can be directly used in the hollow microneedle architecture [1].

Previous studies on the fabrication of hollow microneedles have focused on top-down fabrication approaches using either silicon, metal, or glass. While these approaches have yielded functioning microneedles, their fabrication involves iterated etching or micromachining techniques that ultimately add complexity to the fabrication and limit industrial-scale application [1]. Here, we introduce a new perspective on forming hollow microneedles by fabricating microneedles using a bottom-up approach by using vertically-aligned carbon nanotubes (VA-CNTs). VA-CNTs are first used as a scaffold for forming the shape of the microneedle and then as a fiber component in a CNT-polymer composite to create a high strength material capable of penetrating the skin.

By utilizing standard methods to grow VA-CNTs, including catalyst patterning and thermal chemical vapor deposition (CVD), we directly produce a hollow microneedle. In designing the CNT-polymer composite microneedle, the final product must achieve four mechanical objectives: (i) have high mechanical strength under compression to achieve skin penetration, (ii) conformally coat the VA-CNTs with polymer allowing the microneedle to retain its original shape from catalyst patterning, (iii) anchor microneedles to a common polymer base for easy transfer from the growth substrate to drug delivery platforms, and (iv) maintain an unobstructed hollow cavity for drug delivery.

These objectives are interdependent and thus careful choice and application of polymer must be applied to create a functioning microneedle. Here, we demonstrate our approach using the negative photoresist SU8-2025 (MicroChem, Newton, MA) as a candidate polymer. The high UV absorbance of CNTs allow us to selectively cure the microneedle device. Thus, the polymer base can be cured while leaving the polymer within the microneedle cavity uncured allowing for removal of the polymer in later steps. SU8-2025 is chosen due to its property of forming thick films through spin coating (typically 25μm at 3000 rpm) and high elastic modulus of 3 GPa [2]. A previous study has shown that low viscous SU8, such as SU8-2002 (MicroChem, Newton, MA), can be spin coated on VA-CNTs to create a composite [3]. However, the thin film created by this resist through spin coating (2μm at 3000 rpm) is too fragile for making a structurally supportive base for the microneedles [2]. After proving that the CNT-polymer composite microneedle is mechanically feasible with SU8-2025, future studies will focus on broadening the number of polymers that can be employed in this technique.

## EXPERIMENT

Vertically aligned CNTs are fabricated via thermal CVD on a silicon substrate patterned with catalyst in the shape of hollow circles which act as the microneedle template. To obtain this pattern, the silicon wafer is masked with photoresist patterned through photolithography into the desired microneedle geometry. Alumina and iron of thickness 10nm and 1nm respectively are then deposited on the wafer via electron beam deposition. After deposition, photoresist is stripped from the wafer to finalize the substrate preparation. For the CVD process, ethylene and hydrogen gas is flowed across the substrate at 750°C, 600 torr and 490 sccm and 210 sccm respectively for an hour to achieve VA-CNT growth.

At the end of CVD, the VA-CNTs are patterned into hollow cylinders matching the catalyst pattern of 150 μm outer diameter and inner cavity diameter of 25 μm (Figure 1, Step 1). The height of the microneedles can be varied from 150 μm to 400 μm. Patterned VA-CNTs alone cannot act as a microneedle. VA-CNTs are prone to buckling and have an exceedingly low modulus under compression of only 550 kPa [4]. Additionally, the VA-CNTs must be connected to a common base other than the growth substrate to allow for the microneedles to be transferred onto a device platform for drug delivery.

Starting with the patterned VA-CNTs, we incorporate SU8-2025 to create the composite microneedle. SU8-2025 is dropcasted onto the VA-CNT sample which is then passively wicked into the interspacing of the VA-CNTs. Next, the sample is spin coated at 3000 rpm for 60 seconds to remove the excess material and simultaneously create a base that is thick enough (nominal 25 μm) to allow for the microneedles to be easily removed from the silicon substrate in later steps with minimal damage to the microneedles and the polymer base (Step 2). However, the high viscosity of the SU8-2025 causes the resist to pool in the inner cavity of the microneedle. After spin coating, the sample is soft baked at 95°C for four minutes.

The photoresist is selectively cured under oblique incidence UV light (UV/Visible Light Exposure Chamber, MTI Corp., Richmond, CA) at 30 $mW/cm^2$ for up to 1 minute (Step 3). In this arrangement, the polymer base is cured along with the SU8 embedded within the VA-CNT interspacing. The SU8 in the inner cavity remains uncured due to the oblique incidence of the UV light preventing direct exposure of the inner cavity. Additionally, the VA-CNTs in the CNT-SU8 composite attenuate the amount of UV light reaching the inner cavity through the composite. Following UV exposure, the sample goes through a post exposure bake

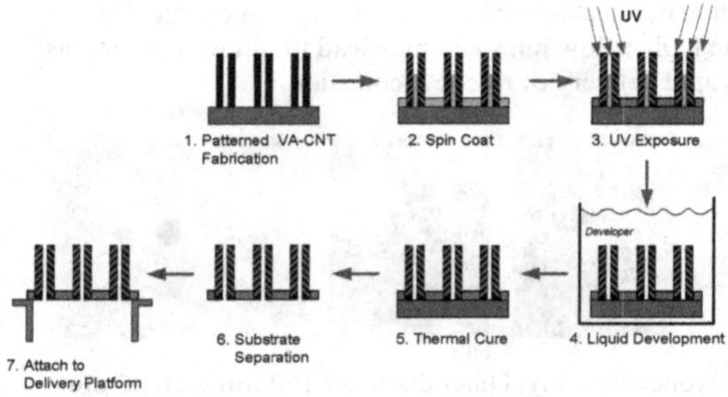


**Figure 1-** Microneedle fabrication process starting with CVD fabrication of patterned VA-CNTs.

for 3 minutes at 95°C. The microneedles are then submerged in SU8 Developer (MicroChem, Newton, MA) to clear the uncured photoresist from the inner cavity (Step 4). The submerged sample is placed on a shaker table set to 150 rpm for 10 minutes. Following development, the sample is rinsed in isopropanol and cured in a vacuum oven at 150 °C for 20 minutes (Step 5).

At this point, the CNT-SU8 composite microneedle is fully formed with a clear inner cavity and a common SU8 base. By taking advantage of the poor adhesion between SU8 and the silicon substrate as well as the thick SU8 base layer, the microneedles can be removed mechanically with a razor blade or tweezers (Step 6). The device is then transferred onto a delivery platform that connects the inner cavity of the microneedles to a liquid reservoir for drug flow (Step 7).

## RESULTS & DISCUSSION

### CNT-SU8 Microneedle

The CNT-SU8 microneedles are spaced on a 1 mm grid to ensure that each needle independently penetrates the skin (Figure 2 a,b). Ideal microneedle height was found to be in the range of 200 μm to 250 μm. Microneedles above this range are more susceptible to buckling failure. Microneedle heights below 200 μm were found to be too short to achieve consistent skin penetration. SU8 conformally coats the microneedle allowing the final product to retain the original pattern defined by catalyst patterning and CVD growth of the VA-CNTs. Comparing the VA-CNT structure to the CNT-SU8 composite structure, we see that the SU8 fully envelops the interspacing between the nanotubes creating a single solid composite structure (Figure 2 c,d). SEM imaging of the underside of the needle confirms that the inner cavity of the needle is clear of polymer after fabrication (Figure 2e).

Liquid flow through the microneedle is demonstrated by connecting a microneedle array to a water filled syringe. Through hand actuation, the microneedle is capable of expelling water into the air at rates of up to 600 μL/min per needle (Figure 3a). Despite the relatively thin polymer base of approximately 25 μm, the base shows no signs of fatigue or cracking under high flow rates. Achieving flow greater than 100μL/min per needle under minimal actuation pressure demonstrates the low hydraulic resistance of the device which ultimately lowers the work needed to flow liquid through the microneedles at any flow rate. For the current microneedle height, we anticipate for *in vitro* drug delivery using delivery rates of about 1 to 10 μL/min per needle and volumes of about 100μL due to the restrictive permeability of the skin. Longer microneedles penetrating deeper into the skin may enable higher flow rates and may lead to future applications of the microneedle as an auto-injector for rapid delivery of rescue medication.

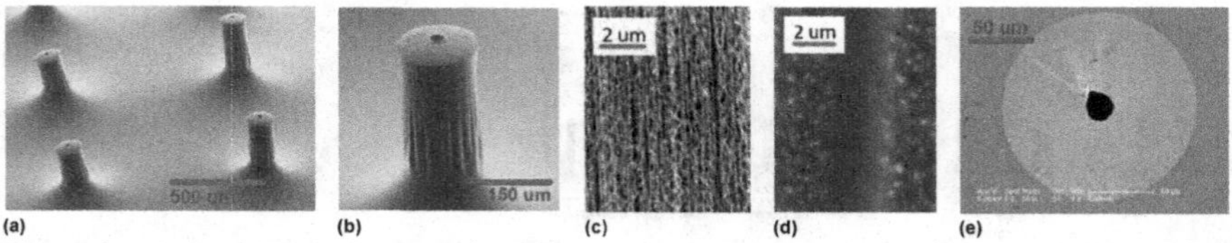


**Figure 2- (a), (b)** CNT-SU8 composite microneedle array. Outer diameter 150 μm with 25 μm inner diameter. Structure of microneedle **(c)** before and **(d)** after SU8 incorporation. **(e)** Underside of microneedle after fabrication showing the inner cavity is clear of polymer.

*In vitro* skin penetration is demonstrated on thin (<0.5 mm) samples of dorsal swine skin prepared by using a dermatome to cut full thickness skin samples. Microneedles were pressed by hand into the swine skin and achieved penetration at tip pressures in the range of 60MPa to 90MPa. Prior to penetration, the microneedles are coated with dry methylene blue powder. Upon contact with the interstitial fluid in the skin, the methylene blue dye is passively released from the needle marking the point of contact with the skin (Figure 3b). The clear pattern of the 2 x 2 microneedle array indicates positive penetration and ultimately demonstrates that the CNT-SU8 composite has sufficient strength to achieve skin penetration. To present, all of the mechanical objectives of the microneedle outlined previously have been achieved. Ongoing work is now looking to build upon these results by characterizing the microneedle's *in vitro* liquid delivery into the skin as well as broadening the method to demonstrate incorporation of other polymers.

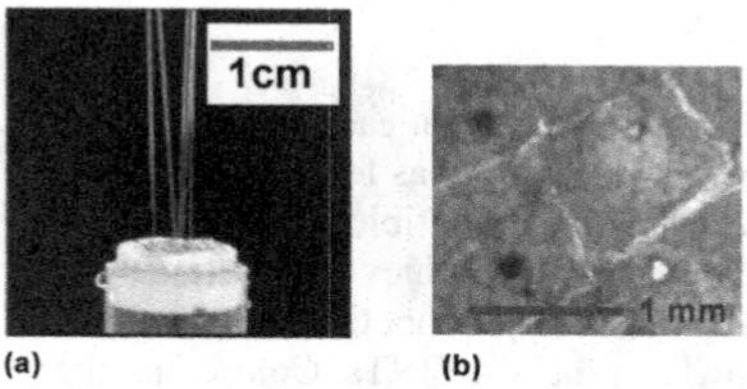


**Figure 3- (a)** Water jets from the microneedle array with exit velocity of about 600 μL/min per needle. **(b)** *In vitro* swine skin penetration marks by microneedles coated in methylene blue dye.

## Fabrication Method Comparison

A trade study was conducted by comparing the laboratory difficulty in fabricating the CNT-SU8 composite microneedle to the current primary approaches for microneedle fabrication including silicon microneedles and micromolding. Fabrication of both silicon and CNT-SU8 microneedles offer similar advantages in terms of geometry customization and parallel processing of large numbers of microneedles. In characterizing the fabrication process for silicon microneedles, reactive ion etching (RIE) is presumed to be the primary method for defining the microneedle shape with additional processes such as wet etching or micromachining used as secondary methods to optimize the microneedle geometry[1, 5-7]. For each process used in the fabrication method, a difficulty factor between 1 through 3 is assigned based on the difficulty of the process on the laboratory scale (Table I).

**Table I**: Trade study of fabrication difficulty for CNT-SU8 composite microneedles and silicon microneedles.

| Techniques | Physical Vapor Deposition | Nanotube Fabrication | Polymer Incorporation | Photolithography | Mask Deposition* | Mask Removal* | Reactive Ion Etching | Wet Etching | Micromachining | Technique Difficulty |
|---|---|---|---|---|---|---|---|---|---|---|
| | Number of Times Process Performed | | | | | | | | | |
| CNT-SU8 Composite | 1 | 1 | 1 | 1 | — | — | — | — | 0-1 | 5-8 |
| Silicon (Dry Etched) | — | — | — | 2-3 | 2-3 | 2-3 | 2-3 | 0-2 | 0-1 | 10-22 |
| Process Difficulty | 1 | 2 | 1 | 1 | 1 | 1 | 2 | 2 | 3 | |

*Metal or Oxide Mask

CNT-SU8 composite has a smaller total technique difficulty than silicon showing that the CNT-polymer method has the potential to be significantly simpler than the silicon approach. This is because creating a hollow cavity in the silicon approach requires iterative etching steps which in turn require iterative processing steps. In contrast, the entire geometry of the CNT-SU8 needle is defined from the sequential steps of catalyst patterning and nanotube fabrication. Another important consideration in the CNT-SU8 approach is that incorporation of SU8 on the VA-CNTs relies on a combination of simple and passive mechanisms such as capillary action and UV exposure that can be easily executed on both the laboratory and industrial level. The primary fabrication challenge for the CNT-polymer microneedle is the fabrication of VA-CNTs. However, the recent increase of commercial fabrication options should lower the overall fabrication difficulty in the long run.

To present, only a hollow cylinder geometry has been considered for our CNT-SU8 microneedle. In comparison with other microneedles, a tapered shape is typically preferred to minimize the tip area as the required penetration force scales linearly with tip area [8]. Previous work has demonstrated that drug release from the top of the microneedle is susceptible to skin occlusion which increases the hydraulic resistance of delivery. A suggested alternative is to release the drug from the side of the microneedle to optimize the delivery by lowering the hydraulic resistance [1, 5]. The fabrication of CNT-SU8 microneedles can be amended to incorporate these more complex geometries if desired. The microneedles can be modified after polymer incorporation via micromachining or before polymer incorporation using techniques such as capillography [3] or focused ion beam milling. Further study of the *in vitro* drug release profile for our current microneedle is necessary before determining if the potential benefit of optimizing the shape of our microneedles outweighs the cost of increased fabrication difficulty.

Micromolding is another common approach to producing hollow microneedles by electroplating micromolds to produce a hollow metallic microneedle. Typical master structures for micromolding are fabricated by micromachining or laser drilling of a bulk material which can be time consuming and ultimately limits the minimum feature size of the needle [1, 9]. However after the master structure is complete, the micromolding process allows for fast fabrication of large numbers of microneedles. Both the silicon and CNT-SU8 needle can achieve much finer feature size and thus may both be potentially used as a master structure for micromolding. A previous study has demonstrated the successful use of CNT-SU8 composites for general micromolding applications [3].

## CONCLUSIONS

A new approach to fabricating hollow microneedles has been shown using vertically-aligned carbon nanotubes and SU8-2025. By taking advantage of self-assembly, VA-CNTs can be simply adapted as a hollow microneedle by incorporating SU8-2025. SU8-2025 allows for the creation of a strong composite with VA-CNTs while simultaneously creating a supportive base for the microneedle array. Initial experiments have shown that the CNT-SU8 needles can penetrate the skin *in vitro* and can structurally support high flow rates of up to 600μL/min per needle. Further studies will investigate expanding the number of polymers that can be used in this architecture as well as *in vitro* characterization of the microneedle's delivery performance.

In comparing the fabrication of CNT-SU8 microneedles with other approaches, we find that the CNT-SU8 fabrication method is potentially simpler than that of silicon microneedles. CNT-SU8 composites may also be incorporated as master structures in micromolds to allow for feature size on the order of several microns. Laboratory scale fabrication represents only a single aspect for comparing microneedles. Future trade studies will need to take into account the performance, cost, and scalability aspects of the CNT-polymer microneedle to properly identify future applications and markets for this technology.

## ACKNOWLEDGMENTS

We acknowledge Zcube s.r.l for their financial support of this work. We also acknowledge the Kavli Nanoscience Institute and the Geology and Planetary Sciences Analytical Facility for their support in running experiments.

## REFERENCES

1. Y.-C. Kim, J.-H. Park and M. R. Prausnitz, Advanced Drug Delivery Reviews **64** (14), 1547-1568 (2012).
2. Microchem, SU-8 2000 Processing Guidelines.
3. M. De Volder, S. H. Tawfick, S. J. Park, D. Copic, Z. Zhao, W. Lu and A. J. Hart, Advanced materials **22** (39), 4384-4389 (2010).
4. L. Ci, J. Suhr, V. Pushparaj, X. Zhang and P. M. Ajayan, Nano letters **8** (9), 2762-2766 (2008).
5. H. J. G. E. Gardeniers, R. Luttge, E. J. W. Berenschot, M. J. De Boer, S. Y. Yeshurun, M. Hefetz, R. van't Oever and A. van den Berg, Microelectromechanical Systems, Journal of **12** (6), 855-862 (2003).
6. B. Ma, S. Liu, Z. Gan, G. Liu, X. Cai, H. Zhang and Z. Yang, Microfluidics and Nanofluidics **2** (5), 417-423 (2006).
7. L. M. Yu, F. E. H. Tay, D. G. Guo, L. Xu and K. L. Yap, Sensors and Actuators A: Physical **151** (1), 17-22 (2009).
8. S. P. Davis, B. J. Landis, Z. H. Adams, M. G. Allen and M. R. Prausnitz, Journal of biomechanics **37** (8), 1155-1163 (2004).
9. J. J. Norman, S. O. Choi, N. T. Tong, A. R. Aiyar, S. R. Patel, M. R. Prausnitz and M. G. Allen, Biomedical microdevices **15** (2), 203-210 (2013).

Mater. Res. Soc. Symp. Proc. Vol. 1569 © 2013 Materials Research Society
DOI: 10.1557/opl.2013.1079

# High Affinity Membranes for Cellulase Enzyme Detection in Subterranean Termites

Arishaun Donald[2], Tariq Taylor[2], Jianjun Miao[3], Robert Linhardt[3], Duane Jackson[2], Juana Mendenhall[1]

[1]Dept. of Chemistry, Morehouse College, 830 Westview Drive, S.W., Atlanta, GA, U.S.A.
[2]Dept. of Psychology, Morehouse College, 830 Westview Dive, S.W. Atlanta, GA, 30314, USA
[3]Dept. of Chemistry and Chemical Biology, Rensselaer Polytechnic Institute, Troy NY, 16801, USA

## ABSTRACT

The United States dependence on fossil fuels has become mandatory over the past few decades. The fuel shortage during the 1970s and after Hurricane Katrina has catalyzed a need for creating alternative energy sources, improving the efficacy of these alternative energy sources, and enhancing energy sustainability. The U.S. Department of Energy has set goals to replace 30% of the liquid petroleum transportation fuel with biofuels and to replace 25% of industrial organic chemicals with biomass-derived chemicals by 2025. In the southeast United States, subterranean termites are prevalent and microbes in their gut degrade wood based materials such as cellulose which produce simple sugars that can be used to produce bioethanol. Upon seasonal change, subterranean termites undergo less enzymatic activity and wood-eating capability limiting the amount of sugars that may be produced. This limited activity sparks an interest to investigate this poorly understood phenomenon of how temperature may affect the enzymatic activity in subterranean termites' guts. In this study, we report the development thermo-responsive biomaterial nanofiber mats containing cellulose to model cellulase activity. Using electrospinning techniques, poly(N-vinylcaprolactam) celluose fiber mats have been prepared via alkaline hydrolysis and labeled with fluorescent tags. Subterranean termites (*reticulitermes species*) were feed fiber mats for 10 consecutive days to assess enzyme mapping and kinetics. Fluorescent microscopy images confirmed spatial and temporal localization of cellulase enzyme throughout the termite gut upon time and temperature change. These novel high affinity enzyme detection membranes show promise towards future biofuel production.

## INTRODUCTION

In the southeastern United States, subterranean termites highly populate the region. The gut of the subterranean termite fosters a natural bioreactor that can be used to digest cellulose from wood products followed by production of simple sugars. Hence, using the termite gut to understand cellulase-cellulose interactions for producing simple sugars towards biofuel production is of particular interest. Therefore by developing affinity fiber membranes using electrospinning techniques, serves as a scaffold to help detect cellulose-cellulase interactions. Previous studies have reported the investigations of cellulase kinetics using cellulose fibrils on hard surfaces such as mica or silicon.[1] By using fluorescent labeled cellulose fibrils, it was noted that cellulase activity could be calculated using binding kinetics of fluorescently labeled cellulase. Towards this end, we have fabricated a temperature-responsive polymer fiber affinity membrane using electrospun fibers. These fibers provided a *smart* substrate that undergoes phase changes upon temperature change which enable attachment and detachment of biological entities. The preparation and characterization of temperature-responsive, poly(N-vinylcaprolactam)[PVCL] and cellulose fiber mats have previously been reported.[2] Herein, we report the use of temperature-responsive PVCL-cellulose fibers as affinity membranes to detect cellulase activity in subterranean termites. We hypothesize that *Reticulitermes sp.* termites will consume PVCL3-CELL fibers due to the addition of carboxylic acid functional groups. Cellulose-cellulase activity was detected using fluorescent microscopy by imaging termites upon (5-(4,6-dichlorotriazinyl)aminofluorescein [DTAF] digestion.

## EXPERIMENT

PVCL was prepared using free radical polymerization (Figure 1) in water and benzene.

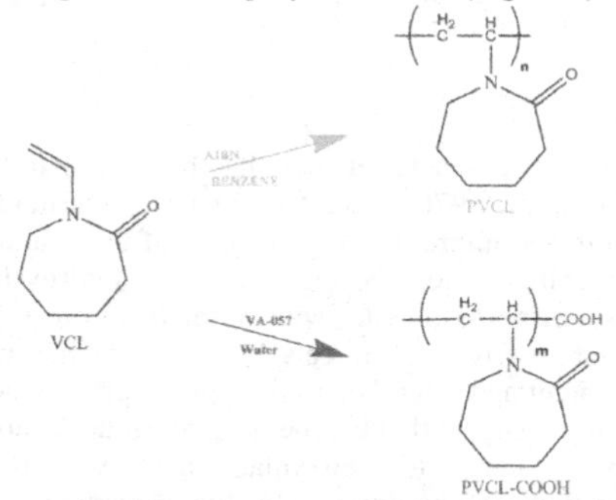


Figure 1. N-vinylcaprolactam polymerization scheme indicating PVCL1 & PVCL2 and PVCL3 (PVCL-COOH)

PVCL polymers were characterized using proton nuclear magnetic resonance (NMR) [Bruker 400 mHz], ATR-FTIR [Nicolet Magna-IR 550 Series II spectrometer], and SEC [Malvern Triple Detector].[2] Three types of electrospun fiber were prepared from a solution of PVCL homopolymers and a cellulose acetate.[2] An 8% (w/v) solution of PVCL-cellulose acetate solution was prepared using dimethylformide (DMF) and PEO and allowed to mix for 15 minutes at 100°C. Fiber mats were characterized using a field emission scanning electron microscope (FE-SEM) JSM-6335 to determine nanofiber size distribution and morphology. PVCL-cellulose acetate fiber mats were hydrolyzed using 0.1 *N* NaOH at 100°C for 2 hours to yield PVCL-cellulose (PVCL-CELL) fibers. Next, fiber mats were tagged with fluorescent molecule (5-(4,6-dichlorotriazinyl)aminofluorescein), DTAF by mixing 5uL in 0.1N NaOH for 24 hours. Simple sugar identification was carried out using para-hydrobenzoic acid hydrazine (PHBH) according to previous methods.[4]

Subterranean termites (*Reticulitermes sp.)* were collected from the campus termite trap and allowed to starve for 48 hours. Termite sample preference, consumption, and mortality rates were determined by using 10 termites per petri dish in an incubator at 27°C in 84% relative humidity. Termites were incubated for 10 days and weighed every 24 hours. Fluorescent microscopy images were taken in 6 hour intervals at 25°C and 37°C to determine spatial and temporal localization of DTAF throughout the termite.

## RESULTS & DISCUSSION

### PVCL-Cellulose Nanofiber

Three PVCL-CELL fibers were prepared via electrospinning and alkaline hydrolysis and characterized using SEM, FTIR, and SEC. SEM images confirm the uniform morphology of PVCL-CELL while the ATR-FTIR image [data not shown] confirms the deacetylation process of PVCL-cellulase acetate.[2] The molecular weights of each PVCL sample (Table 1) affected the morphology of the electrospun fiber mats (Figure 2).

Table 1. PVCL Sample Chart

| Sample | Solvent | Mw | PDI |
|---|---|---|---|
| PVCL1 | Benzene | 38,000 | 2.24 |
| PVCL 2 | Benzene | 49,000 | 2.49 |
| PVCL 3 (PVCL-COOH) | Water | 948,000 | 4.93 |

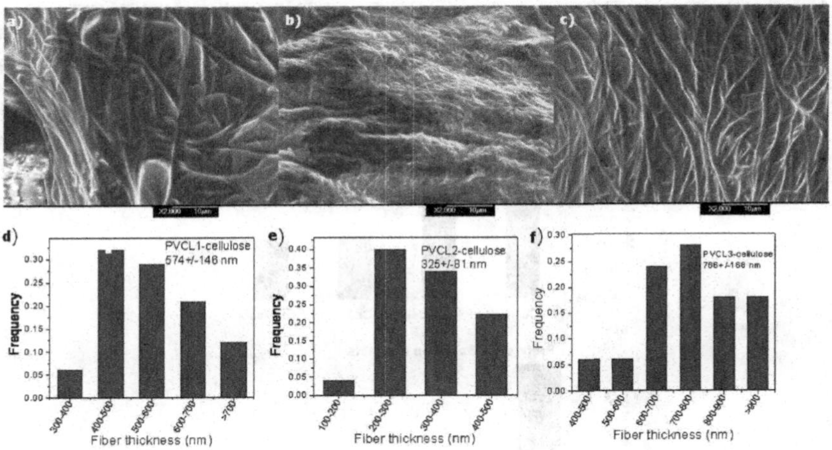


Figure 2. SEM images of a) PVCL1-CELL, b) PVCL2-CELL, and c) PVCL3-CELL fibers (scale bar is 10 μm) with respective size distribution d,e, and f).

Homopolymers depicting PVCL1, PVCL2, and PVCL3 fiber morphology and size distributions are described elsewhere.[2] The PVCL-CELL composite fiber morphology and size dimension changed after alkaline hydrolysis. The fiber size distribution for PVCL1-CELL ranged from 574 ± 146nm, whereas PVCL2-CELL fibers were smaller having 325 ± 81nm, and PVCL3-CELL fibers were the largest in size having a size distribution of 766 ± 168nm.

## Termite Investigations

Termite preference, consumption, and mortality rates were determined after seven days. Although experiments were conducted at >80% humidity, PVCL-CELL fibers were not analyzed for the water retention. It was also noted that *Reticulitermes sp.* preferred the sample PVCL1-CELL in comparison to PVCL2-CELL and PVCL3-CELL. Conversely , termites prefer PVCL1-CELL over the control filter paper (Figure 3). The average mass per termite increased for PVCL1-CELL fibers versus the PVCL3-CELL and the filter paper control, anova analysis confirmed there was not a difference among the groups (PVCL 1, 2 & 3 and filter paper) in regard to mass or weight. Additionally, termite contact preference was higher on the PVCL1-CELL samples in comparison to the PVCL2-CELL and PVCL3-CELL samples. Filter paper controls were not used in the preference students in an effort to elucidate the preference to three novel samples PVCL-CELL membranes. Filter paper was not presented during preference test because termites are raised on eating filter paper. It was also determined that termite viability was higher when termites digested PVCL1-CELL samples in comparison to all other samples including the filter paper. The Chi-square goodness-of-fit test determined that $P<.005$. This confirmed that termites definitely preferred PVCL 1, when correlating preference with the number of contacts made. Fluorescence imaging confirmed the digestion of PVCL-CELL fibers conjugated with DTAF. Figure 4, confirms cellulase digestion of PVCL1-CELL fibers up to 204 hours after digestion. This suggests that *Reticulitermes sp.* subterranean termites prefer consuming PVCL1-CELL fiber samples conjugated with DTAF molecules in comparison to other PVCL-CELL samples (images not shown).

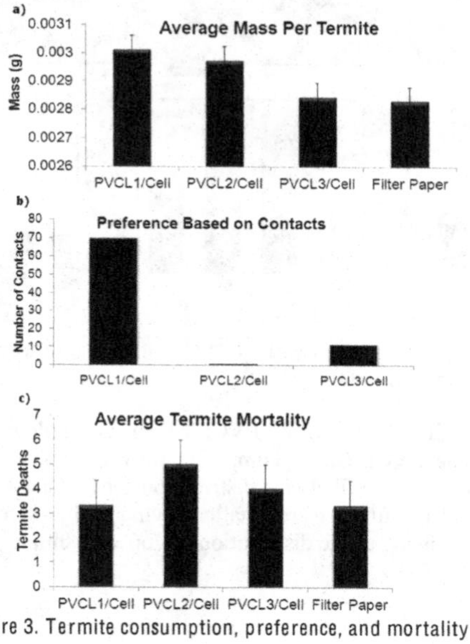


Figure 3. Termite consumption, preference, and mortality data

## CONCLUSIONS

In this study, *Reticulitermes sp.* subterranean termites were fed electrospun PVCL-CELL fiber mats conjugated with the fluorophore DTAF (Figure 4). Fluorescence microscopy was used to visualize cellulase degradation of PVCL-CELLULOSE fiber mats throughout the termite indicating cellulase activity is dependent upon time and temperature. This study determined that subterranean termites preference and viability was higher for the PVCL1-CELL fiber mat over the filter paper controls and other PVCL-CELL substrates.

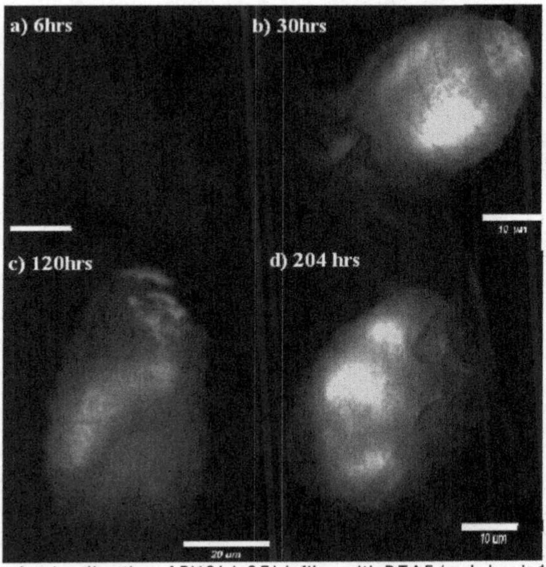


Figure 4. Termite gut showing digestion of PVCL1-CELL fiber with DTAF (scale bar is 10 and 20 μm).

## ACKNOWLEDGMENTS

The authors would like to acknowledge funding provided by the NSF PUI Grant DMR-0642573 and NSF HBCU-UP Morehouse - Wide Initiative for Sustainable Energy.

## REFERENCES

1. Moran-Mirabel, J., Santhanam,N., Corgie, S., Craighead, H., Walker, L. *Biotech & Bioeng* 101, 1130, (2008).
2. Webster, M., Miao, J., Lynch, B., Green, D., Jones-Sawyer, R., Linhardt, R., Mendenhall, J. *Macromol Mater Eng* 298, 447-453 (2013).
3. Herbert, W., Chanzy, H., Husum, TL, Schulein, M., Ernst, S. *Biomacromol* 4, 481-487, (2003)
4. Moretti, R., Thorson, J. *Analytical Biochemistry*, 377, 251-258, (2008).

Mater. Res. Soc. Symp. Proc. Vol. 1569 © 2013 Materials Research Society
DOI: 10.1557/opl.2013.1022

# Application of Hypothetical Cathepsin-like Protein from *Nematostella vectensis* and Its Mutant Silicatein-like Cathepsin for Biosilica Production

Mi-Ran Ki, Ki Ha Min and Seung Pil Pack
Department of Biotechnology and Bioinformatics, Korea University, 2511 Sejong-Ro, Sejong city 339-700, Korea

## ABSTRACT

Silicatein is general catalyst for synthesis of silica structure in siliceous sponges. However, the advent of biomimetic silicification by this recombinant version is limited by its poor yield. To overcome this limitation, we employed a cathepsin L as an alternative to silicatein. Cathepsin L has high sequence identity and similarity with silicatein alpha except cysteine other than serine residues at the active site. Here, we expressed recombinant hypothetical cathepsin-like protein (CAT) from *Nematostella vectensis,* displaying not only protease activity but also silica condensing activity. To increase the silica forming activity, some residues including cysteine in active site were changed into silicatein conserved residues. The mutant silicatein-like cathepsin (SLC) revealed increased protein stability in comparison with that of CAT when expressed in *E. coli.* The silica forming activity of SLC was comparable to that of SIL. SLC produced silica particles of size less than 50 nm which were increased to 200~300 nm in the presence of a structure-directing agent, Triton X-100. Protein immobilization by SLC-mediated silicification was performed using bovine carbonic anhydrase under ambient conditions. Immobilized protein retained its enzymatic activity for a longer time and was reused up to several times. In conclusion, CAT from *Nematostella vectensis* was evolved to a more soluble and available biosilica forming protein that can be applied for various silica-based materials.

## INTRODUCTION

Silicates have a wide range of industrial uses and its nontoxic and highly biocompatible characteristics. Silification is widespread in the biological world including bacteria, single-celled protists, plants, invertebrates and vertebrates wherein gigatons of silica is formed every year in the condition of undersaturated silica concentration, around neutral pH and ambient temperature [1]. Silicatein has been discovered and isolated from spicules, skeletal element of *Demosponge* and are general catalysts for synthesis of silica structure in siliceous sponges. Biosilicification by silicatein is safer and lower energy catalytic pathway for nanomaterial synthesis and furthermore, it can deposit non-biological metal oxide other than silica including titanium oxide, gallium oxide and zirconium oxide at surfaces [2, 3]. Silicatein have been expressed both in eukaryotic and prokaryotic systems, resulting in low level and inclusion body formation when expressed in *E. coli* [4]. Therefore the prerequisite for the utilization of this fascinating biosilicification machinery is to obtain the sufficient amount of enzymatically active recombinant protein. Silicatein has high sequence similarity with cathepsin L protein [5]. The main differences of silicatein from cathepsin L are substitution of serine for cysteine in the catalytic triad and the serine-rich cluster that is regarded as a template in biosilicification on the protein surface [5, 6]. Based on high sequence similarity, cathepsin L might have silica condensing activity by computer-aided molecular modeling and site-directed mutagenesis. Fairhead et al. demonstrated this concept by using human cathepsin L [7]. However, human cathepsin L itself cannot deposit silica and the mutant of it had different property to precipitate silica from silicatein did [7]. We searched the cathepsin L from *Nematostella vectensis* which is higher sequence similarity with silicatein than human's. We demonstrated this protein and its mutant have silica condensing activity as good as silicatein did and can be used for enzyme immobilization by cathepsin-like protein mediated biosilicification.

## EXPERIMENTAL DETAILS

The cDNA sequences of cathepsin L-like protein (CAT) of *Nematostella vectensis* and its mutant protein, silicatein-like cathepsin (SLC) were optimized for expression in *E. coli* and synthesized. The each DNA fragment was inserted to pET42b+ vector to create pET-CAT or pET-SLC. For silicatein expression, we used a vector named pET-SIL previously constructed. The expressing vector was transformed into *E. coli* BL21 (DE3). These recombinant proteins were finally purified using NTA-His tag resin (Thermo Fisher Scientific) according to the manufacturer's instructions. The silica deposition catalyzed by recombinant proteins was started by adding 100 μl of 1M TEOS to 200 μl of 25 mM Tris-HCl (pH 6.8) buffer containing 2 μg of corresponding protein based on the previously described in the literature [5]. In order to observe the effect of agent for directed self-assembly, 1 % Triton X-100 was added to the reaction mixture for silica deposition. Silica deposition was measured according to the modified molybdenum blue method previously described [8]. The morphology of synthesized silica was analyzed using Variable pressure field emission scanning electron microscope (VP-FE-SEM) (KBSI, Chuncheon, Korea). All values are presented as the mean ± SD. Statistical analyses were determined using one-way analysis of the variance (ANOVA) followed by LSD post hoc test for multiple comparisons. The value of statistical significance was set at $*p<0.05$. All statistical analyses were performed with the SPSS 18.0 software program (IBM, USA).

## DISCUSSION

### Protease activity of CAT and SLC

Enzymatic activities of recombinant proteins as a protease were assessed using the substrate Z-FR-AMC as shown in Figure 1. Expectedly, CAT displayed the highest protease activity among CAT, SLC and SIL.

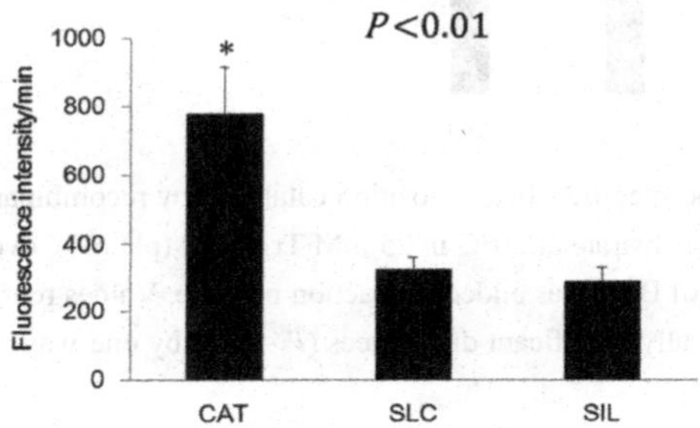


**Figure 1**. The protease activities of recombinant proteins against the substrate Z-FR-AMC. The enzymatic activity was monitored every 1 min over a period of 60 min at $\lambda_{ex}$ 380 nm and $\lambda_{em}$ 460 nm in a fluorescence microplate reader. Values represent mean ± SD. Asterisks indicate statistically significant differences ($P < 0.05$ by one-way ANOVA).

### Silica deposition activity

SIL has dual function as an enzymatic catalyst and organic template for biosilicification. Based on cathepsin L-like structure, a mechanism has been proposed for polymerization of TEOS in which a protein-O-Si intermediate is formed by nucleophilic attack of catalytic serine oxygen on the substrate silicon center. This intermediate bonding is then hydrolyzed by addition of water to form a silanol (Si-OH), followed by subsequent condensation occurring to build up the silica framework [5]. CAT and SLC revealed significantly higher silica condensing activity than bovine serum albumin (BSA) did, which was comparable to that of SIL (Figure 2).

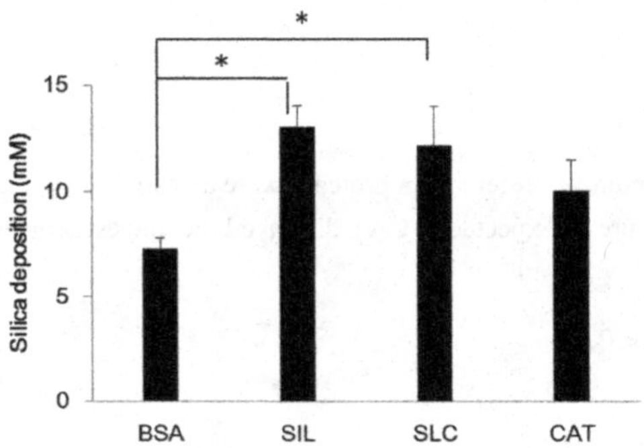


**Figure 2**. Silica condensing activity. The specific silica deposition catalyzed by recombinant proteins was carried out using TEOS as substrate at 20°C in 25 mM Tris-HCl (pH 6.8), overnight. In control assay, the equivalent amount of BSA was added to reaction mixture. Values represent mean ± SD. Asterisks indicate statistically significant differences ($P < 0.05$ by one-way ANOVA).

Unlike human cathepsin L, *Nematostella* cathepsin-like protein expressed in *E. coli*, in our study, displayed silica condensing activity from TEOS in 25 mM Tris-HCl, pH 6.8. BSA has known to have ability to precipitate silica due to a high number of positive charged residues on the surface [9]. Unlike these, SIL catalyzes silica condensation from silicon alkoxide substrates such as TEOS as mentioned above. Likewise, CAT and SLC appeared to precipitate silica from TEOS. A higher sequence similarity of CAT with SIL (53 %) as compared to that of human cathepsin L (50 %) may allow CAT to deposit silica. However, the SLC which has 65 % sequence similarity with SIL did not display a significantly increased silica deposition activity as compared to that of CAT. This suggests that the catalytic sites are more important than other sites to precipitate silica. However, CAT readily degraded after 2 h incubation in the presence of 0.1 mM IPTG, probably due to its autocatalytic activity [10], whereas SLC kept its expression levels as shown in Figure 3. In addition, a soluble expression level of SLC was higher than that of SIL, indicating that SLC can be an attractive substitute for SIL.

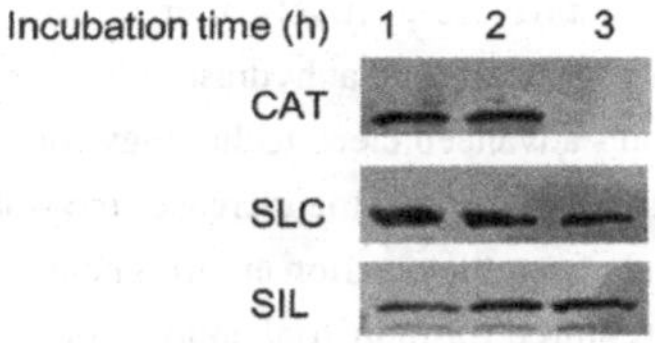


**Figure 3**. Comparison of levels of soluble expression among recombinant proteins depending on incubation times. The recombinant proteins with His-tag on PVDF membrane were probed with anti-mouse His-tag antibody.

SLC produced silica particles of size less than 50 nm which were increased to 200~300 nm in the presence of a structure-directing agent, Triton X-100 (Figure 4). This data indicates that silica precipitates formed from SLC can be directed self-assemble in the presence of Triton X-100. In addition, without protein or BSA-mediated silica precipitates yielded to no silica particles as shown in Figure 4B even in the presence of Triton X-100 (data not shown). Faster rate of silica condensation by SLC than no catalyst or BSA on micelle structure mediated by Triton X-100 may give rise to form a larger size of silica particles.

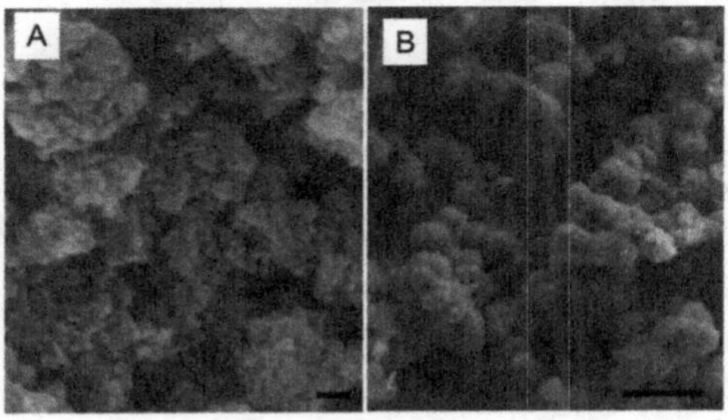


**Figure 4.** SEM micrographs of silica precipitates formed from recombinant SLC without (A) and with 1% Triton X-100 (B). Scale bar is 100 nm (A) and 1 μm (B), respectively.

**BCA immobilization by SLC-mediated biosilicification**

Bovine carbonic anhydrase (BCA) was efficiently immobilized by simple adding it to the reaction mixture during SLC-mediated silicification with TEOS as a substrate at near-neutral pH in an aqueous solution at ambient temperature. Immobilized BCA by SLC-mediated silicification retained its enzymatic activity for a longer time compared to that of BSA-mediated by displaying significantly decreased activity at the fourth day and the first day, respectively after immobilization and was reused up to several times (Figure 5). Carbonic anhydrase (CA) has been focused on as a mean of biocatalyst for $CO_2$ sequestration - advanced clean technology for capture of excess $CO_2$ from the atmosphere [11]. Immobilizing the enzyme increases the stability and reusability of enzyme and makes it possible to recover from the reaction environment. Therefore, the advent of immobilized CA techniques is required for industrial application. Furthermore, the enzyme immobilization by SLC-mediated biosilicification provides a very simple and efficient alternative method for enzyme immobilization under eco-friendly condition.

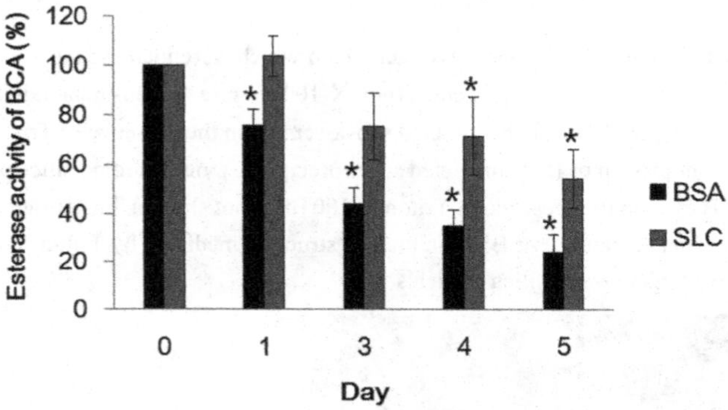


**Figure 5.** BCA immobilization by BSA or SLC-mediated biosilicification and stability of immobilized BCA. The rate of hydrolysis of *p*-nitrophenylacetate into *p*-nitrophenol by BCA was determined spectrophotometrically at 25°C in 50 mM Tris-$SO_4$ buffer (pH 7.6) at 348 nm according to previously described [12]. Values represent mean ± SD. Asterisks indicate statistically significant differences ($P < 0.05$ by one-way ANOVA).

## CONCLUSIONS

Hypothetical cathepsin-like protein from *Nematostella vectensis* was expressed as soluble form in *E. coli* and it was capable of silica deposition from TEOS. This protein was evolved to a more stable and available biosilica forming protein by changing some residues into conserved ones in silicatein. This mutant-mediated silicification conferred a very simple and efficient method for enzyme immobilization, displaying an increase in enzyme stability and reusability.

Therefore, cathepsin L mutant can be an alternative method of producing various silica-based materials simply under eco-friendly conditions.


## ACKNOWLEDGMENTS

This work was supported by the Marine Biomaterials Research Center grant from Marine Biotechnology Program funded by the Ministry of Land, Transport and Maritime Affairs, Korea.


## REFERENCES


1. Perry, C. C. (2003). Silicification: The Processes by Which Organisms Capture and Mineralize Silica. *Reviews in Mineralogy and Geochemistry*, 54(1), 291-327..
2. Andre, R., Tahir, M. N., Link, T., Jochum, F. D., Kolb, U., Theato, P., Tremel, W. (2011). Chemical mimicry: hierarchical 1D TiO2@ZrO2 core-shell structures reminiscent of sponge spicules by the synergistic effect of silicatein-alpha and silintaphin-1. *Langmuir*, 27(9), 5464-5471.
3. Brutchey, R. L., & Morse, D. E. (2008). Silicatein and the translation of its molecular mechanism of biosilicification into low temperature nanomaterial synthesis. *Chem Rev*, 108(11), 4915-4934..
4. Cha, J. N., Shimizu, K., Zhou, Y., Christiansen, S. C., Chmelka, B. F., Stucky, G. D., & Morse, D. E. (1999). Silicatein filaments and subunits from a marine sponge direct the polymerization of silica and silicones in vitro. *Proc Natl Acad Sci U S A*, 96(2), 361-365.
5. Shimizu, K., Cha, J., Stucky, G. D., & Morse, D. E. (1998). Silicatein alpha: cathepsin L-like protein in sponge biosilica. *Proc Natl Acad Sci U S A*, 95(11), 6234-6238..
6. Sumerel, J. L., & Morse, D. E. (2003). Biotechnological advances in biosilicification. *Prog Mol Subcell Biol*, 33, 225-247.
7. Fairhead, M., Johnson, K. A., Kowatz, T., McMahon, S. A., Carter, L. G., Oke, M., van der Walle, C. F. (2008). Crystal structure and silica condensing activities of silicatein alpha-cathepsin L chimeras. *Chem Commun (Camb)*,(15), 1765-1767.
8. Ki, M. R., Yeo, K. B., & Pack, S. P. (2012). Surface immobilization of protein via biosilification catalyzed by silicatein fused to glutathione S-transferase (GST). *Bioprocess Biosyst Eng*. [Epub ahead of print].
9. Coradin, Thibaud, Coupé, Aurélie, & Livage, Jacques. (2003). Interactions of bovine serum albumin and lysozyme with sodium silicate solutions. *Colloids. Surf., B*, 29(2-3), 189-196.

10. Menard, R., Carmona, E., Takebe, S., Dufour, E., Plouffe, C., Mason, P., & Mort, J. S. (1998). Autocatalytic processing of recombinant human procathepsin L. Contribution of both intermolecular and unimolecular events in the processing of procathepsin L in vitro. *J Biol Chem*, 273(8), 4478-4484.
11. Trachtenberg, M. C., Bao, L. (2005). $CO_2$ capture: Enzyme vs. Amine. Fourth Annual Conference On Carbon Capture and Sequestration DOE/NETL, May 2-5 (2005).
12. Ki, M. R., Kanth, B. K., Min, K. H., Lee, J., Pack, S.P. (2012) Increased expression level and catalytic activity of internally-duplicated carbonic anhydrase from *Dunaliella* species by reconstitution of two separate domains. *Process Biochem*. 47, 1423-1427.

Mater. Res. Soc. Symp. Proc. Vol. 1569 © 2013 Materials Research Society
DOI: 10.1557/opl.2013.801

# Tuning Drug Release Profile from Nanoporous Gold Thin Films

Ozge Kurtulus[1] and Erkin Seker[2]

Departments of [1]Chemical Engineering and Materials Science and [2]Electrical and Computer Engineering, University of California - Davis, Davis, CA 95616, USA

## ABSTRACT

Nanoporous gold (np-Au) with its high surface area to volume ratio, tunable pore morphology, ease of surface modification with well-studied thiol chemistry, as well as integration with conventional microfabrication techniques is a promising candidate for controlled drug delivery studies. While it has been demonstrated that np-Au can retain and release drugs, release mechanisms and governing parameters are unclear. This paper reports on the effect of film thickness and morphology on the molecular release from np-Au films.

## INTRODUCTION

Delivering drugs with high spatio-temporal resolution and precise dosage is a major requirement for effective management of disorders. Therefore, there is a need for controlled drug delivery platforms, in which advanced materials play an instrumental role in controlling drug delivery. Recent drug delivery platforms utilize several biomaterials including polymers as eroding templates [1] and inorganic materials as non–eroding templates such as nanoporous anodic aluminum oxide (AAO) and anodic titanium oxide (ATO) [2]. While controlled drug delivery has been established with polymers and ceramics, porous metals, particularly nanoporous gold, has not been explored. Np-Au, which can be produced by dealloying, which consists of selective dissolution of silver from a silver-rich gold alloy, exhibits several advantages such as enabling iontophoretic release of drug molecules and functioning as a high-fidelity neural electrode coating [3]. In addition, its tunable properties, such as pore morphology, provide an opportunity to investigate effect of nanoporosity on drug release profile [4]. The objective of this paper is to demonstrate the effect of film thickness and morphology on the loading capacity of np-Au films and consequent release profile of a small-molecule drug surrogate.

## EXPERIMENTAL DETAILS

### Fabrication:

Nanoporous gold films (3 mm x 3 mm) were patterned on glass coverslips through silicone stencil masks. Prior to film deposition, the coverslips were cleaned for 15 minutes in piranha solution (mixture of hydrogen peroxide and sulfuric acid), rinsed in deionized (DI) water, and dried with nitrogen. Direct-current sputtering system was used to deposit the metal stacks onto the glass coverslips through a laser-cut stencil mask. Following a cleaning step of 50 W Ar plasma treatment, metal films were created by sequential deposition of 15 nm-thick chrome adhesive layer, 50 nm-thick gold seed layer, and co-sputtered $Au_{0.3}Ag_{0.7}$ alloy layer. By varying the deposition time, we obtained two different film thicknesses – 366 nm and 772 nm. The samples were dealloyed in 65% nitric acid for 15 minutes at 65 °C to produce the nanoporous films. Samples were then rinsed and stored in DI water for a week while changing the water every other day. Before use, samples were dried with nitrogen.

**Characterization:**

Scanning electron microscopy (SEM) was used for characterizing film morphology. Top and cross-sectional views of the np-Au films were captured by SEM. Film thickness, average pore area, and percent void/crack coverage in the np-Au films were extracted by digital image analysis of scanning electron micrographs using ImageJ. Briefly, a threshold gray value was chosen to segment the pores and voids. Size of a small void was used as a cut-off point, where smaller features were counted as pores while larger features were counted as voids.

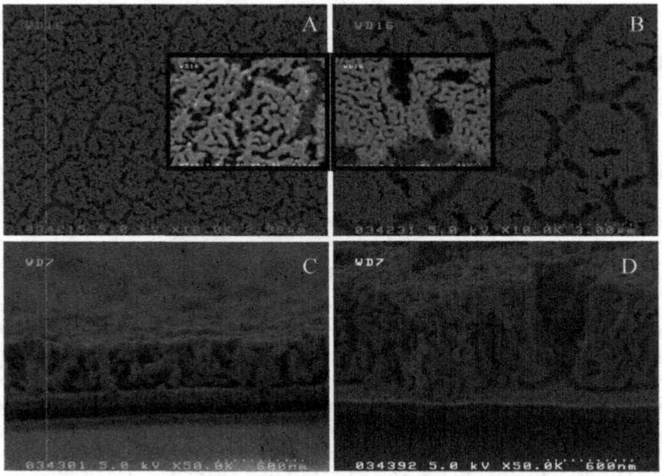


**Figure 1. A:** Top view of 366 nm-thick sample, **B:** Top view of 772-nm thick sample, **C:** Cross-section of 366 nm-thick sample, **D:** Cross-section of 772-nm thick sample. **Insets:** High-magnification images of nanoporous islands separated by voids.

**Drug surrogate release:**

In order to evaluate the performance of np-Au films for retaining and releasing small-molecule drugs, we utilized fluorescein molecules (MW: 376.27 g/mol) as drug surrogates. The samples (np-Au patterns on glass cover slips) with different np-Au film thicknesses (366 nm and 772 nm) were treated with plasma (Harrick Plasma Cleaner) at 10 W for 40 seconds to clean and increase the wettability of surfaces, particularly that of the gold network. All chips were incubated in 10 mM fluorescein solution (Sigma) overnight. Following several rinsing steps, they were placed in a centrifuge tube filled with 250 μL DI water to initiate the molecular release. At specific time points, 5 μL sample solutions were taken from the elution tube and mixed with the same amount of 50 mM NaOH (Sigma) solution in order to enhance the fluorescence intensity. Samples were characterized at an emission wavelength of 515 nm using a fluorospectrometer (NanoDrop 3300).

## RESULTS AND DISCUSSION

In this study, we investigated the effect of the film thickness and morphology on the release profile of fluorescein molecules. Figure 2 illustrates that the cumulative amount of fluorescein (presented as fluorescein concentration in the 250 μL elution tube) released by the end of fifth day is higher for the thicker sample. The released amount for the thinner sample was 0.45 ± 0.02 μM, while 0.60 ± 0.04 μM for the thicker film. Surprisingly, the ratio of the released amounts from the thick and thin films (1.33 ± 0.03) was less than the ratio of film thicknesses (2.11 ± 0.82). This suggests that the loading and release capacity of the nanoporous films examined here does not scale linearly with the film thickness.

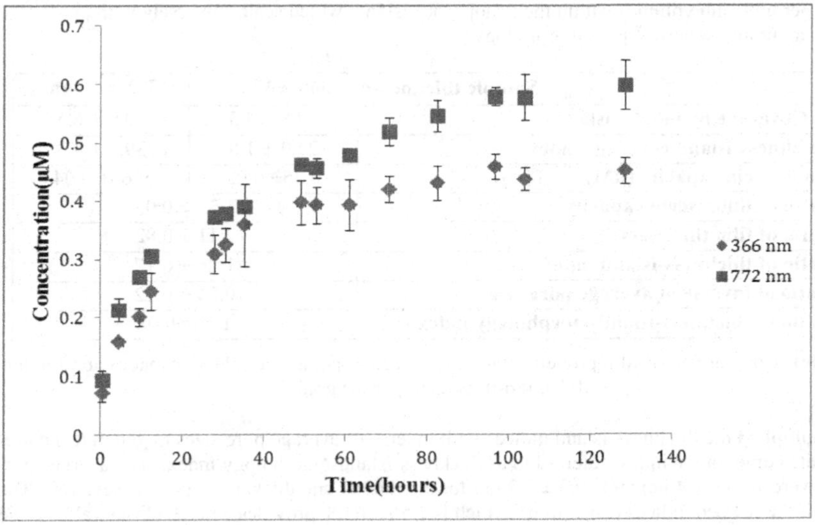


**Figure 2.** Release from 366 nm and 772 nm-thick samples

One reason for this discrepancy may be due to the difference in the tortuosity of thin and thick films, which would result in a non-linear alteration in the diffusion length since it can pose hindrance to molecules traveling through the pores. For pores with comparable size to the molecules, diffusivity of molecules is reduced and transport through the pores becomes hindered, where molecule-wall and molecule-molecule interactions dominate the release [5]. The degree of these interactions can alter the diffusion path in combination with the tortuosity. On the other hand, for pores with sizes greater than the size of loaded molecules, the mass transport mechanism governing the release is bulk diffusion which can be modeled with Fick`s law of diffusion. Therefore, we hypothesized that fluorescein loaded into the voids (mudcrack-like large areas in the film) are removed during the rinsing step, and the main source of the release are the nanoporous islands, where morphological differences and tortuosity play a role in the observed release profile. For simplicity's sake, we first examined a case with an assumption that the pore morphology within the islands is similar for both film thicknesses. This led us to establishing indices (referred to as thickness-island coverage index) to compare the loading capacity of the porous islands for the two thicknesses by simply multiplying film thickness and the percent area occupied by the islands in the SEM image. Thickness-island coverage indices for 366 nm and 772 nm np-Au films were 21.2 ± 1.8 and 39.3 ± 4.3 respectively. The ratio (thick:thin) of these indices was 1.86 ± 0.17, which better approximates the ratio of fluorescein (1.33 ± 0.03) released from the two films, but still falls short of fully explaining the difference. This suggests that the assumption of similar pore morphology for the two film thicknesses is not accurate. We therefore expanded definition of the index to include the effect of specific surface area (surface area of pores per unit film volume) within the nanoporous island, which scales inversely with pore radius assuming spherical pore morphology.

| **Sample thickness:** | **366 ± 17 nm** | **772 ± 19 nm** |
|---|---|---|
| **% Coverage by porous islands** | 57.8 ± 3.3 | 50.9 ± 6.5 |
| **Thickness-island coverage index** | 21.2 ± 1.8 | 39.3 ± 4.3 |
| **Fluorescein capacity (μM)** | 0.45± 0.02 | 0.60± 0.04 |
| **Ratio of fluorescein capacity** | 1.33 ± 0.03 | |
| **Ratio of film thickness** | 2.11 ± 0.82 | |
| **Ratio of thickness-island index** | 1.86 ± 0.17 | |
| **Ratio of inverse of average pore size** | 0.84 ± 0.12 | |
| **Ratio of thickness-island-morphology index** | 1.56 ±0.08 | |

**Table1.** Comparison of film properties and fluorescein loading and release capacity for 366 nm and 772 nm-thick nanoporous gold films

We multiplied the thickness-island indices with inverse of average pore sizes to generate a more accurate comparative index, referred to as thickness-island-morphology index. The average pore sizes were 16.3 ± 1.4 nm and 19.3 ± 1.2 nm for the thinner and thicker samples respectively. The ratio of the revised indices (thick:thin), which is 1.56 ± 0.08, provides a much closer value to the ratio of fluorescein capacity for the two thicknesses. This result is in agreement with the notion that release from the porous islands is largely dictated by molecule-surface interactions, such as adsorption and desorption of molecules at the pore walls. Specific surface area, which is a function of pore morphology, therefore plays a dominant role in defining the release profile. These observations highlight that controlling film thickness and pore morphology are essential for accurately tuning molecular release.

## CONCLUSIONS

We reported our preliminary findings on the effect of film properties on molecular release from np-Au films. The combination of different film thickness, presence of voids, and nanoporous islands dictated the fluorescein loading capacity. We are currently investigating effects of rationally-tuned pore distributions on molecular release profiles. We expect that np-Au coatings will create opportunities for fundamental studies of molecular release as well as advancement of biomedical device coatings.

**REFERENCES**


1. M. Goldberg, R. Langer, and X.Q. Jia, Journal of Biomaterials Science-Polymer Edition **18**(3), 241 (2007).
2. E. Gultepe, D. Nagesha, S. Sridhar, and M. Amiji. Advanced drug delivery reviews **62**(3), 307 ( 2010).
3. E. Seker, Y. Berdichevsky, M. Begley, M. Reed, K. Staley and M. Yarmush, Nanotechnology **21**, 125504 (2010).
4. E. Seker, Y. Berdichevsky, K. J. Staley and M. L. Yarmush, Adv Healthc Mater. **1**(2), 174 (2012).
5. P. Dechadilok and W.M. Deen, Industrial & Engineering Chemistry Research **45**(21), 6953 (2006).

# AUTHOR INDEX

# SUBJECT INDEX